Thermal Comfort Perception

Kristian Fabbri

Thermal Comfort Perception

A Questionnaire Approach Focusing on Children

Second Edition

Springer

Kristian Fabbri
Architect and Adjunct Professor
Department of Architecture
University of Bologna
Cesena (FC), Italy

ISBN 978-3-031-52609-1 ISBN 978-3-031-52610-7 (eBook)
https://doi.org/10.1007/978-3-031-52610-7

This Springer imprint is published by the registered company Springer Nature Switzerland AG
The registered company address is: Gewerbestrasse 11, 6330 Cham, Switzerland

Paper in this product is recyclable.

I dedicate this to my mother, Piera Mainetti,
as always, as everything

Contents

Part I
Thermal Comfort and Children Thermal Cognitive Knowledge

Chapter 1
Introduction

Abstract This chapter describes the purpose and content of the book, and acknowledgement.

1.1 Why This Book?

The study of Thermal Comfort provides useful suggestions for medicine and architecture, combining the features of the physical environment, buildings and/or built environment, with the physiological and health needs of the human body. Energy exchanges that take place between the human body and the environment are perceived and encoded by Thermal Sensation.

The difficulty lies in the evaluation of Thermal Comfort, because it is a judgment that depends not only on the energy exchanges with the body-environment, but also includes the psychological, emotional, cultural and social aspects of people. The feeling of comfort is an intrinsically objective condition.

In the seventies, Povl Ole Fanger proposed a method of statistical investigation, and introduced the indexes of feeling: from the responses of a questionnaire, Fanger extrapolated an index that may be the judgment on Thermal Comfort, once known for the size of the physical environment and the physiological issues relating to the subject. The questionnaire directly asks subjects what their perception of comfort is, and this is one of the ways to study comfort.

Since the first studies in the seventies, the discipline has evolved, trying to assess the Adaptive Thermal Comfort, and the perception of the different subjects.

The First Part of the book describes the history of comfort, physiological variables that affect the human body, the environment and the comfort indexes required by law, and finally, we talk about how to write and administer questionnaires to assess the perceived comfort.

Thermal Comfort studies the phenomena at the human scale, and seeks to identify common criteria to describe the perception of the different types of human beings that vary by gender (male/female), culture and age (children, adults, the elderly).

K. Fabbri, *Thermal Comfort Perception*, https://doi.org/10.1007/978-3-031-52610-7_1

In the case of the phenomena at the "human-scale", it is the perception of the phenomenon by the individual or by a group (and their dynamics) that determines the actions. In the case of comfort, the actions are linked to energy use, to the expected quality and perception of an attractive environment and the psychological effects of the conditions of comfort or discomfort. The perceptive aspect can only be assessed by directly asking individual subjects and interpreting the results of a statistical basis, looking for the correlation between the physical and the physiological model.

The discipline still presents margins of uncertainty in the application of standards, developing new standards also in situations of different microclimates, in addition to those of moderate and extreme, in stressful situations, in vehicles, etc.

Another area of uncertainty concerns the concept of the art of comfort by different users such as children, the sick, the elderly, etc.

Children perceive and describe the world (first to themselves and then to others) differently than adults do. In administering questionnaires to children, we must also pay attention to the meaning of the questions. For their part, the description of the world depends on the Child Development Psychology.

This book deals with the study of comfort through the use of questionnaires. Comfort is a well-established area of research, so much so that technical standards ISO and ASHRAE were issued for standardized evaluation.

The goal of this book is to add a wedge of knowledge to this subject area, add a small contribution to the discipline, focusing on the comfort of a particular category: children.

The idea for the book came to mind following a series of research studies carried out at nurseries and schools. In Part Two we report the results of the research and the criteria used to obtain the results that were comparable and consistent with the discipline, as well as with the requirements of the legislation as described in Part One.

During the research we learned many things about children, on the construction of questionnaires, on pedagogy and Child Development Psychology and so forth. In addition, it has been a lot of fun hearing children's responses to our strange questions.

In many situations, the children were surprised that we asked such obvious and trivial questions, and together we tried to find the words to express their judgment (in some cases, already the fact of defining themselves as individuals with their own opinions was a novelty). And together with us, other figures learned something: children, teachers and pedagogues, and if there had been the designers of buildings (architects, engineers, etc.) they too would have certainly learned concepts useful for design.

The book will appeal to students who already know about comfort, but also to professionals in other subject areas, educationalists and doctors of psychology. For this reason, it is seeking to bring together a summary of the concepts that affect Child Development Psychology and Pedagogy. These concepts may seem minor to those who already know them, but they are useful notions for those working in other disciplines, as it was for me.

The goal is to provide a useful tool for those who design buildings and facilities (from an architectural point of view, in HVAC engineering, building management, etc.) and items of furniture (interior design, furniture designers, etc.); for educationalists and experts in Environmental Psychology.

1.2 New Edition Upgrade

Since the initial publication of this book, many things have changed. In recent years, research on the relationship between thermal comfort and the school has increased, and I am happy to have, albeit in a small part, made a contribution.

The novelties proposed in this edition concern the new role assumed by the school outside the school building: from Distance Learning to Outdoor Education.

The Coronavirus pandemic event (or COVID-19 pandemic, 2020–2023), in different ways from country to country, has modified the classroom by moving the teaching activity from the school building to other places, in particular at the students' home or room or in the open spaces. In order to ensure a safe return to the classroom, it was necessary to adapt the environments so as to increase the Air-Change per Hour (ACH).

Children, young people, teachers, school directors and all school staff have seen the number and type of places where it is possible to do education expand: even one's own room at home can be school. Existing and experimental or occasional experiences, such as Distance Learning or outdoor lessons, became common experiences during the months of the pandemic (Chap. 7). In this regard, the flipped classroom experiences are interesting (Abeysekera and Dawson 2015; O'Flaherty and Phillips 2015; Kurbanoğlu and Akkoyunlu 2017; Subramaniam and Muniandy 2019; Hew et al. 2021) where the classroom is the place of active confrontation and not only of passive learning, with ex-cathedra lessons.

In my opinion, a distinction has been made between the class and the classroom, a sort of new deal that I describe in the third part of the book.

In recent years, themes such as Cognitive Thermal Comfort by children have expanded outside the classroom: their classroom, as mentioned, and the outside, the city, the Built Environment and Nature. The relationship with nature has changed, both adults and children ask for a greater number of hours outdoors and in contact with nature.

My research interest has shifted towards new themes, related to the relationship between thermal comfort, children, learning/education and school buildings, in particular: (a) the sensory perception of architecture (Chap. 10), given by the relationship between the senses, emotions and characteristics of indoor classrooms; and (b) the relationship between Outdoor Education and Outdoor Thermal Comfort.

In the educational field there have been, for years, consolidated experiences in the field of what is called Outdoor Education. The scientific literature in the pedagogical field, reports a great deal of research, experience and practice in this regard. To these experiences, briefly described in Chap. 11, book adds to the research related to Outdoor Thermal Comfort, the Heat Stress Index or Thermal Stress Indexes in natural or urban outdoor environment and presents some reflections on how Outdoor Thermal Comfort can be managed during Outdoor Education.

1.3 How to Read This Book

The book is divided into three parts:

In the First Part are shown the concepts and variables that interest the human body and the indoor environment for the evaluation of comfort, following that are the provisions of the technical regulations (if any) and the scientific literature. In particular:

- a brief description of the history of Thermal Comfort, a concept born in the twentieth century, from the first experiences until future prospects;
- the ergonomics of the thermal environment, and the variables that describe the physiological behavior of the human body on Thermal Comfort and the Indoor Air Quality (IAQ);
- the sensation indexes of Thermal Comfort and IAQ;
- the criteria to properly build a questionnaire to assess subjective judgment;
- a summary of the relationships between concepts in Thermal Comfort and Child Development Psychology;
- finally, an upgrade about how school thermal comfort perception has changed during the last eight years and after the Coronavirus pandemic (COVID-19), especially about the role of Distance Learning (sometimes also called e-learning) and of Indoor Air Quality in the classroom and its ventilation.

In the Second Part, are the case studies:

- the Nursery School (Kindergarten) "*Diana*" in Reggio Emilia (children aged 3–5 years old);
- the Elementary School "*CEIS*" in Rimini (children aged 6–10 years old);
- the Middle School "*Tito Maccio Plauto*" in Cesena (children aged 11–14 years old);
- for each case study described, the criterion followed for administration of the questionnaires and the results of the research.

In the new Third Part, I propose a New Deal about school teaching and buildings:

- the role of the senses and emotions given by the characteristics of the classroom, during the lessons and the relationship between Architecture and Sensitivity;
- the relationship between Outdoor Education, a well-established discipline, and Outdoor Thermal Comfort.

The classroom is no longer an anonymous place to learn, but rather, it plays a role in determining the '*classroom climate*' in the double meaning of physical place (classroom environment) and relationship between people (class environment). In addition, the classroom and its thermal comfort is no longer and only limited to the indoors: one can also learn outdoors in the courtyard, in the city, in parks or in the forests.

Acknowledgements The writing of this book has involved several people. I thank the colleagues of the Architecture Department of the University of Bologna, Professors Boeri A., Antonini E., Gaspari J., Longo D., Ugolini A., for their availability, therefore I thank the student-researchers, now graduates and architects, for the passion and meticulous execution of our experiments and I thank the educational facilities for their willingness to carry out the thesis and research.

A special thanks to Silvia Magnani and our chatting about teaching, clothes and Outdoor Education.

Finally, I thank the reader, the research and writing of this book was a useful learning experience, I hope that the reader will find in this book the same enthusiasm of those who carried out the research.

www.kristianfabbri.com.

References

Abeysekera L, Dawson P (2015) Motivation and cognitive load in the flipped classroom: definition, rationale and a call for research. High Educ Res Dev 34:1–14. https://doi.org/10.1080/07294360.2014.934336

Hew KF, Bai S, Huang W et al (2021) On the use of flipped classroom across various disciplines: insights from a second-order meta-analysis. Australas J Educ Technol 37:132–151. https://doi.org/10.14742/ajet.6475

Kurbanoğlu S, Akkoyunlu B (2017) Information literacy and flipped learning. In: Pathways into information literacy and communities of practice. Elsevier, pp 53–84

O'Flaherty J, Phillips C (2015) The use of flipped classrooms in higher education: a scoping review. Internet High Educ 25:85–95. https://doi.org/10.1016/j.iheduc.2015.02.002

Subramaniam SR, Muniandy B (2019) The effect of flipped classroom on students' engagement. Technol Knowl Learn 24:355–372. https://doi.org/10.1007/s10758-017-9343-y

Chapter 2
A Brief History of Thermal Comfort: From Effective Temperature to Adaptive Thermal Comfort

Abstract The study of Thermal Comfort was born in the early decades of the twentieth century, with the studies of A. P. Gagge to resolve particular problems due to stressful situations in the workplace. The postwar period and in the seventies, with P. O. Fanger and other researchers, marked the study as a real discipline. This chapter is a brief history of the discipline that studies Thermal Comfort.

2.1 The Discipline of Comfort

The discipline of Well-Being (or Comfort), Thermal Comfort, and the concept of a comfortable environment itself, was born in the twentieth century, when it became possible to directly—control—the microclimate of the indoor environment: houses, vehicles, etc.

In previous centuries indoor comfort conditions were controlled by adaptive processes related to behavior and clothing, in addition to the use of fireplaces or stoves to control the temperature. Not being able to act on the comfort of the rooms, it was not useful to study the parameters that could influence comfort. In addition to studying comfort, it was necessary to model the building as an open system and apply the laws of thermodynamics, a discipline born in the second half of the nineteenth century.

In the twentieth century, the architectural theories (Mouvement modern, Functionalism, Bauhaus, Le Corbusier with Le Modulor, De Stijl, CIAM, International Style, etc.) and technical manuals, put man at the center, as an individual with a physical dimension, founded an interest in the design and construction of residential buildings.

Thanks to heating and air conditioning systems like those invented by Willis Carrier, it becomes possible for individuals to adjust the characteristics of their own indoor environments, and consequently to demand the best indoor comfort conditions.

K. Fabbri, *Thermal Comfort Perception*, https://doi.org/10.1007/978-3-031-52610-7_2

The history of comfort evolved with other inventions, in the military, when it became necessary to ensure that the military personnel could continue to work on ships and airplanes even in environments with high temperatures.

Comfort is the result of the interaction of physical exchanges, physiological, psychological, social and cultural rights, it depends on the architecture, the clothing, the eating habits and the climate.

The history of the discipline that deals with studying comfort, especially Thermal Comfort, is recent. Before beginning to describe the tools, the physical dimensions and indexes to evaluate comfort, a brief history of Thermal Comfort is given: a first embodiment of physical-physiological, up to a greater level of detail, Adaptive Thermal Comfort, and the study of Thermal Comfort for specific types of subjects, such as in the case of children, and also the evaluation of perceptual cognitive aspects, are the subject of this book.

2.2 The Beginnings A. P. Gagge and the Military Requirements

In 1936, A. Pharo Gagge (1908–1993) of J. B. Pierce Laboratory of Hygiene of New Haven Connecticut, in the article "The linearity criterion as applied to partitional calorimetry" (Gagge 1936), in which, by processing the experimental data on the relationship of the human body and the environment, shows the expression of the first principle of thermodynamics for the human body. The model consists of two nodes: the "core", or inside of the human body that produces energy through the metabolic activity and the mechanical work (muscles), and the "skin" that exchange energy and matter outward. Gagge was not the only one in the thirties, to deal with these issues (Bedford 1936), but he was the first dedicated to finding an application of the principles of thermodynamics to energy exchanges between man and his environment.

The "*Two node model*" proposed by Gagge provides that the sum of the thermal exchanges due to metabolic activity, the body's energy, evaporation, radiation and conduction are zero.

In this way, he writes the Human Body Heat Balance, where the variation of the accumulated heat of the body to keep its temperature at 37 °C is given by the sum of the metabolic energy (as measured by the oxygen consumed) the dispersions due to evaporation and sweat, the energy flow through the body boundary by: conduction, conventions and thermal radiation.

The J. B. Pierce Laboratory was founded in 1933, adjacent to the medical school at Yale University, with a mission to promote research to increase the knowledge and the advancement of human health and comfort. The will to establish such an institution is due to the "John B. Pierce Foundation", founded in 1924, as a result of the legacy of John Bartlett Pierce (1844–1917) businessman and founder of the "*American Radiator Company*", with the purpose of promoting research, education, technical and scientific work in the field of heating, ventilation and health care, to increase the knowledge to improve the health and comfort of men and their homes.

John B. Pierce was born in 1824 and began his career in a shop selling stoves in Buffalo. In 1892, he founded a factory for the production of boilers, heating systems and radiators, which was successful and became one of the most important industries of heating systems in the United States. In 1917, following his death, having no children, he left a fund of more than a million dollars to friends and employees, who decided to establish a foundation and then the institute, in memory of J. B. Pierce.

The experience of J. B. Pierce, like T. A. Edison or W. H. Carrier, highlights the link between the industrial and entrepreneurial activities and the activities of research, particularly in the United States.

Space heating is related to the health of the population, both these issues included the business of J. B. Pierce and his philanthropic interests, which led to the foundation of J. B. Pierce Laboratories. In this laboratory, since the thirties, with the pioneering works of C. Winslow, L. P. Herrington and A. P. Gagge (Winslow et al. 1937) the physical and physiological principles to understand and measure the thermal exchanges between body and human environments were defined.

The research in this area continues along two lines:

An inherent thermodynamic study of the physiological processes, continuing the first approaches of the studies of von Helmholtz.

While the other focuses on the relationship between the human body and the environment, and the indexes of well-being, that depend on physical factors, and physiological behavior.

The biophysicist A. P. Gagge was born in 1908 in Columbus, Ohio, and, after graduating in physics at the University of Virginia, obtained a doctorate in physics in 1933 at Yale, where he worked with the J. B. Pierce Laboratory. Here the concepts of thermodynamics were applied to physiology, through a series of experiments which measured the behavior of the human body. In 1936, the article was published (Gagge 1936) which elaborates the "*Two node model*" temperature control system, the equation of heat balance of the human body.

In the years of World War II, he worked in the medical aviation laboratories in the Wright Patterson Air Force Base, where he developed the equipment to ensure respiration at an altitude of 43,000 ft. for airline pilots, and he continued to work with aviation up until 1963.

The studies of A. P. Gagge helped define the field of study of the energy exchanges between the human body and the environment, the applications of which have repercussions in the field of health and safety in the workplace, in the military, in space exploration and in the design of buildings.

In the seventies, the Danish physiologist Povl Ole Fanger (1934–2006), following a series of experiments and tests that allowed him to define the indexes of comfort and well-being, revised the equation of the "two-node model", setting equal to zero the variation of the amount of energy necessary to the human body at 37 °C, and expresses energy exchanges of the human body with a double equation.

In severe cold or hot workplace environments how can you evaluate if the conditions are due to discomfort or abnormal working conditions, as long as they are tolerable?

On these issues, researchers began to carry out the first study in the twenties, in the United States and the United Kingdom. The first studies on comfort, based on empirical rules, were developed by researchers F. C. Houghten and C. P. Yaglou, who simulated different conditions in the ASHVE (American Society of Heating and Ventilating Engineers) research laboratories of Pittsburgh to locate the comfort zone.

In 1923, the article "Determination of the comfort zone" was published in the journal of ASHVE (Houghten and Yaglou 1923a) and the study "Determining equal comfort lines" was published, again in 1923 (Houghten and Yaglou 1923b) in which comfort lines were proposed, on the psychrometric chart (or ASHRAE) of the moist air, which corresponds to the empirical index called "Effective Temperature (ET)", corresponding to the correspondence between the temperature of the real environment and the temperature of a notional environment in which there is no temperature difference between the ambient air and that of the walls, there are no air currents and the relative humidity corresponds to 100%.

In summary, the equivalent temperature of an environment corresponds to the same temperature there would be in an environment where the temperature is uniform, the air is stationary and the moisture content corresponds to 100%, and therefore the human body cannot exchange energy with the environment. The logic is that of an analogy between the variables of a specific environment and any conditions of a standard environment, for example, the actual temperature of an environment which is at 22 °C with relative humidity of 50% and air speed of 0.2 m/s, is equal to the temperature that the subject would receive in an environment where the relative humidity is equal to 100% and the air is stationary, which corresponds to the actual temperature of about 19.6 °C (Houghten and Yaglou 1953). Standard ASHRAE 55 (ASHRAE 2010) adopts 'air speed' while Standard ISO 7730 (ISO 2005) adopts 'air velocity'.

The equation does not take into account the variables linked to the person, and in subsequent ASHVE studies by means of experimental tests, in 1932, following the studies of researchers H. M. Vernon and C. G. Warner "The influence of the humidity of the air on capacity of work at high temperatures" (Vernon and Warner 1932), it was decided to include the air velocity in the diagrams of wellness.

The analytical study was elaborated in the hygiene laboratory of the J. B. Pierce Institute, by A. P. Gagge and others in 1971, with the article "An effective temperature scale based on a simple model of human physiological regulatory response" (Gagge et al. 1972) which introduces the "Effective Temperature Scale" which takes into account the clothing, activity and radiation exchange, expressed through a series of nomograms.

In the seventies there were several studies on comfort (Rohles and Levins 1971; Rohles and Johnson 1972; Givoni and Pandolf 1973; Gagge and Nishi 1976) and in the eighties (Bell 1981; Collins and Hoinville 1980).

The actual temperature is an index based on the empirical basis of an analogy between the real environment and the standard environment, among these, in 1957, as a result of the studies in the US military centers, Yaglou and Minard, in "Control of heat causalities at military training centers" (Yaglou and Minard 1957) developed the WBGT (Wet Bulb Globe temperature) an indicator that combines the effect of temperature, relative humidity, heat exchange by radiation and solar radiation, and is used to determine the extent of exposure to heat conditions.

Along the same lines, other temperature indicators were introduced for "extreme" climatic conditions in any case different from the standard conditions of the environments and buildings in climatic condition temperatures.

These include the Equatorial Comfort Index (ECI), developed in 1959, based on 393 observations, from C. G. Webb in "An analysis of some observations of thermal comfort in an equatorial climate" (Webb 1959) that corresponds to the response of a subject perfectly acclimatized in the equatorial climatic conditions, and the Tropical Summer Index (TSI) developed in the eighties by the Central Building Research Institute Roorkee (India) to evaluate the welfare conditions in countries where the relative humidity is greater than 50%.

2.3 The Revolution of Povl Ole Fanger: Evaluating the Thermal Sensation

The sixties and seventies are fertile with studies on the subject, and in addition to the studies of the Pierce Laboratory by Gagge and other American scholars who focused on the indexes of thermal stress and the approach to engineering, probably derived from the role ASHRAE had and the air-conditioned building in the United States.

In parallel, a European approach to the problem grew, which focused on the evaluation of the "feel-good" parameter. The first to set the research in this direction was Povl Ole Fanger (1934–2006), physiologist of the Technical University of Denmark, the forefather on the study of the welfare of confined spaces. Fanger focused on the relationship between the physical parameters of an environment and the physiological parameters of people, and the perception of well-being expressed by the people themselves.

The research began in the sixties at the Laboratory of Heating and Air Conditioning of the Technical University of Denmark and also at the Institute for Environmental Research at Kansas State University. In 1965, after five years of study, Fanger published the article "Calculation of Thermal Comfort: Introduction of a basic comfort equation" (Fanger 1967) which proposes a rating scale of *perceived sense of well-being.*

Following successive trials and research, in 1970 he published the book "Thermal Comfort" (Fanger 1970), which defines the contents of a new discipline: the study of the condition of comfort and well-being in indoor environments.

The conceptual leap introduced by Fanger, compared to previous studies, is in the introduction of the rating/judgment scale from the people themselves. Based on the feedback and the votes of thermal sensation expressed by people, Fanger elaborates an equation that relates the physical physiological environmental parameters, and indexes of thermal sensation.

The research Fanger starts from studies on energy exchanges between the human body and the environment, and the equation of balance of body heat (Heat Balance Equation) which defines the conditions of heat, that is the range of well-being, and a series of Comfort Diagrams, which correlates metabolic rate, clothes, air temperature, mean radiant temperature, air velocity and Relative Humidity (see comfort lines in Figs. 2.1, 2.2, 2.3 and 2.4).

In research, Fanger submits 128 subjects, Danish students, half male and half female, college-age (a mean age of 23 years) in an experiment, repeated with another 128 subjects with a mean age of 68 years. The fact is that Danish students were not affected by the experimental activity but, as in any experiment, the boundary conditions of the experiment must be defined given the role that culture and habits of the conditions of well-being can have.

The experiment took place in the autumn of 1968 in the climatic chamber (Environmental chamber) of the Technical University of Denmark, and included the participation of 128 subjects, with measurements, height, weight and body area, with the formula of DuBois (DuBois and DuBois 1916), and wearing a T-shirt and cotton trousers, with the T-shirt out of the trousers, all wear cotton underwear and wool socks without shoes, so as to have a uniform insulation clothing value of 0.6 clo.

The climatic chamber is a cube of 2.8 $\times$ 5.6 m with a height of 2.8 m, within which one can monitor and record the air temperature, humidity, the mean radiant temperature with a digital system and the air speed with a dedicated air conditioning system. The conditions of illumination (150 lx) and acoustics (45 dB) are kept constant, and the air is filtered and exchanged so as to avoid the formation of dust or odors.

The test includes eight set-point conditions kept constant for three hours so as not to create the feeling of variation of the climatic conditions within the test chamber. 32 tests are conducted lasting three hours each, in the afternoon or evening, during which subjects respond to a pre-questionnaire, which includes questions related to the confirmation that they have slept well the previous day and if they have eaten well, in order to verify that there aren't uncontrollable factors of interference.

Before entering the climatic chamber subjects wait for about thirty minutes in a pre-chamber where the body temperature is measured orally, the purpose of the experiment is explained along with the method of scoring. Climatic chamber, results and photo of a similar experiment reported in Figs. 2.5, 2.6, 2.7 and 2.8.

Once inside the climatic chamber, subjects are seated and have their "quiet" activity, reading, studying or similar, or "quiet conversation", to avoid any metabolic effect of controversy and psychological or verbal fights, during they which mustn't exchange views on the climatic conditions of the environment.

After about a half-hour they are asked to complete the questionnaire, giving a score to the ambient temperature conditions: cold, cool, slightly cool, neutral, slightly warm, warm, hot.

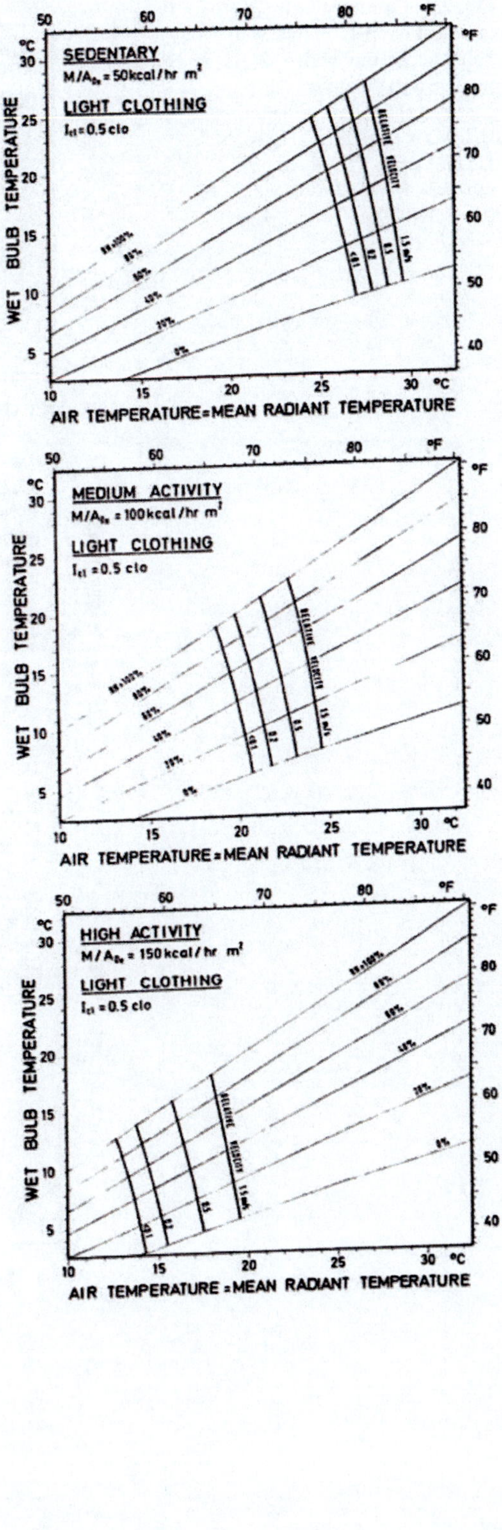

Fig. 2.1 Comfort lines (ambient temperature vs. wet bulb temperature with relative air velocity as the parameter) for person with LIGHT CLOTHING (I_{cl} = 0.5 cl, f_{cl} = 1.1) at three different activity levels (Fanger 1970)

Fig. 2.2 Comfort lines (ambient temperature vs. wet bulb temperature with relative air velocity as the parameter) for person with HEAVY CLOTHING (I_{cl} = 1.5 cl, f_{cl} = 1.2) at three different activity levels (Fanger 1970)

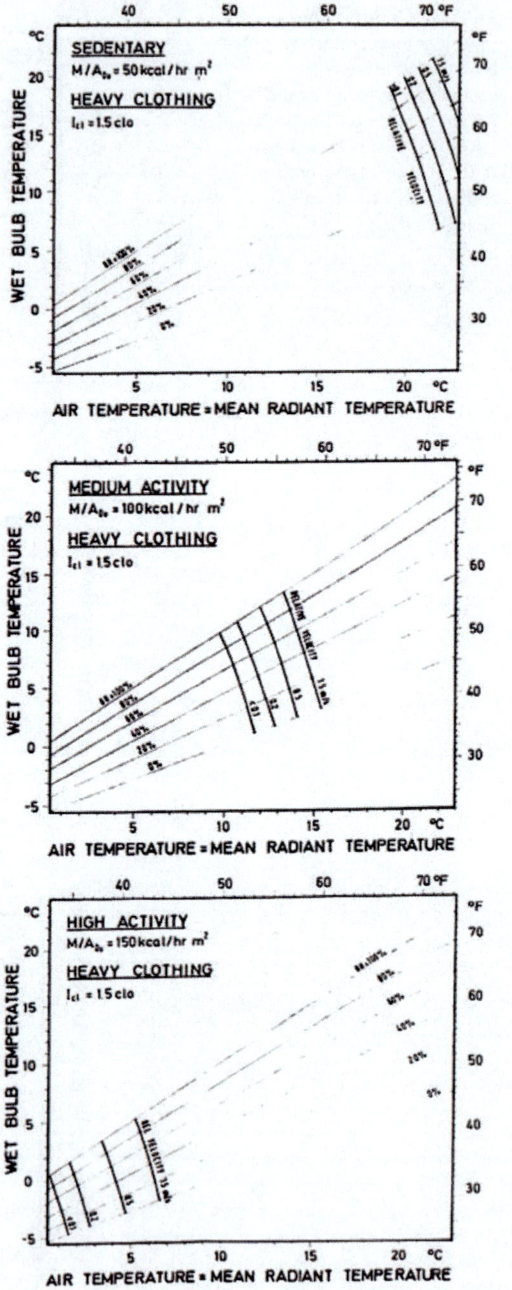

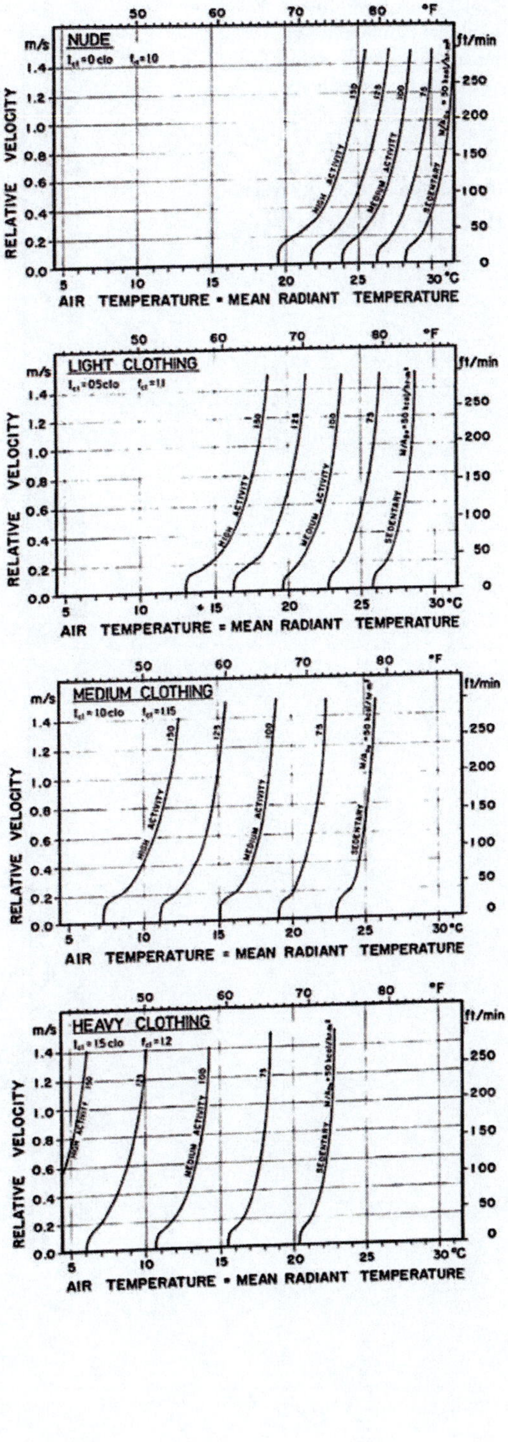

Fig. 2.3 Comfort lines (ambient temperature vs. wet bulb temperature with relative air velocity as the parameter) at 4 different clo-value levels (rh = 50%) (Fanger 1970)

Fig. 2.4 Comfort lines (ambient temperature vs. mean radiant temperature with relative air velocity as the parameter) for person with MEDIUM CLOTHING (I_{cl} = 1.0 cl, f_{cl} = 1.15) at three different activity levels (rh = 50%) (Fanger 1970)

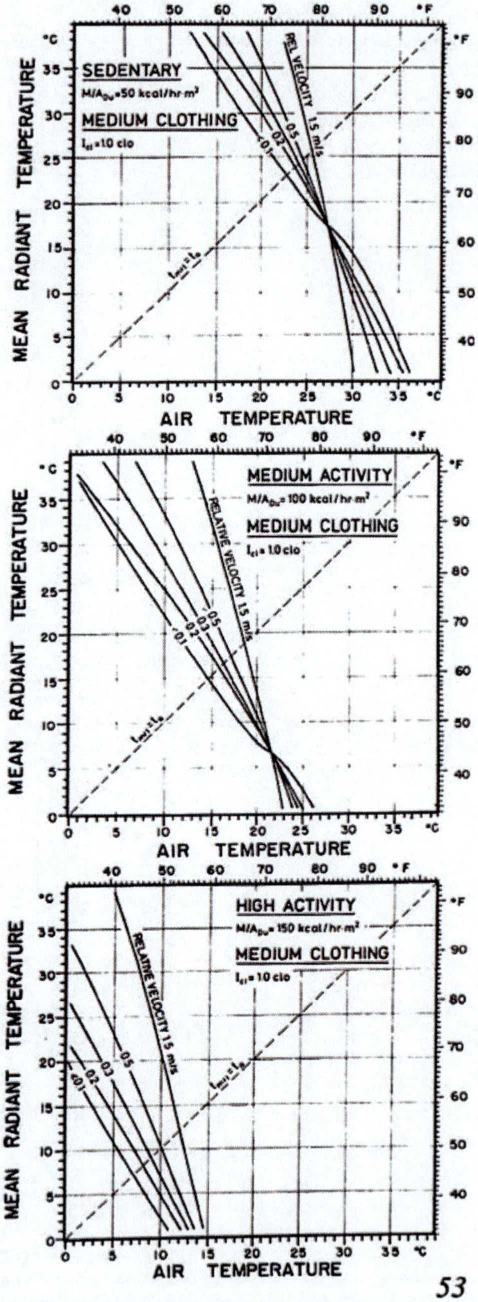

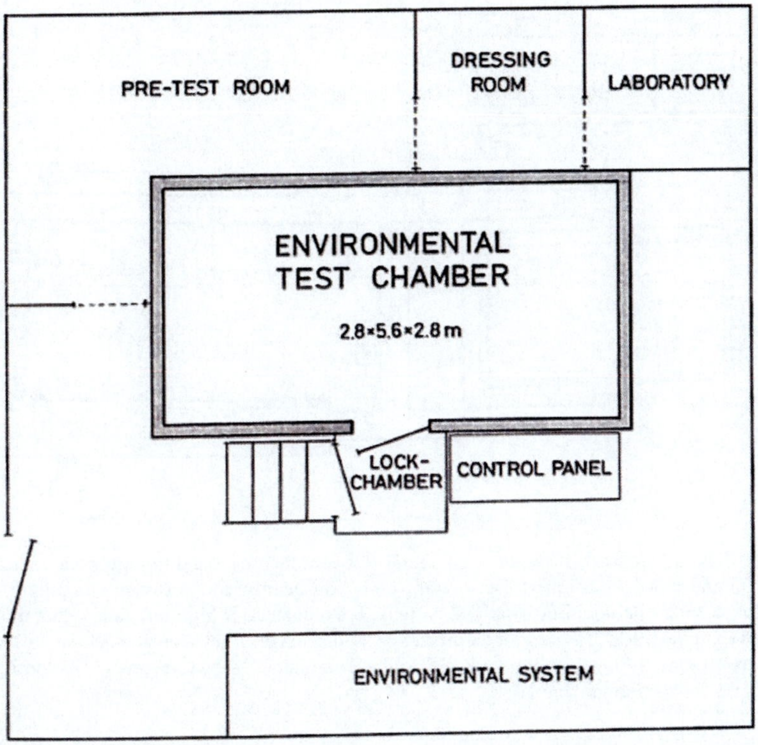

Fig. 2.5 Fanger experiment: environmental test chamber at the Technical University of Denmark (Fanger 1970)

The scoring is repeated every half hour for a total of six scores for the subject. The subjects are weighed in sitting position before the first questionnaire and after the sixth questionnaire with a precision balance so as to determine the weight loss due to evaporation. During the test they may drink but not eat, and the amount of drinks is measured. After the sixth questionnaire a score is given on the temperature that the subject deems as comfortable during the experiment, to which subjects respond more or less indicating the same value.

In the experimentation, Fanger aims to eliminate any factors that might disturb the evaluation scale of judgment, or at least to evaluate their impact, and it is interesting to learn, along with the evaluation of a parameter with many variables, what the sense of comfort is, that can be measured once the criteria and the assessment scale are determined. At the end of the test, a questionnaire is completed on dietary habits, sleep and the menstrual cycle, so as to identify any abnormalities or factors that may have influenced the test.

The same Fanger presents a series of considerations about the geographical location; the equation for calculating the predicted mean score on the well-being of an environment is valid for people who live in temperate climates, as there are no

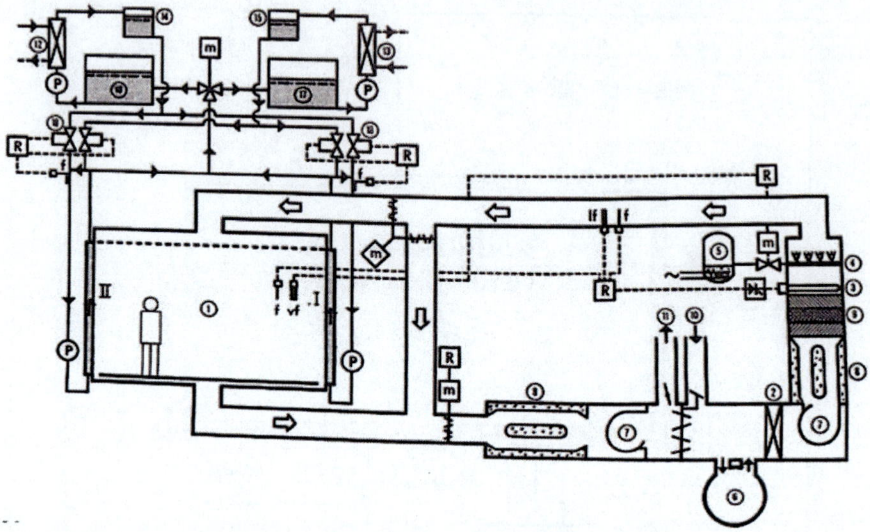

Fig. 2.6 Scheme diagram of the environmental test chamber, air conditioning system and water system for end walls. *1* Chamber; *2* air-cooling coil; *3* air-heating coil; *4* steam humidifier; *5* steam generator; *6* rotary dehumidification unit; *7* fans; *8* attenuators; *9* high-efficiency dust filters and activated charcoal filters; *10* outdoor air intake; *11* air discharge; *12* heat exchanger (steam); *13* heat exchanger (freon); *14* heat accumulator; *15* cold accumulator; *16* heat receiver; *17* cold receiver; *18* solenoid valves (Fanger 1970)

significant variations between the different age groups, except that older people and women prefer slightly warmer environments, although this difference is irrelevant to our study, as they are not being investigated, but rather, children are our subjects.

The comfort index introduced by Fanger is the Predicted Mean Vote (PMV) that allows one to express the score that a person gives to an environment, from the measurement of the physical parameters of the environment: air temperature, mean radiant temperature, air speed and humidity, and from the metabolic rate and clothing of the subject itself.

The index allows one to find an area of well-being bounded by the values of the physical parameters of the environment, which can also be reported on the ASHRAE Psychrometric Chart, within which the environment is considered in terms of comfort. The practical use enables one to define the conditions for set-point of the environments and the variables of the heating systems in buildings for collective use, such as cinemas, theaters, hospitals, shopping centers, etc.

The PMV expresses the opinion of the people, but does not assess what is the acceptability of the conditions of comfort, even under conditions in which the score is positive, for example, with a value of PMV equal to—0.3 ("slightly cool" with respect to the neutral feeling) it was not able to assess whether it is a condition in which the majority of people consider as an acceptable one.

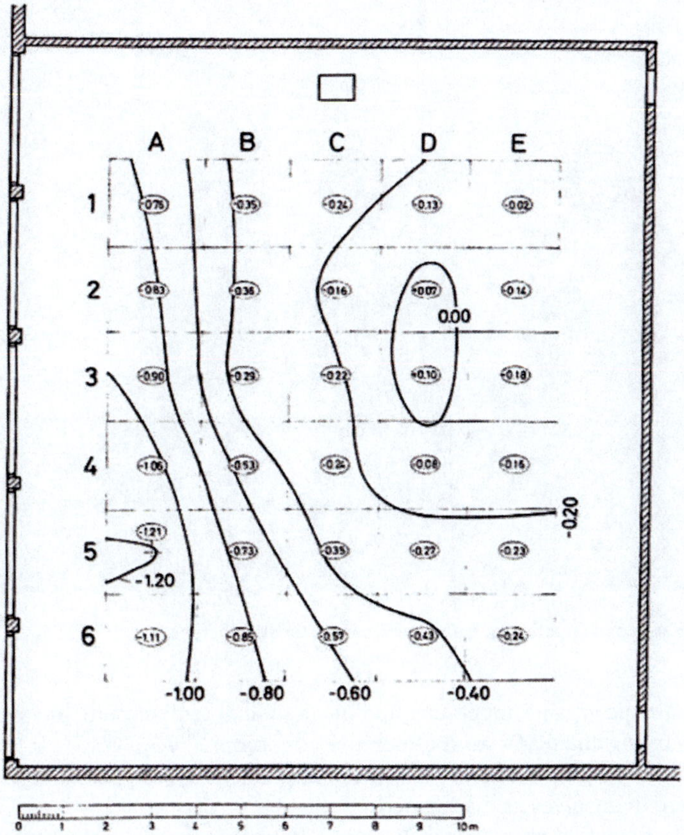

Fig. 2.7 Assessment of the thermal environment in the lecture hall. Measurements are taken in the center of each square and the corresponding PMV values (Fanger 1970)

Following these considerations, Fanger proposes an index for the evaluation of the conditions of non-comfort (or discomfort) to an environment, expressed as a Predicted Percentage of Dissatisfied (PPD). The PPD index expresses the percentage of people in those conditions of metabolism, clothing and physical parameters of the environment, expressing, however, a negative judgment, who in fact complain; even when an environment is assessed by most people as neutral, it is believed that there are however 5% of people who consider this condition as unsatisfactory.

In the case of a room where there are 20 people, even if you are in conditions of comfort "neutral", at least one person will complain because it is either too hot or too cold, consider that this is what happens in the compartment of a train or in the waiting room of a doctor to confirm, at least empirically, this report.

The PPD-PMV Diagram allows one to check what percentage of people are dissatisfied (PPD) to vary the judgment on the feeling of comfort PMV, which in turn always depends on the same physical para meters of the environment, relative

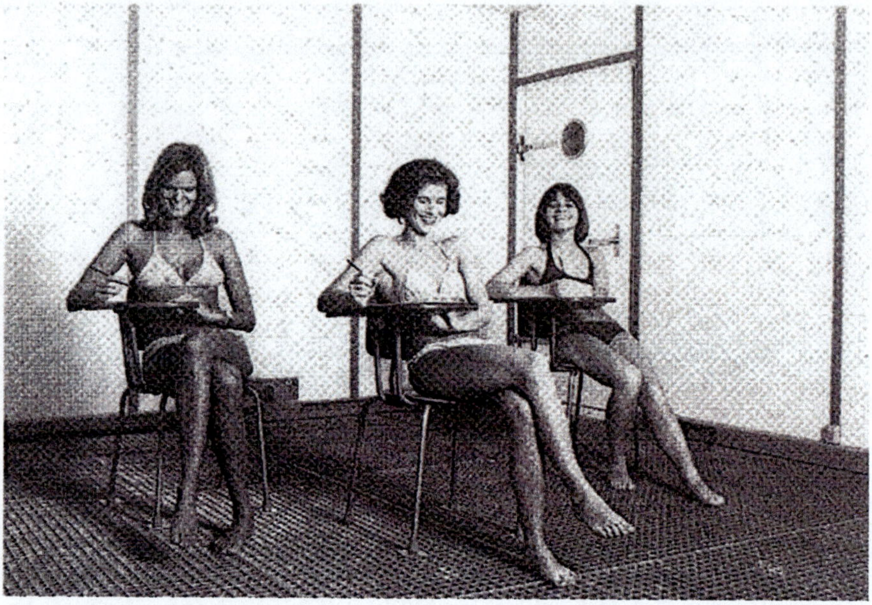

Fig. 2.8 Example of experiment with subjects in a climatic chamber (by B. W. Olesen)

humidity, air speed, air temperature and mean radiant temperature, metabolism and clothes. Varying the PMV to the extreme conditions, “very cold” or “very hot”, increase, according to a logarithmic law, or exponentially, the percentage of people who declare themselves as dissatisfied.

The dissatisfaction of the microclimate of an environment is expressed with other indicators, stress indexes and local discomfort indexes, due to non-uniformity of the parameters in the environments, also considering the Heat Stress Indexes (Epstein and Moran 2006).

The Thermal Stress Index (TSI) was introduced by Israeli architect Baruch Givoni (1932–2019), a graduate of the Faculty of Architecture at the Institute of Technology in Israel, who then specialized with a Master and Doctorate in Hygiene and Public Health at the School of Medicine of the University of Jerusalem (1963) that in the study “Estimation of the effect of climate on man: developing a new thermal index” (Givoni 1963a, b) then took up in the book “Man, climate and architecture” in 1969 (Givoni 1969), which expresses the amount of heat transferred from the human body through perspiration to maintain a given condition of comfort. When the index is high this means that the body expends more thermal energy than is normally required (under stress) to stay in conditions of well-being.

The text of Baruch Givoni in 1969 followed another crucial work in the architecture and construction industry, the book by architect Victor Olgyay (1910–1970), “Design with Climate” (Olgyay 1963), that constitutes the cornerstone of architecture called “bioclimatic”. Olgyay studies the relationship between the architectural form of the buildings and the related climate, albeit in an empirical way in the exchange of energy that the building has with the context.

From these two books, and the book by Edward Mazria "The Passive Solar Energy Book" of 1979 (Mazria 1979), develops that part of architectural design, solar architecture, passive architecture (or passive buildings), green architecture, bioclimatic architecture.

Other indicators of discomfort are introduced, in the sixties, by Tennenbaum in the article "The physiological significance of the cumulative discomfort index" of 1961 (Tennenbaum et al. 1961; Sohar et al. 1962) and other research centers in which students of the same Fanger, such as Bjarne Olesen Wilkens (1947–present), an engineer from the Technical University of Denmark continuing the approach of the master, studied indexes of local discomfort and air quality, the sector which is now active.

The indexes of local discomfort depend on the uneven distribution of environmental parameters, in Tennenbaum et al. (1961) in confined spaces such as homes, hospital or hotel rooms, schools and buildings in general, but also trains, planes and automobiles. The causes of discomfort can be due to the temperature difference between the floor and the ceiling, the said radial asymmetry and leads to a non-homogeneous distribution of the surface temperature of the skin between the ankle and head, and the uncomfortable feeling of having hot/cold in the head, also in this case it is possible to express a percentage of dissatisfied in relation to the difference in temperature between the ankles and the head, for example, if the difference is between 5 and 8 °C to at least half of the subjects one declares as dissatisfied.

First studies and the evolution brought by P. O. Fanger in the discipline, the scientific literature has deepened the physiological basis of comfort, the human body and of the human dynamic thermoregulatory system, thermal comfort models and techniques, several studies on the adaptive approach, how resumed in a review paper van Hoof (2008), Djongyang et al. (2010), Frontczak and Wargocki (2011), and Mishra and Ramgopal (2013).

2.4 The Adaptive Thermal Comfort

In last decade of XX century, the last frontier in the study of Thermal Comfort is the study of Adaptive Thermal Comfort (Olesen and Parsons 2002; Brager and de Dear 1998; Schweiker et al. 2012; Humphreys and Hancock 2007; Halawa and van Hoof 2012), an approach that takes into account the variations that the individual shall adopt, in any case, even in a condition of neutral PMV, to feel in comfortable condition. This approach takes into account the dynamic variation of internal and external environmental conditions, and the individual.

The latest version of ASHRAE 55-2020 includes a separate adaptive mode for determining acceptable thermal conditions in occupant-controlled naturally conditioned spaces and local discomfort upgrade.

Already in 2008, J. van Hoff asked questions concerning the forty years of Fanger's thermal comfort model (van Hoof 2008) with a useful literature review. The observations concern the adjusting of the input model(s) and validity of the application intervals in climatic conditions different from those of the west and tropical climates, such as Jakarta (Karyono 2000). J. van Hoff also shows an application of the PMV model for semi-outdoor environments used by Potter and de Dear (2000) and the need to extend the models to different genders, ages and disablements. In 2008, van Hoof concluded by saying,

> The current PMV model is used for assessing and predicting thermal comfort in indoor spaces, often offices (...) The many input parameters of the comfort equation make PMV an index with great potential for the computerized 21st century in which occupants have increased demands regarding the indoor environment. In combination with means for individual control, engineers should be able to create indoor environments that provide nearly 100% acceptability and comfort to all, based on the work of Fanger.

A new perspective concerns the interaction with Building Automation and Control systems (BACs) for Heating, Ventilation, and Air-Conditioning systems (HVACs). An unusual approach is that of Kim and Ham (2023) that evaluates personal thermal comfort based on facial expression.

The other field of study concerns the extension of the entities and individuals, not only healthy adult men and women, but also sick people, the elderly, children, and not only the mental perception, but also with the cognitive information and with results that can be useful for design.

Finally, climate change will require us to assess how thermal comfort will change in air-conditioned and non-air-conditioned indoor environments. Presumably, there will be an adaptation to extreme conditions for the human body in conditions of heat stress, in relation to circadian rhythm, particularly during the night hours.

This book is a contribution in this direction.

2.5 History of Outdoor Thermal Comfort Indexes

The evaluation of the Outdoor Thermal Comfort indexes and Heat Stress Indexes is also worthy of consideration. Although the object of the evaluation concerns the evaluation of thermal comfort, or energy exchanges between the human body and the environment and the related sensations, unlike indoor studies that derive from studies on the physiology of the human body and architecture, in the outdoor field most of these indexes are developed within the studies on climate and bioclimatology. In addition, they are indexes that express the heat stress of the human body in particular conditions (e.g. pedestrian comfort, or tourist comfort, or during heat waves) and discomfort. The models refer to the physiology of the human body, energy balance equations and empirical evaluations that relate outdoor climate data with the possible judgment of the individual. The indexes are proposed by the National Weather Bureaus themselves of the different countries, or similar institutions, with the aim of identifying alert thresholds for extreme weather conditions, in particular, the Heat Wave or Cold Wave.

The first index was proposed by E. C. Thom, a researcher working at the 'Office of Climatology' of the United States Weather Bureau. Thom sought a useful tool to evaluate the discomfort of people and, based on experiences in the field of indoor air conditioning, in 1959 published an article (Thom 1959) proposing the *Discomfort Index* wherein two variables are coordinated by means of an empirical formula: outdoor air temperature and relative humidity. The Discomfort Index was an immediate success in the sectors of climatology, epidemiology and urban climate. Despite the simplicity of the model and the absence of field observations, as P. O. Fanger had done, the Thom Index is still in use among the risk indicators.

In 1963, Baruch Givoni, based on his experience in indoor environments, proposed the Thermal Stress Index (TSI), which would be adapted in 2007 to evaluate *Body Heat Stress* (Erell et al. 2014), or the risk of collapse of the body due to the excess of energy exchanged between the human body and the outdoor environment. This is a different approach, in which the limit is defined not by the thermal sensation based on climatic variables, but by the amount of energy exchanged by the human body, with the limit of collapse for Heat Stroke.

Almost a decade after Earl Thom, Robert G. Steadman, a researcher at the Textile Research Center of Texas Tech University in the United States, in evaluating the effect of clothing and convective exchanges due to wind on people, proposes the *Indices of Wind-Chill of Clothed Persons*.

In the introduction to the 1971 article (Steadman 1971), he reports that

> Many environmental factors besides temperature have a bearing on subjective feelings of warmth or cold. More than 70 publications (Macpherson 1962) deal with attempts to combine these factors into a single index. Two of the most common indices of the combined effects of wind and low temperature are based on the work of Siple and Passel (1945) the windchill index (kcal m^{-2} hr^{-1}). (…) The present publication deals with some of the limitations and applies Siple's approach to extend the concept of windchill to clothed persons.

It is interesting to note that already in 1962, four years after the proposal of the Thom index, there was a literature review on outdoor indexes (Macpherson 1962).

For the record, already in the first post-war period Siple and Passel (1945) proposed a model based only on freezing temperatures, similar to the work of R. G. Steadman in the Textile Research Center.

In 1984, Steadman (1984) goes a step further and proposes the *Apparent Temperature* for the evaluation of the pedestrian exposition.

In 1979, on behalf of the Canadian Atmospheric Environmental Service (AES), Masterton and Richardson (1979) proposed the *Humidex* index, which makes it possible to measure the discomfort associated with humid conditions as is the case of the Canadian province of Ontario, characterized by a humid continental climate, with high temperatures and high humidity rates, in particular, southern Ontario with humid summers ranging from warm to hot as well as mild to cold winters. The Humidex index, measuring the level of discomfort in muggy conditions (high humidity and high temperature), is based on the Dew Point Temperature and would be adopted by the AES of Canada and other countries in other similar climatic zones. It should be noted that, for the first time, the article refers to computing using '*a computer program*'.

In the late nineties, Peter Höppe, researcher at LRZ University of Munich, Germany, on the basis of the MEMI model (Munich Energy-balance Model for Individuals) developed in the same university, proposes the *Physiological Equivalent Temperature* (Mayer and Höppe 1987; Höppe 1997, 1999). The PET redefines outdoor comfort studies because it is a non-empirical index that considers the conditions of people's clothing and activity. This is the first index similar to the Predicted Mean Value although, in this case, not validated and always refers to an '*ideal reference room*' as in the Equivalent Temperatures (see Chap. 11).

In 1990, Rothfusz (1990) for the Scientific Services Division (NWS Southern Region Headquarters) proposed the *Heat Index* (HI).

The next step came in 1995, when Robert D. Brown and Terry K. Gillespie of the University of Guelph, Ontario, Canada, published the book '*Microclimatic landscape design: creating thermal comfort and energy efficiency*' (Brown and Gillespie 1995). The book describes the different variables that affect the microclimate, according to an approach aimed at '*landscape design to improve the comfort in outdoor space*' and in which they adapt the thermal comfort index of indoor environments in outdoor environments, thanks to the COFMA model, then finalized as COMFA+.

At the end of the millennium, in 1999, perhaps in anticipation of the evolution of the climatic context due to Climate Change, the European research project Cooperation in Science and Technical development COST Action 730 addresses thermal comfort in urban areas.

The intergovernmental program European Cooperation in Science and Technology (COST) has existed since 1971 and funds the creation of an interdisciplinary research network that brings researchers and innovators together to investigate a topic of their choice for four years. The COST Action 730 '*Towards a Universal Thermal Climate Index UTCI for Assessing the Thermal Environment of the Human Being*' (www.cost.eu/actions/730/) (Jendritzky and Nübler 1981) is a research project launched in 2005 and concluded in 2009 involving 19 European partners plus Australia and New Zealand, and "*The final outcome was presented in a symposium with WMO. A big advantage is the fact that WMO clearly decided to publish the outcome of COST Action 730 as a WMO Guideline which assures the perception of UTCI as a suitable standard on an international level.*"

The result of the research is the Urban Thermal Comfort Index (UTCI), for which there is a specific website (www.utci.org). The UTCI index adopts the assumptions of the PET by analyzing in detail the model of the human body and the body energy balance with the outdoor environment, in different World Climate Zones. The peculiarity of the UTCI index, in fact, is that it can be adopted in different climatic, urban and rural contexts, and Köppen climate classification (Kottek et al. 2006), hot-dry or hot-humid or cold climates, etc. Together with the PET, it is the most widespread index for the evaluation of outdoor thermal comfort and urban bioclimate.

The twenty-first century sees the expansion of research and new indexes are added, for the most part according to the empirical approach of the Thom index and referring to the different climate zones and not to specific cities or territories. In 2004, the European Project RURSO adopted the *Actual Sensation Vote* (ASV) based on an onsite monitoring questionnaire (Nikolopoulou et al. 2001).

In 2005, Emmanuel (2005) proposed the *Relative Strain Index* for evaluation of the comfort of pedestrian standard urban environments in warm-humid cities (Sri Lanka).

In 2006, it was the turn of the *Environmental Stress Index* (ESI) developed in the military to assess stress in hot-dry and hot-humid climates (Moran and Epstein 2006). In 2011, another thermal index was introduced, the *Universal Effective Temperature* (ETU) (Nagano and Horikoshi 2011). In 2012, *Perceived Temperature* (PT) proposed by Staiger et al. (2012) and adopted by the German Weather Service (Burton et al. 2009), is applied in urban bioclimatology and for the design of bioclimatic maps on a regional and urban scale, to evaluate the impacts of climate change on thermal perception and on people's health.

It is interesting to note that, like climate change, the search for outdoor indexes is subsequent to that of indoor comfort indexes and has developed over the last thirty years.

In my opinion, there is a lack of robust validation of data based on the actual sensations of people, as in Fanger's experiment (if such a thing is even possible) and standardization based on the different applications: weather alerts, bioclimatology, design of nature and open space by landscapers, architects and urban planners. Future studies could investigate and further explore these aspects.

References

ASHRAE (2010) ASHRAE 55 thermal environmental conditions for human occupancy

Bedford T (1936) The warmth factor in comfort at work: a physiological study of heating and ventilation. Industrial health research board no. 76. HMSO, London

Bell PA (1981) Physiological, comfort, performance, and social effects of heat stress. J Soc Issues 37:71–94

Brager GS, de Dear R (1998) Thermal adaptation in the built environment: a literature review. Energy Build 27:83–96

Brown RD, Gillespie TJ (1995) Microclimatic landscape design: creating thermal comfort and energy efficiency. Wiley, New York

Burton I, Ebi KL, McGregor G (2009) Biometeorology for adaptation to climate variability and change. Biometeorology for adaptation to climate variability and change. Springer Netherlands, Dordrecht, pp 1–5

Collins KJ, Hoinville E (1980) Temperature requirements in old age. Build Serv Eng Res Technol 1(4):165–172

Djongyang N, Tchinda R, Njomo D (2010) Thermal comfort: a review paper. Renew Sustain Energy Rev 14:626–2640

DuBois D, DuBois EU (1916) Tenth paper a formula to estimate the approximate surface area if height and weight be known. Arch Intern Med XVII(6–2):863–871

Emmanuel R (2005) Thermal comfort implications of urbanization in a warm-humid city: the Colombo Metropolitan Region (CMR), Sri Lanka. Build Environ 40:1591–1601. https://doi.org/10.1016/j.buildenv.2004.12.004

Epstein Y, Moran DS (2006) Thermal comfort and the heat stress indices. Ind Health 44:388–398
Erell E, Pearlmutter D, Boneh D, Kutiel PB (2014) Effect of high-albedo materials on pedestrian heat stress in urban street canyons. Urban Clim 10:367–386. https://doi.org/10.1016/j.uclim.2013.10.005
Fanger PO (1967) Calculations of thermal comfort: introduction of a basic comfort equation. ASHRAE Trans 73:1–4
Fanger PO (1970) Thermal comfort-analysis and applications in environmental engineering. Danish Technical Press, Copenhagen
Frontczak M, Wargocki M (2011) Literature survey on how different factors influence human comfort in indoor environments. Build Environ 46:922–937
Gagge AP (1936) The linearity criterion as applied to partitional calorimetry. Am J Physiol 31:656–668
Gagge AP, Nishi Y (1976) Physical indices of the thermal environment. ASHRAE J 18:47–51
Gagge AP, Stolwijk JAJ, Nishi Y (1972) An effective temperature scale based on a simple model of human physiological regulatory response. ASHRAE Trans 77(part 1):247–262
Givoni B (1963a) Estimation of the effect of climate on man; development of a new thermal index. Technion, Israel Institute of Technology Building Research Station
Givoni B (1963b) The effect of climate on man: development of a new thermal index. Research report to UNESCO. Building Research Station, Technion, Haifa
Givoni B (1969) Man, climate and architecture. Elsevier Publishing Company Limited, New York
Givoni B, Pandolf RR (1973) Predicting heart rate response to work, environment and clothing. J Appl Physiol 34:201–204
Halawa E, van Hoof J (2012) The adaptive approach to thermal comfort: a critical overview. Energy Build 51:101–110
Höppe P (1997) Aspects of human biometerology in past, present and future. Int J Biometeorol 40:19–23. https://doi.org/10.1007/BF02439406
Höppe P (1999) The physiological equivalent temperature—a universal index for the biometeorological assessment of the thermal environment. Int J Biometeorol 43:71–75. https://doi.org/10.1007/s004840050118
Houghten FC, Yaglou CP (1923a) Determination of the comfort zone. ASHVE Trans 29
Houghten FC, Yaglou CP (1923b) Determining equal comfort lines. J Am Soc Heat Vent Eng 29:165–176
Houghten FC, Yaglou CP (1953) Determining equal comfort lines. Transl Am Soc Heat Vent Eng 29:165–176
Humphreys MA, Hancock M (2007) Do people like to feel 'neutral'? Exploring the variation of the desired thermal sensation on the ASHRAE scale. Energy Build 39:867–874
ISO (2005) ISO 7730 ergonomics of the thermal environment—analytical de-termination and interpretation of thermal comfort using calculation of the PMV and PPD indices and local thermal comfort criteria
Jendritzky G, Nübler W (1981) A model analysing the urban thermal environment in physiologically significant terms. Arch Meteorol Geophys Bioclimatol Ser B 29:313–326. https://doi.org/10.1007/BF02263308
Karyono TH (2000) Report on thermal comfort and building energy studies in Jakarta—Indonesia. Build Environ 35:77–90. https://doi.org/10.1016/S0360-1323(98)00066-3
Kim BY, Ham Y (2023) Personal thermal comfort modeling based on facial expression. J Build Eng 75:106956. https://doi.org/10.1016/j.jobe.2023.106956
Kottek M, Grieser J, Beck C, Rudolf B, Rubel F (2006) World map of the Koppen–Geiger climate classification updated. Meteorol Z 15:259–263
Macpherson RK (1962) The assessment of the thermal environment. A review. Occup Environ Med 19:151–164. https://doi.org/10.1136/oem.19.3.151
Masterton JM, Richardson FA (1979) Humidex: a method of quantifying human discomfort due to excessive heat and humidity. Atmospheric Environment Service, Canada

Mayer H, Höppe P (1987) Thermal comfort of man in different urban environments. Theor Appl Climatol 38:43–49. https://doi.org/10.1007/BF00866252
Mazria E (1979) Passive solar energy book. Rodale Press, Emmaus, PA
Mishra AK, Ramgopal M (2013) Field studies on human thermal comfort—an overview. Build Environ 64:94–106
Moran DS, Epstein Y (2006) Evaluation of the environmental stress index (ESI) for hot/dry and hot/wet climates. Ind Health 44:399–403. https://doi.org/10.2486/indhealth.44.399
Nagano K, Horikoshi T (2011) Development of outdoor thermal index indicating universal and separate effects on human thermal comfort. Int J Biometeorol 55:219–227. https://doi.org/10.1007/s00484-010-0327-z
Nikolopoulou M, Baker N, Steemers K (2001) Thermal comfort in outdoor urban spaces: understanding the human parameter. Sol Energy 70:227–235. https://doi.org/10.1016/S0038-092X(00)00093-1
Olesen BW, Parsons KC (2002) Introduction to thermal comfort standards and to the proposed new version of EN ISO 7730. Energy Build 34:537–548
Olgyay V (1963) Design with climate: bioclimatic approach to architectural regionalism. Princeton University Press, Princeton, NJ
Potter JC, de Dear RJ (2000) Field study to calibrate outdoor thermal comfort index. In: Biometeorology and urban climatology at the turn of the millennium, pp 315–320
Rohles F, Johnson MA (1972) Thermal comfort in the elderly. ASHRAE Trans 78(Part I):131–137
Rohles FJ, Levins R (1971) The nature of thermal comfort for sedentary man. ASHRAE Trans 77:239–246
Rothfusz LP (1990) The heat index "equation" (or, more than you ever wanted to know about heat index)
Schweiker M, Brasche S, Bischof W, Hawighorst M, Voss K, Wagner A (2012) Development and validation of a methodology to challenge the adaptive comfort model. Build Environ 49:336–347
Siple PA, Passel CG (1945) Measurement of dry atmospheric cooling in subfreezing temperature. Proc Am Philos Soc 89:177–199
Sohar E, Tennenbaum DJ, Robinson N (1962) A comparison of the cumulative discomfort index (Cum DI) and cumulative effective temperature (Cum ET), as obtained by meteorological data. In: Tromp SW (ed) Biometeorology. Pergamon Press, Oxford, pp 395–400
Staiger H, Laschewski G, Grätz A (2012) The perceived temperature—a versatile index for the assessment of the human thermal environment. Part A: scientific basics. Int J Biometeorol 56:165–176. https://doi.org/10.1007/s00484-011-0409-6
Steadman RG (1971) Indices of windchill of clothed persons. American Meteorological Society
Steadman RG (1984) A universal scale of apparent temperature. American Meteorological Society
Tennenbaum J, Sohar E, Adar R, Gilat T (1961) The physiological significance of the cumulative discomfort index. Harefuah 60:315–319
Thom EC (1959) The discomfort index. Weatherwise 12:57–61. https://doi.org/10.1080/00431672.1959.9926960
van Hoof J (2008) Forty years of Fanger's model of thermal comfort: comfort for all? Indoor Air 18:182–201. https://doi.org/10.1111/j.1600-0668.2007.00516.x
Vernon HM, Warner CG (1932) The influence of the humidity of the air on capacity for work at high temperatures. J Hyg (Lond) 32(3):431–463
Webb CG (1959) An analysis of some observation of thermal comfort in an equatorial climate. Br J Ind Med 16(III):297–310
Winslow C-EA, Herrington LP, Gagge AP (1937) Physiological reaction of the human body to various atmospheric humidities. Contribution no. 16 from the John B. Pierce Laboratory of Hygiene, New Haven, CT, pp 288–299
Yaglou CP, Minard D (1957) Control of heat causalities at military training centers. Am Med Assoc Arch Ind Health 16:302–316

Chapter 3
Ergonomics of the Thermal Environment. Human Body and Clothes

Abstract The Human Body constantly exchanges energy and mass with the environment, to ensure its own metabolism. The study of the physiology of the human body used to determine the terms of the Heat Balance Equation. This chapter describes the indexes that are used to determine: the Heat Balance Equation, the Clothes resistance and how the human body perceives the Indoor Air Quality.

3.1 What About Thermal Comfort

The condition of comfort of an environment depends on the interaction between multiple factors physical, physiological and psychological. The World Health Organization (WHO 1999) defines the comfort such as:

> Health is a state of complete physical, mental and social well being and not merely the absence of disease or infirmity. (WHO 1999)

And the ASHRAE 55 standard defines comfort

> that condition of mind which expresses satisfaction with the thermal environment and is assessed by subjective evaluation. (ASHRAE 55 2004)

The study of comfort must therefore take into account the variables:

- *Physiological* due to metabolic activity of the Human Body and its modelling of physiological processes;
- *Physical*, of environmental, indoor or outdoor, in which we find the human body affecting exchanges between the human body and environment. Frontczak and Wargocki (2011) report a literature survey on the different factors that affect indoor comfort environments;
- *Psychological and Cultural*, not directly measurable, and that have an influence on the perception of the feeling of comfort, or perceptual coding of what happens in the exchange between the human body (physiology) and environment (physical).

K. Fabbri, *Thermal Comfort Perception*, https://doi.org/10.1007/978-3-031-52610-7_3

Thermal Comfort is the result of a *process* due to exchanges of mass and energy between the person and the environment (modelled according to the laws of thermodynamics) and of the Cognitive Encoding (or Knowledge Code) of the person regarding this exchange.

Indeed, Lucrezio in "De rerum natura" wrote that

> Eyes cannot know the nature of things

emphasizing that it is the reason, and not in accordance with, the study of reality; and then subsequently Poincarè

> This tool, which relates to all, that of which we use instinctively, is nothing more than our body. It is in reaction to our body that we situate foreign objects, and the only spatial relationships of these objects that we can represent are the relationships with our body. It is our body that we need, so to speak, as a coordinate system. (Poincarè 1905, cited in Bellone 2011, p. 92)
>
> questo strumento, al quale rapportiamo tutto, quello del quale ci serviamo istintivamente, altro non è che il nostro corpo. È in reazione al nostro corpo che situiamo gli oggetti esteri, e le uniche relazioni spaziali di questi oggetti che ci possono rappresentare sono le relazioni con il nostro corpo. È il nostro corpo che ci serve, per così dire, da sistema di assi coordinati. (Poincarè 1908, cited in Bellone 2011, p. 92)

Regarding the Cognitive Encoding (Knowledge Code) the theme includes the study on the construction of the world through the sense organs, related to neurological studies, or philosophical epistemology, which is beyond the areas of the study of comfort and of this book. The scientific literature on Thermal Comfort limited to the study of the *Related Windchill Factor* directly to the subject.

The study of comfort is part of the discipline of *Ergonomics*, defined

> The scientific study of the efficiency of man in his working environment. (Oxford English Dictionary)

and as:

> the scientific discipline concerned with the understanding of the interactions among human and other elements of a system, and the profession that applies theory, principles, data and methods to design in order to optimize human well-being and overall system performance. (International Ergonomic Association)

The general principles of ergonomics put the man in the center, all the design process of products, systems and services take into account the human scale, the characteristics (geometry, physiological, behavioral, psychological, dynamics, etc.) of the human body in relation to the *Environmental Contest*, understood as the physical attributes of the organizational context and social.

Ergonomics defined as directing the design process, understood as a series of activities of continuing correction of deficiencies. The process shall take account of the human tasks and interaction.

The ergonomic approach is, therefore, oriented towards the designer, among them also the architects and anyone who designs buildings and Heating Ventilation and Air-Condition Systems. In these cases, we speak of "*Ergonomics of the thermal environment*" (ISO 11339) which has as its principle the

> Ergonomics investigation of the thermal environments involve and understanding a number of underlying concepts and principles concerning human response to thermal environment methods.

The Ergonomics of thermal environment and Comfort have similar definitions, the first related to the engineering aspects, while the second refers to a discipline that considers not only the human body, but also other personal interactions, psychological and perceptual, concerning the feeling and judgment that the subject expresses with respect to the environment.

In turn, the subject can have distinct characteristics. Standards refer to a healthy adult man or woman who expresses a positive opinion in terms of thermal neutrality.

Finally Thermal Comfort takes into account the processes of acclimatization

> physiological change, occurring within the lifetime of an organism, which reduces the strain caused by experimentally induced stressful changes in particular climatic factors, (Glossary of terms for thermal physiology 1987)

- to ensure Homeothermy

> The pattern of temperature regulation in a tachymetabolic species in which the cyclic variation in core temperature, either nychthemerally or seasonally, is maintained within arbitrarily defined limits (± 2 °C) despite much larger variations in ambient temperature. (Glossary of terms for thermal physiology 1987)

To study comfort it is necessary to know the size that defines the Thermal Comfort, those relating to the Human Body and Clothes (this chapter) and the indices of thermal sensation (next chapter).

This chapter describes the modelling of physiological aspects that determine the energy exchanges between Human Body and Environment, according to the Comfort Equation.

3.2 The Human Body (Physiology)

Physiology is the discipline that studies the functions of living organisms, the conditions and the laws that determine and regulate vital phenomena. In particular for the Human Body, the physiological systems that allow you to perceive the outside world are:

- The Thermo-Regulatory System of the human body, such as exchange receptors between the human body and profile, which can ensure the proper metabolic activity;
- Respiratory system, larynx, pharynx and lungs, those receptors of air quality and pollution load;
- The human eye, which is the receptor for sense of sight, shapes and colors, but also the quality of air and temperature;
- The human ear, which the sound receptor and of the sense of balance.

To each of these organs corresponds an assessment of Comfort, which helps to make the judgment on the conditions of well-being. The four categories of comfort are:

- *Thermal Comfort*, as regards the preservation of the conditions of homeothermy of the human body in the variation of the physiological activities and physical conditions of the environment, and therefore the design of buildings and HVAC (Heating, Ventilation and Air-Conditioning);
- *Indoor Air Quality* (IAQ), which concerns the concentrations of pollutants, bacteria, viruses, etc. in the air compared to the risk of irritation, allergies, illnesses, etc.;
- *Lighting Comfort*, which concerns the correct lighting and vision of the environment, and therefore the natural and artificial lighting;
- *Acoustic Comfort*, which covers both the absence of noise, listening to music, conferences, etc.

Due to external stress factors the human body can be the subject to disorder and diseases, this happens for several factors borne from the nervous system, the immune system and the endocrine system. Factors of Disease/Disorder may include: the external stress factor, concerning the human senses (heat/cold, smell, noise, etc.), the effects of external conditions (fatigue, irritation, etc.) or psychological effects (anxiety, depression, etc.). Other factors have a direct impact on the nervous system, endocrine and immune systems, causing damage such as: (slight) irritation, allergy, (acute) legionella, infectious diseases, etc. (chronic) infections, cancer and other debilitating diseases.

Levels of Disease/Disorder for each sense organ may vary from state of Discomfort, Systemic effect Allergic or Stinging, Infectious, until the conditions of Toxic Chronic Effect (Bluyssen 2009).

For example, the skin, which sense organ receptors thermo-regulatory system, can have feelings of Discomfort if the environment is warm/cold or sweat/draught, may be the subject of allergies such as dermatitis, red-skin, itchy, or have chronic conditions such as radiation-related disorders. Similarly, the respiratory system (respiratory tract) based on the IAQ can have conditions of discomfort (cough, shortness of breath), allergic (asthma, bronchitis), infection (bacterial, viral or fungal) or toxic chronic effect (damage and/or tumors).

In conclusion, the human body interprets and reacts to the outside world according to its own sense organs. In the case of thermal comfort, the sense organ is given by the receptors of thermoregulation, the skin, the activity of respiration and metabolic processes that maintain the internal temperature of the human body always at about 37 °C. The human body temperature can decrease or increase in relation to physical activity, the thermal exchanges, and under conditions of disease or disorder.

Some lines of research, in the medical field, focus on the relationship between the physiological aspects (e.g. Yao et al. 2009; ASHRAE 1993; Huizenga et al. 2001; Jones and Ogawa 1992) the relationship between breathing, heart rate, etc. In this book mentioned only Human Thermal Physiological models.

3.2.1 Human Thermal Physiological Models

The study of the Thermal Comfort model of the human body modelled from the numerical point of view. There are several thermal Human physiological models (Cheng et al. 2012), which have evolved over time

- One-node model, empirical, proposed in 1971 by Givoni and Goldman (1971);
- Two-node model, proposed in 1971 by Gagge et al. (1971) revised in 1977 by Azer and Hsu (1977);
- 40-layer finite difference skin model of 1991 by de Dear and Ring (1990);
- Multi-node thermal model, of 1999 by Fiala et al. (1999, 2010);
- Multi-node thermal model and UC Berkeley Comfort Model, in 2001 described by Huizenga et al. (2001).

The model has been adopted by some authors according to the psychological aspects, in Psychological thermal comfort model:

- Dynamic Thermal Stimulus Mode and whole-body thermal state, of 1993 by de Dear et al. (1993);
- Transient thermal sensation prediction model, of 2003 by Guan et al. (2003);
- Thermal Comfort Zone by Equivalent Temperature (t_{eq}) of 2003 by Nilsson and Holmér (2003);
- UC Berkley Thermal Comfort, of 2005 by Arens et al. (2006).

The numerical approach of thermo-physiological models, in relation to the exchange of energy with the indoor environment, was investigated by Schellen et al. (2013) who propose the model ThermaSEM (van Marken Lichtenbelt et al. 2007), in which the human body is modeled as a set of 18 cylinders (trunk, neck, arm, forearm, etc.) and one sphere (head) to each from which is applied the mass balance (Fig. 3.1).

3.2.2 Human Sense and Thermal-Regulatory

The human senses enable us to perceive the outside world and ensure the conditions of equilibrium between the human body and exterior, as regards the activities of movement, controlled by the muscles and voluntary movements, sense of balance, receptors of the cardiovascular circulation, and receptors of digestion.

The main organ tasked with the transposition of the outside world is the brain through which the information comes from the Nervous System, Endocrine System and Immune System.

The sense organ of the Thermo-regulation system is the skin (skin).

The Human Body boundary is the skin that allows you to feel the "*pressure sensed*", due to mechanical pressure (touch) and the light (light), the "*temperature*

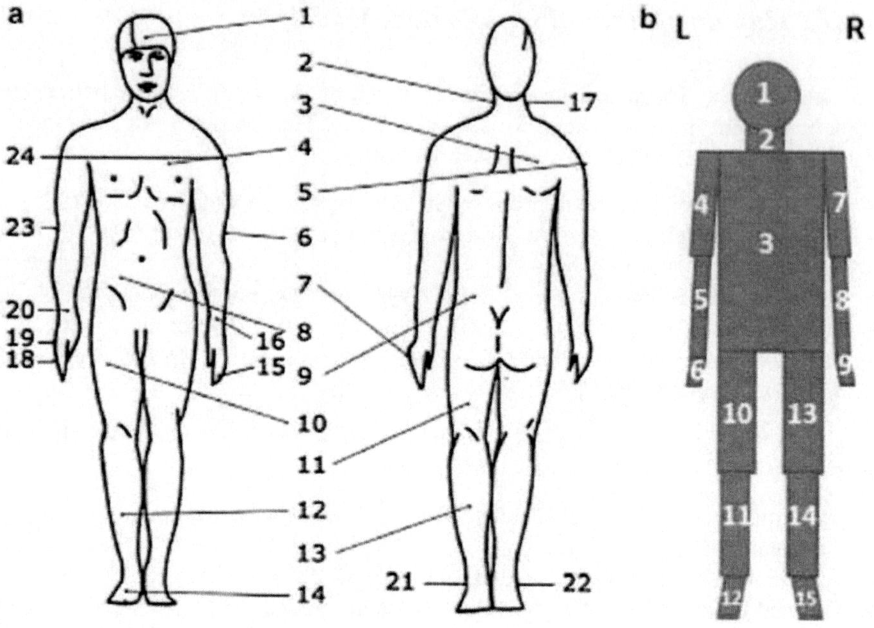

Fig. 3.1 **a** Measurement sites skin temperature. **b** Schematic representation of body parts to access local thermal sensation

sensitivity senses" (hot/cold) and the "pain sense" (nocicpetor). The skin is constituted from *Epidermis*, consists of dead cells in direct contact with the exterior, and the *Dermis*, the living part with nerve receptors and glands.

The Pain Senses (nocicpetor) are neuron terminals that allow the brain to signal the notions of pain, due, for example, to the cutting of the skin (cutaneous pain), the deep tissues (somatic pain) or the internal organs (visceral pain).

The *Temperature Senses* distinguish the receptors of hot (warm), for the temperatures ranging from 38 to 43 °C and cold (cold), for temperatures between 15 and 34 °C. The cold receptors can sense the decrease of 0.004 °C per second and the heat receptors that have a speed of perception of temperature of 0.001 °C per second.

The *Eccrine Gland* (or Thermoregulatory Sweat Gland) is composed of 99% water and 1% from the electrolyte, amino acids, and other vitamins which are distributed throughout the body and allow secretions of sweat, or lipid materials, triglyceride and cause a bad smell (smell of sweat).

The mechanism of heat balance and temperature control is regulated by the *Hypothalamus*, which is activated: by vascular dilatation and sweat in the case of excess heat, or by vascular narrowing and shivering in cold. The activation criterion of the hypothalamus follows the equation:

$$\text{Energy Supply} + \text{Energy produced (body)} = \text{Energy Removed}$$

The Human Senses that interest the IAQ, are:

- the *Human Nose*, that thanks to the *Olfactory Epithelium*, located inside the nasal cavity, allows recognition of odors and irritation due to alterations of the nasal mucosa; and
- the *Human Respiratory Tract*, which includes the paranasal cavities (sinuses), the pharynx (throat) and the oral cavity, or the part of the respiratory system in direct contact with the outside, through which air exchange takes place between the internal and external; and then the exchange of mass—*water vapour, air, and gas molecules—and energy—according to the temperature of the air entering and exiting the body and the latent heat of vaporization of the breath.*

The system for thermoregulation allows the temperature of the human body to maintain at about 37 °C with change of ± 0.5 °C, and reacts by increasing the temperature, in the event given intense metabolic activity such as infections, influences and other febrile phenomena. The temperature control can be performed in two ways:

- *Vasomotor*, against the heat and the cold, activated by thermal receptors;
- *Behavioral*, with the increase or decrease in metabolic activity, through mechanical work, clothing, etc.

The Thermoreceptors of the human body act like thermostats and can operate according to the principle "On/Off": the temperature follows a corrective action instantaneous.

In the case of excess heat (Heatstroke) when the internal body temperature exceeds 40 °C is because it is no longer able to transfer heat to the outside, the temperature controllers "detach" the correction systems so as to cause fainting, which is a condition in which is guaranteed "only" the basal metabolic rate (if the internal temperature exceeds 41 °C there can be permanent damage, and over 42 °C death can occur).

In the case of the phenomena of freezing the process of vasoconstriction reduces the temperature of the skin and the core of the human body, if this temperature falls below 33 °C, the body stops react by trying to raise the temperature by producing shivering, below this temperature loss of consciousness, and if the body temperature falls below 25 °C death. In the process of freezing the production of the mechanical work of the muscles, is such as to produce a sensation of heat, which can become so unbearable that leads to the elimination of clothing accelerating freezing.

To the On/Off control system you add other control strategies, proportional type (P), by modulating the reaction in relation to the temperature variation, proportional and integral (P + I) or proportional, integral and derivative (P + I + D) where the amount of adjustment is continuous and free of inertia and acts not based on the value of the initial temperature, it acts in time to ensure the future value, and then follows the evolution of the phenomenon, and its speed with which it evolves.

The corrective actions to increase or decrease the temperature refer to the process of increase/decrease of temperature with respect to the condition of the set-point, the same mechanism is used in thermostats with P + I control, the heating system is activated and provides energy until reaching a temperature exceeding that set as a set-point in the environment, ensuring the desired temperature in relation to the inertia of the ambient air.

The control system of the type *Behavioral*, is the production/heat dissipation through the *Mechanical Work* of the muscles.

For example, to counteract the sensation of cold when the vasoconstriction is not sufficient, the human body acts producing the work of involuntary muscle, like *shivering* which consists in the activation of muscles and given that the mechanical work of the shivering is null there is an increase of thermal energy within the human body.

In addition it is possible to add to the mechanical work not zero, producing work with the human body, for example by rubbing the palms of the hands against one another, or hopping on the spot, actions that result in the production of kinetic energy to the outside of the human body, result in an increase of internal energy of the human body.

Thermoregulation against heat consists of the production of sweat, in the reduction of motor activity, also due to the feeling of fatigue, and increased ventilation by breathing.

The exchange of energy can be increased with the production of sweat, through the sweat glands, which can produce up to 2 or 3 L of sweat per hour. The sweat, being composed mostly of water, form a surface film to the liquid state that with the increase of skin temperature and air that touches the body, involves the transfer of energy through the process of vaporization of the water vapor, about 2.43 kJ per g of steam produced.

The energy exchange through steam is also achieved through breathing, which allows an exchange of heat by evaporation of about 40 g per hour corresponding to about 28 W, for this reason, when it is hot, we tend to pant.

The study of the well-being conditions for cold workplaces, such as refrigerators, or hot or severely-hot workplaces, such as foundries, steel plants or the like, takes into account the dynamic behavior of the body over time, worktime is determined by the time intervals during which the thermoregulatory vasomotor functions ensure equilibrium conditions in the energy balance of the human body.

3.2.3 The Core Temperature and Skin Temperature Model

In the *Core-Skin model* the human body is divided into two:

- An inner zone maintained at a constant internal temperature of 37 °C ± 0.5 °C, said to be "Core-Temperature",
- And an outside area of the skin (Skin-Temperature) that can vary from 45 °C, for the area of the forehead when you have a fever, up to 17 °C, again for the front and to the 4 °C for the fingers of the hand.

Below these temperatures the temperature control system is unable to ensure the blood flow to increase the temperature and starts the freezing process. The minimum and maximum temperatures that the human body can endure depend on two factors:

- The outside temperature, which varies with a range between – 5 °C below which there is the freezing and 45 °C above which there is heat stroke, the external temperature of the skin can vary from a maximum 45 °C (fever of the forehead) to a minimum of 4 °C (fingers) or 17 °C (forehead);
- The relative humidity, which expresses as a percentage of the content of water vapor in the air which must be between 40 and 70% for the proper exchange through the evapotranspiration, at below 40% there is the feeling of dry mouth and skin and the risk of dehydration. Above the 70% up to 80–90% increases the feeling of heat or cold and the body tries to increase the exchange of heat through sweating but finds the obstacle of external air is already saturated, and thus increases the feeling of warmth.

3.3 The Modelling of the Human Body: Body Heat Balance Equation

The human body can be modeled as an open system that exchanges energy through the boundary to maintain an internal temperature of about 37 °C.

The heat produced by the human body is due to the basal metabolism, the basis that allows the body to be kept alive, and from muscle metabolism or by the mechanical work of the human body.

The energy balance of the human body consists of an open system, exchanging energy and matter (food, breathing, evapotranspiration) with the outside through its boundaries (*Human Body Area, Skin*) (Fig. 3.2).

The objective of the thermoregulation system of the body is to maintain an essential constant internal body temperature.

The HEAT BALANCE EQUATION, in the formulation by Fanger (1970) is given the following formula:

$$H - E_d - E_{sw} - E_{re} - L = K = R + C \tag{3.1}$$

where

- H is the internal heat production in the Human Body
- E_d is the heat loss by water vapor diffusion through the skin
- E_{sw} is the heat loss by evaporation of sweat from the surface of the skin
- E_{re} is the latent respiration heat loss
- L is the dry respiration heat loss

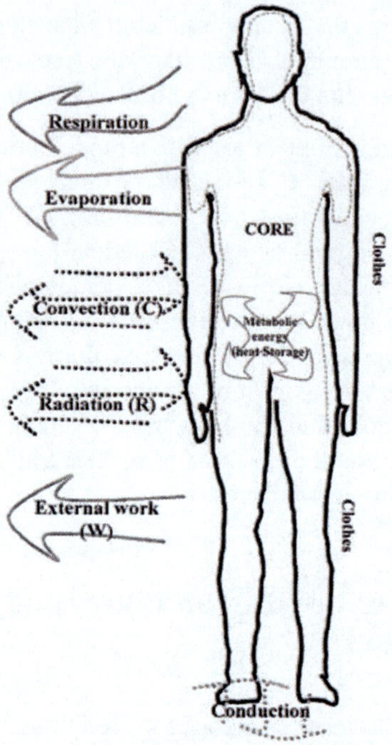

Fig. 3.2 Schematic energy exchanges of the human body. Remaking design

K is the heat transfer from the skin to the outer surface of the clothed body (conduction through the clothing)
R is the heat loss by radiation from the outer surface of the clothed body
C is the heat loss by convection from the outer surface of the clothed body.

The formula consists of a "double equation", to the right and to the left of the term (K), to the left heat exchanges due to the metabolic activity internal to the human body (Core-Skin), the thermal exchanges between the right boundary of the human body and environment (Skin-environment).

The following paragraphs describe the meaning of each term in the equation.

The formulation of Fanger's heat Balance Equation can be written explaining the metabolic activity, as in the formula of Annex of ISO 11399, following formula:

$$M - W = C + R + E + C_{res} + E_{res} + S \tag{3.2}$$

where

M is the metabolic power minus external work achieved (W)
W is the external work achieved
C is the heat losses from the skin by convection
R is radiation
E is evaporation (sweat)
C_{res} is the heat losses due to respiration by convection
E_{res} is the heat losses due to respiration by evaporation
S is the rate of heat storage.

The *Glossary of terms for thermal physiology* (1987) report another definite of the *Body Heat Balance Equation,* as:

> A mathematical expression that describes the net rate at which a subject generates and exchanges heat with its environment. (First Law of Thermodynamics)

The dimension of the equation and its terms respectively, are in Watts (W) but are also expressed in relation to unit area of body surface (W/m^2) or to unit body weight (W/kg):

$$S = M - W - E - C - K - R \tag{3.3}$$

where

S storage of heat in body (positive = increase in body heat content; negative = decrease in body heat content).
M metabolic energy transformation (always positive in a living organism).
W work, positive (= useful work accomplished; negative = mechanical work absorbed by body: work, negative).
E evaporative heat transfer (positive = evaporative heat loss; negative = evaporative heat gain).
C convective heat transfer (positive = transfer to the environment; negative = transfer into the body).
K conductive heat transfer (positive = transfer to the environment; negative = transfer into the body).
R radiant heat exchange (positive = heat transfer to the environment; negative = heat absorption by the body).

The equation is designed to describe the prevailing physiological situation of a homeothermic (e.g. human) subject.

In all equations the variables are expressed in Watts per square meter of Body Surface Area (A_{Du}) measured in W/m^2.

3.3.1 Internal Heat Production—Metabolism (met)

The metabolic activity of the human body depends on the state of health, age, sex, and type of diet, and is divided into:

- Basal energy metabolism, calorie requirements necessary for the maintenance of vital functions in conditions of rest or thermal neutrality (temperature 29–31 °C when naked, 25–29 °C if clothed);
- Metabolism related to body position (lying, sitting, standing);
- Metabolism linked to the type of work/activities carried out (rest, physical activity light, medium, heavy, etc. …) (Fig. 3.3).

The *Internal Heat Production* in the Human Body is a magnitude that depends on *Human Body Metabolism* (M) and of *External Mechanical Power* (W), eexpressed per unit surface area of the human body.

The metabolism is given by

$$M = H + W \tag{3.4}$$

The External Mechanical Power (W) refers to the effective mechanical power external work useful work energy spent in overcoming external mechanical forces on the body, for most activities it can be neglected (definition by ISO 13731), if it can be expressed according to the External Mechanical Efficiency η:

$$\eta = \frac{W}{M} \tag{3.5}$$

therefore

$$H = M \cdot (1 - \eta) \tag{3.6}$$

The value of η is, in most situations, equal to zero, it is considered only in the case of intense physical activity such as uphill races, carpentry works, and heavy activity.

The mechanical work due to physical activity performed by the subject is work when it "exits" from the human body system, that is, work that the human body carries

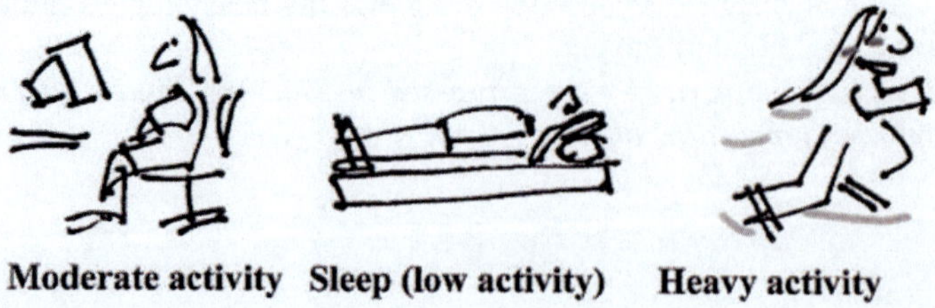

Fig. 3.3 Metabolic activity

out is positive, such as running or lifting an object, while if it is the work required to ensure the conditions for energy metabolism, the work environment carried out by man, this is negative work.

The External Mechanical Efficiency (η) expresses the efficiency of the human body, which is typically zero, or equal to 20% or for each unit of metabolic energy supplied is obtained 0.20 units of mechanical work.

The relationship is not linear, for example, if an *Ergometer*, the kind of medical exercise bikes used for strength testing and to measure the activity of the human body aerobic or anaerobic (Conconi test), work is performed equal to 20 W, the relationship of the lightest bikes, corresponding to the freewheeling, metabolic energy required is 200 W; but if you must perform work of 40 W, the metabolism is 260 W with a yield of 15%.

A man of about 80 kg and 1.80 m tall, that accomplishes 20 W of mechanical work, develops metabolic energy of 200 W compared to 100 W in the state at rest.

Metabolic Rate

The metabolic activity (M), or Metabolic Rate is expressed in W/m^2, and is referred to as m^2 of human body surface, or directly in ***met***.

The value of the Metabolic Rate of ***met*** is provided directly by the Standard, as shown in the following tables, or can be calculated in accordance with the ISO 8996 "*Ergonomic of the thermal environment—Determination of metabolic rate.*" In the case of elderly or younger people the values can be reduced.

The method of calculation and accuracy are 4:

- *Level 1, screening*: a simple and fast method to determine the Metabolic Rate according to the activity carried out, according to occupation, or according to the kind of activity (e.g. Table 3.1);

Table 3.1 Table of metabolic rate (by ISO 7730 Annex B—metabolic rates of different activities)

Activity	Metabolic rates	
	W/m^2	Met
Reclining	46	0.8
Seated, relaxed	58	1.0
Sedentary activity (office, dwelling, school, laboratory)	70	1.2
Standing, light activity (shopping, laboratory, light industry)	93	1.6
Standing, medium activity (shop assistant, domestic work, machine work)	116	2
Walking on level ground		
2 km/h	110	1.9
3 km/h	140	2.4
4 km/h	165	2.8
5 km/h	200	3.4

- *Level 2, observation,* which takes into account the baseline metabolic rate and metabolic rate for the body posture, the calculation of the metabolism takes account of the values tabulated for various scenarios of activities (e.g. Table 3.2);

Table 3.2 Classification of metabolic rate by category (by ISO 8996 Annex A—Table A.2)

Class	Average metabolic rate (with range in brackets)		Examples
	W/m^2	W	
0 Resting	65 (55–70)	115 (100–125)	Resting, sitting at ease
1 Low metabolic rate	100 (70–130)	180 (125–235)	Light manual work (writing, typing, drawing, sewing, book-keeping); hand and arm work (small bench tools, inspection, assembly or sorting of light materials); arm and leg work (driving vehicle in normal conditions, operating foot switch or pedal). Standing drilling (small parts); milling machine (small parts); coil winding; small armature winding; machining with low power tools; casual walking (speed up to 2.5 km h^{-1})
2 Moderate metabolic rate	165 (130–200)	295 (235–360)	Sustained hand and arm work (hammering in nails, filing); arm and leg work (off-road operation of lorries, tractors or construction equipment); arm and trunk work (work with pneumatic hammer, tractor assembly, plastering, intermittent handling of moderately heavy material, weeding, hoeing, picking fruits or vegetables, pushing or pulling lightweight carts or wheelbarrows, walking at a speed of 2.5–5.5 km h^{-1}, forging)
3 High metabolic rate	230 (200–260)	415 (360–465)	Intense arm and trunk work; carrying heavy material; shoveling; sledgehammer work; sawing; planning or chiseling hard wood; hand mowing; digging; walking at a speed of 5.5–7 km h^{-1}. Pushing or pulling heavily loaded hand carts or wheelbarrows; chipping castings; concrete block laying
4 Very high metabolic rate	290 (> 260)	520 (> 465)	Very intense activity at fast to maximum pace; working with an axe; intense shoveling or digging; climbing stairs, ramp or ladder; walking quickly with small steps; running; walking at a speed greater than 7 km h^{-1}

- *Level 3, analysis*, the metabolic rate is determined from Heart Rate recording over a representative period;
- *Level 4, expertise*, the more sophisticated level that takes account of the oxygen consumption measured over a short period (10–20 min), or a doubly labelled water method at characterizing the average metabolic rate over much longer periods (1–2 weeks), and finally the most accurate method: direct calorimetry method.

The relationship between Heart Rate (HR) and Metabolic Rate is given by the equation:

$$HR = \mathrm{HR}_0 + RM \cdot (M - M_0) \tag{3.7}$$

where

M is the metabolic rate, in W/m^2
M_0 is the metabolic rate at rest, in W/m^2
HR_0 is the hear rate at rest, under neutral thermal condition
RM is the increase in heart rate per unit of metabolic rate, and it is given by formula:

$$RM = \frac{HR_{\max} - 0}{MWC - M_0} \tag{3.8}$$

where

HR_{max} is equal to do $HR_{\max} = 205 - 0.62 \cdot A$
MWC is the Maximum Working Capacity, following

$$MWC_{(man)} = (41.7 - 0.22 \cdot \mathrm{A}) \cdot P^{0.66} \quad (\mathrm{W/m^2}) \tag{3.9}$$

$$MWC_{(woman)} = (35.0 - 0.22 \cdot \mathrm{A}) \cdot P^{0.66} \quad (\mathrm{W/m^2}) \tag{3.10}$$

where A = Age in years and P = weight in kg.

The Level 4 method for determining the Metabolic Rate from Oxygen Consumption Rate is a more accurate method and that allows you to refer to the individual (or to an individual average as reported in the case studies in Part 2).

The calculation of the Metabolic Rate (M) is given by the following formula

$$M = \mathrm{EE} \cdot V_{O_2} \cdot \left(\frac{1}{A_{Du}}\right) \quad (\mathrm{W/m^2}) \tag{3.11}$$

where

A_{Du} is the body surface area, by the DuBois Formula, follow formula (3.14)
EE is the Energetic Equivalent, in Watt hours per liter of oxygen (Wh/l_{O_2}) follow formula (3.13)
V_{O_2} is the oxygen consumption rate, in liters of oxygen per hour

RQ is the respiratory quotient, follow formula (3.12).

The respiratory quotient is the ratio of carbon dioxide production to oxygen consumption as measured from analysis of expired gases, and it is given by formula:

$$RQ = \frac{V_{CO_2}}{V_{O_2}} \quad (-) \tag{3.12}$$

V_{CO_2} is the carbon dioxide consumption rate, in liters of carbon dioxide per hour.

The Energetic Equivalent (EE) for oxygen is the energy produced by the consumption of 1 L of oxygen at 0 °C and 101.3 kPa, and it is given by formula:

$$EE = (0.23 \cdot RQ + 0.77) \cdot 5.88 \quad \left(\mathrm{Wh/l_{O_2}}\right) \tag{3.13}$$

Dubois Area

The area of Dubois (A_{Du}) is used to determine, empirically, the body surface, by the formula:

$$A_{Du} = 0.202 \cdot \mathrm{W}^{0.425} \cdot H^{0.725} \quad (\mathrm{m}^2) \tag{3.14}$$

where

W is the body weight, in kilograms (kg)
H is the body height, in meters (m).

The formula of DuBois or Dubois and Dubois, was proposed in 1916 in the article "*A formula to estimate the approximate surface area if height and weight be known*" in the journal Archives of Internal Medicine, by Delfield Dubois (1882–1959) by Eugene Floyd Dubois (1882–1959) American physicist who worked in the study of metabolism, he worked at the hospital in New York and was a professor of medicine at Cornel Medical College.

The formula of Dubois is not the only one that enable you to express this calculation but it is the one used in the evaluation of the metabolism in the determination of comfort conditions and thermal comfort, studies that are based on statistical evaluations, and the formula of Dubois and Dubois allows only data related only to the weight and height of the subjects tested.

The Oxygen and Carbon Dioxide Consumption Rate

The values of the oxygen consumption rate (V_{O_2}) and the carbon dioxide consumption rate (V_{CO_2}) are calculated based on the volume flow rate (V_{ex}) given by the formula:

$$V_{ex} = \frac{V_{ex,STPD}}{\tau} \tag{3.15}$$

where

τ is the test duration (e.g. 1 h)
$V_{ex,STPD}$ is the expired volume at STPD (Standard Temperature and Pressure, Dry), and is given by the formula

$$V_{ex,STPD} = V_{ex,ATPD} \cdot f \tag{3.16}$$

where

$V_{ex,ATPD}$ is the expired volume at ATPD (Atmospheric Temperature and Pressure, Dry)
f is STPD reduction factor, by formula:

$$f = \frac{273 \cdot (p - p_{H_2O})}{(273 + \theta) \cdot 101.3} \tag{3.17}$$

where

p is the measure atmospheric pressure (kPa)
p_{H_2O} is the pressure of saturated water vapor (kPa)
θ is the temperature of the expired air (°C).

When you have measured the above data, also using the values provided by the standard you can calculate V_{O_2} and V_{CO_2} using the formulas

$$V_{O_2} = V_{ex} \cdot (0.209 - F_{O_2}) \tag{3.18}$$

$$V_{CO_2} = V_{ex} \cdot (F_{CO_2} - 0.0003) \tag{3.19}$$

where

F_{O_2} is the fraction of oxygen in the expired air (standard value 0.162)
F_{CO_2} is the fraction of carbon dioxide in the expired air (standard value 0.042).

Tables 3.1, 3.2, 3.3 and 3.4 report some examples of metabolic rate values, following ISO 7730.

In summary, the unit of measurement of the metabolism is the met, defined as:

met: an assigned unit of measurement to designate "sitting-resting" metabolic rate of man, the formula is:

$$1\,\text{met} = 58.15\,\text{W/m}^2 \tag{3.20}$$

Table 3.3 Metabolic rate for specific activities (by ISO 8996 Annex B—Table B.3)

Activity	W/m²
Sleeping	40
Reclining	45
At rest, sitting	55
At rest, standing	70
Walking on the level, even path, solid	
1. Without load at 2 km h^{-1}	110
At 3 km h^{-1}	140
At 4 km h^{-1}	165
At 5 km h^{-1}	200
2. With load 10 kg, 4 km h^{-1}	185
30 kg, 4 km h^{-1}	250
Walking uphill, even path, solid	
1. Without load 5° inclination, 4 km h^{-1}	180
15° inclination, 3 km h^{-1}	210
25° inclination, 3 km h^{-1}	300
2. With load of 20 kg 15° inclination, 4 km h^{-1}	270
25° inclination, 4 km h^{-1}	410
Walking downhill at 5 km h^{-1}, without load 5° inclination	135
15° inclination	140
25° inclination	180
Ladder at 70°, climbing at a rate of 11.2 m min^{-1}	
Without load	290
With a 20 kg load	360
Pushing or pulling a tip-wagon, 3.6 km h^{-1}, even path, solid	
Pushing force: 12 kg	290
Pulling force: 16 kg	375
Pushing a wheelbarrow, even path, 4.5 km h^{-1}, rubber tyres, 100 kg load	230
Filing iron 42 file strokes/min	100
60 file strokes/min	190
Work with a hammer, 2 hands, mass of the hammer 4.4 kg, 15 strokes/min	290
Carpentry work hand sawing	220
Machine sawing	100
Hand planning	300
Brick-laying, 5 bricks/min	170
Screw driving	100
Digging a trench	290
Sedentary activity (office, dwelling, school, laboratory)	70

(continued)

Table 3.3 (continued)

Activity	W/m^2
Standing, light activity (shopping, laboratory, light industry)	95
Standing, medium activity (shop assistant, domestic work, machine work)	115
Work on a machine tool	
Light (adjusting, assembling)	100
Medium (loading)	140
Heavy	210
Work with a hand tool	
Light (light polishing)	100
Medium (polishing)	160
Heavy (heavy drilling)	230

Table 3.4 Relationship between metabolic rate (in W/m^2) and heart rate (in beats per min), predicted as a function of the age and the weight of the subject (for women and men) (by ISO 8996 Annex C—Table C.1)

Age (years)	Weight (kg) (HR is heart rate)				
	50	60	70	80	90
Women					
20	2.9 × HR – 150	3.4 × HR – 181	3.8 × HR – 210	4.2 × HR – 237	4.5 × HR – 263
30	2.8 × HR – 143	3.3 × HR – 173	3.7 × HR – 201	4.0 × HR – 228	4.4 × HR – 254
40	2.7 × HR – 136	3.1 × HR – 162	3.5 × HR – 192	3.9 × HR – 208	4.3 × HR – 244
50	2.6 × HR – 127	3.0 × HR – 155	3.4 × HR – 182	3.7 × HR – 207	4.1 × HR – 232
60	2.5 × HR – 117	2.9 × HR – 145	3.3 × HR – 170	3.6 × HR – 195	3.9 × HR – 219
Men					
20	3.7 × HR – 201	4.2 × HR – 238	4.7 × HR – 273	5.2 × HR – 307	5.6 × HR – 339
30	3.6 × HR – 197	4.1 × HR – 233	4.6 × HR – 268	5.1 × HR – 301	5.5 × HR – 333
40	3.5 × HR – 192	4.0 × HR – 228	4.5 × HR – 262	5.0 × HR – 295	5.4 × HR – 326
50	3.4 × HR – 186	4.0 × HR – 222	4.4 × HR – 256	4.9 × HR – 288	5.3 × HR – 319
60	3.4 × HR – 180	3.9 × HR – 215	4.5 × HR – 249	4.8 × HR – 280	5.1 × HR – 311

It is an empirical unit of measurement to express the metabolic rate of a man whose clothing has an insulative value of 1 clo when he is sitting at rest in comfortable indoor surroundings (21 °C). (Glossary of terms for thermal physiology 1987)

3.3.2 *Heat Loss by Skin Diffusion (E_d)*

The Heat Loss by Skin Diffusion (Ed), or *Evaporative Heat Transfer* (E) expresses the amount of energy lost through skin evaporation, a process not linked to thermoregulation. The amount of energy exchanged depends on the amount of water vapor dispersed through sweat.

The calculation done with the formula:

$$E_d = \lambda_m \cdot A_{Du} \cdot (p_s - p_a) \quad \text{(W)} \tag{3.21}$$

where

λ_m is heat of vaporization of water at 35 °C.

Latent heat of vaporization λ is the quantity of heat absorbed (or released) by a volatile substance (fluid) per unit mass in the process of its reversible change of state by evaporation (or condensation) under isobaric and isothermal equilibrium conditions (J/g). For water: λ = 2490.9 − 2.34 T, with T = temperature of water in ~ e.g., 1 g of water, when changing from fluid to vapor at T = 34~ absorbs 2411.3 J (Glossary of terms for thermal physiology, 1987), it given by formula:

$$E_d = 3.05 \times 10^{-3} \cdot (p_s - p_a) \quad (\text{W/m}^2) \tag{3.22}$$

where

p_a is vapor pressure at air temperature (Pa)
p_s is saturated vapor pressure at skin temperature (Pa), that can be considered equal to:

$$p_s = 256 \cdot t_s - 3373 \quad \text{(Pa)} \tag{3.23}$$

then

$$E_d = 3.05 \times 10^{-3} \cdot (256 \cdot t_s - 3373 - p_a) \quad (\text{W/m}^2) \tag{3.24}$$

The Heat Loss by Skin Diffusion (or *Evaporative Heat Transfer*) also expresses the components of thermoregulation due to adjustment through sweating thermal and thermal tachypnea, and behavioral thermoeffectors, like saliva spreading, swallowing, and other behavioral modes of surface wetting.

At this, in some formulations of the Heat Balance equation, plus the amount of *Heat Loss by Evaporation of Sweat Secretion (E_{sw})*, which can range from 0 W/m^2 until 400 W/m^2 in the case of very hard work. In the case of moderate work, it accounts for about 25% of the total heat loss.

3.3.3 Latent Respiration (E_{re}) and Dry Respiration Heat Loss (L)

The Latent Respiration Heat Loss (E_{re} or E_{res}) depends on the amount of energy and mass of water vapor transferred through the inhalation and exhalation of air, and evaporation due to the mucosal lining of the respiratory tract. The calculation considers the humidity ratio of the air expiration/inhalation, with the formula:

$$E_{re} = V \cdot (W_{ex} - W_a) \cdot \lambda_m \quad (\mathrm{W}) \tag{3.25}$$

where

V is pulmonary ventilation (kg/h), equal to $V = 0.006 \cdot M$ (where M is metabolism)
W_{ex} is humidity ratio of the expiration air (kg_{water}/kg_{air})
W_a is humidity ratio of the inspiration air (kg_{water}/kg_{air})
λ_m is heat of vaporization of water at 35 °C

that, given a value of

$$(W_{ex} - W_a) = 0.029 - 0.00066 \cdot p_a \tag{3.26}$$

Therefore, the formula can be written:

$$E_{re} = 1.72 \times 10^{-5} \cdot M \cdot (5867 - RH \cdot p_a) \quad (\mathrm{W/m^2}) \tag{3.27}$$

where

RH is Relative Humidity (%).

The *Dry Respiration Heat Loss* (L) is due to the temperature difference between expired and inspired air (Sensible Heat) and is described by the formula:

$$L = V \cdot c_p \cdot (t_{ex} - t_a) = 0.0014 \cdot M \cdot (t_{ex} - t_a) \quad (\mathrm{W/m^2}) \tag{3.28}$$

where

V is pulmonary ventilation (kg/h), equal to $V = 0.006 \cdot M$ (where M is metabolism)

c_p is specific heat of dry air at constant pressure
t_a is temperature of the inspiration air (°C)
t_{ex} is temperature of the expiration air (°C)

$$t_{ex} = 32.6 + 0.066 \cdot t_a + 32 \cdot W_a \quad (°C) \tag{3.29}$$

in conclusion it can be written:

$$L = 0.0014 \cdot M \cdot (34 - t_a) \quad (W/m^2) \tag{3.30}$$

3.3.4 Heat Conduction Through the Clothing (K)

Fanger's equation of Thermal Comfort of considers energy exchanges of the human body and the environment, in the first part of the equation, and the exchanges between the outer surface of the body and the environment (conduction, convection, radiation), in the second part of the equation. Between the two modes of exchange is between (K) the coefficient due to the clothing, which includes the heat exchange between the body and clothing and between the clothing and the environment. In this way it is possible to express, in a single value, the quantity of heat exchanged through the clothing which depends on the temperature difference between the skin and the outer layer of clothing.

The Heat Conduction through the Clothing (K) or *Transfer from the skin to the outer surface of the clothed body*, expresses the transfer of heat by conduction through clothing, between the temperatures of the skin and the external environment.

K measures the conduction of the passage of heat through the clothing, and is described by the formula:

$$K = \frac{t_{sk} - t_{cl}}{0.155 \cdot I_{cl}} \quad (W/m^2) \tag{3.31}$$

where

t_{sk} is mean skin temperature (°C)
t_{cl} is clothing surface temperature (°C)
I_{cl} is total Thermal Resistance from the skin to the other surface of the clothed body (clo).

The thermal resistance I_{cl} is calculated with the formula:

$$I_{cl} = \frac{R_{cl}}{0.18} \quad (clo) \tag{3.32}$$

where

R_{cl} is the total heat transfer resistance from skin to the other surface of the clothed body, measured in m^2 °C/W.

The value of I_{cl} depends on the characteristics of individual items of clothes and clothing, the characteristics of the textile material, type of fiber (wool, cotton, silk, raylon, etc.).

The criteria of measurement and the values are defined by ISO 9920 for single items of clothing or all clothing, as described in the following paragraphs.

The definition of Thermal insulation, clothing (I_{cl}) is *the intrinsic insulation of a clothing assembly.*

> The effective insulation of clothing is (I_{cl} + I_a) where I_a the reciprocal of the thermal conductance of the ambient environment. (I_{cl} + I_a) is usually measured as the temperature gradient from the surface of a heated man-sized manikin to the ambient air divided by the heat production per unit area of manikin surface (m^2 °C/W). The value is expressed in "clo" units. (Glossary of terms for thermal physiology, 1987)

Energy exchanges through the boundaries are due to the phenomena of heat exchange by convection and radiation between the surface of the body and the external environment, and in part for conduction.

In assessing the exchange through the boundaries of the human body it is always necessary to refer to the "*outer surface of the clothed body*", as it is the area that exchanges energy by Radiation (R) and by Convection (C).

3.3.5 *Heat Loss by Radiation (R)*

The heat exchange by radiation between the body and the environment by radiation is due to the temperature difference between the outer surface of the clothing (t_{cl} in °C) and the mean radiant temperature of the walls (t_{mr} in °C), described with the *Law of Stefan–Boltzmann*:

$$Q_{1\to 2} = A_1 \cdot \varepsilon_1 \cdot \sigma_0 \cdot \left(T_1^4 - T_2^4\right) \quad (W) \tag{3.33}$$

where

A_1 is body 1 area, in m^2
ε_1 is emittance of body 1 versus body 2 (a-dimensional)
σ_0 is the Stefan–Boltzmann constant = 5.67×10^{-8} W/m^2 K^4
$\left(T_1^4 - T_2^4\right)$ is the difference of temperature in Kelvin, between body 1 and 2, measure in Kelvin (K).

In the case of the Human Body, we can consider the area of the human body much smaller than the area of the cavity in which it is inserted (the room), and in this case it is possible to report the heat exchange by radiation only to the smallest, which is that of the Human Body.

In this case, reference should be made to the effective area (A_{eff}) of the Human Body and the difference between the temperature of the changing room (t_{cl}) and the mean radiant temperature of the walls (t_{mr}). The above formula then becomes:

$$R = A_{eff} \cdot \varepsilon_1 \cdot \sigma_0 \cdot \left[(t_{cl} + 273)^4 - (t_{mr} + 273)^4\right] \quad \text{(W)} \tag{3.34}$$

where

ε is used 0.9 value (average emittance of clothes)
A_{eff} is Effective Radiation Area, by formula

$$A_{eff} = A_{Du} \cdot f_{cl} \cdot f_{eff} \quad (\text{m}^2) \tag{3.35}$$

where

A_{Du} is the Human Body Area, by DuBois formula, measured in m^2;
f_{cl} is the ratio of the surface are of the clothed body to the surface area of the nude human body, which depend on type of clothing (a-dimensional);
f_{eff} is the ratio of the effective radiation area of the clothed body to the surface are of the clothed body (by pre-calculated value or by tables e.g. 0.696 for sedentary body posture, or 0.725 for standing posture).

The mean radiant temperature is calculated as

$$t_{mr} = \sum_n t_n \cdot F_{p-n} \quad (^\circ\text{C}) \tag{3.36}$$

where

t_n is isotherm temperature of room surface *n* (walls, floor, ceiling, etc.)
F_{p-n} is the angle factor between person (*p*) and surface *n* (by tables)

and is measured with the Globo-Thermometer.

The final version of the formula is:

$$R = 3.96 \times 10^{-8} \cdot f_c \cdot A_{Du} \cdot \left[(t_{cl} + 273)^4 - (t_{mr} + 273)^4\right] \quad \text{(W)} \tag{3.37}$$

3.3.6 Heat Loss by Convection (C)

The Heat Loss by Convection from the outer surface of the clothed body (C) is described in the formula

$$C = A_{Du} \cdot f_{cl} \cdot h_c \cdot (t_{cl} - t_a) \quad (\text{W/m}^2) \tag{3.38}$$

where

A_{Du} is the Human Body Area, by DuBois formula (m^2)
f_{cl} is the ratio of the surface are of the clothed body to the surface are of the nude human body, which depend on type of clothing (a-dimensional)
t_{cl} is clothing surface temperature (°C)
t_a is air temperature (°C)
h_{cl} is the *convective heat transfer* coefficient (W/m^2 °C) or (W/m^2 K), between surface and clothing, and the room air, which depends on the type of convection:

Low air velocity "*still air*" (natural convection), by

$$h_c = 2.38 \cdot (t_{cl} - t_a)^{0.25} \quad (\text{W/m}^2\,\text{K}) \tag{3.39}$$

Speed air velocity (forced convection), by

$$h_c = 12.1 \cdot \sqrt{v_{air}} \quad (\text{W/m}^2\,\text{K}) \tag{3.40}$$

where

v_{air} is air velocity (m/s) if $v_{air} < 2.6$ m/s

that can be calculated with the formula:

$$v_{air} = v_a + 0.005 \cdot \left(\frac{M}{A_{Du}} - 58.15\right) \quad (\text{m/s}) \tag{3.41}$$

v_a is air velocity (m/s).

3.3.7 Heat Balance Equation (Thermal Comfort Equation)

In the Heat Balance Equation, if we put all the terms in the (3.1) initial equation dividing the whole for the Human Body Dubois Area (A_{Du}), the final formula is:

$$\begin{aligned}&(M-W)-\left[3.05\times 10^{-3}\cdot(256\cdot t_s-3373-p_a)\right]-E_{sw}\\&-1.72\times 10^{-5}\cdot M\cdot(2587-RH\cdot p_a)-0.0014\cdot M\cdot(34-t_a)\\&=\frac{(t_{sk}-t_{cl})}{(0.155\cdot I_{cl})}=2.96\times 10^{-8}\cdot f_{cl}\cdot A_{Du}\cdot\left[(t_{cl}+273)^4-(t_{mr}+273)^4\right]\\&+A_{Du}\cdot f_{cl}\cdot h_{cl}\cdot(t_{cl}-t_a)\end{aligned} \quad (3.42)$$

In the Heat Balance Equation there remain some unknowns: the t_s (skin temperature) and the E_{sw} (Heat Loss by respiration) both depend on the activity level (H) and thus can be expressed, as an average value, in function of the metabolism (M) with the equations:

$$t_s=35.7-0.027\cdot(M-W)\quad(°C) \quad (3.43)$$

$$E_{sw}=0.42\cdot[(M-W)-58.15]\quad(W) \quad (3.44)$$

Both equations can be written as a function of the Area of DuBois (Bartal et al. 2012) with the following formulas

$$t_s=35.7-0.032\cdot\frac{M}{A_{Du}}\quad(°C) \quad (3.45)$$

$$E_{sw}=0.42\cdot\left[\left(\frac{M}{A_{Du}}\right)\cdot\left(\frac{1}{\eta}\right)-58.15\right]\quad(W) \quad (3.46)$$

Substituting these values in the formula (3.42), all referring to the unit area of the Human Body (m^2) you obtain the *Thermal Comfort Equation*:

$$\begin{aligned}&(M-W)-\left\{3.05\times 10^{-3}\cdot[256\cdot(35.7-0.027\cdot(M-W))]-3373-p_a\right\}\\&-\{0.42\cdot[(M-W)-58.15]\}-\left\{1.72\times 10^{-5}\cdot M\cdot(2587-RH\cdot p_a)\right\}\\&-\{0.0014\cdot M\cdot(34-t_a)\}=\left\{\frac{[35.7-0.032\cdot(M-W)]-t_{cl}}{(0.155\cdot I_{cl})}\right\}\\&=3.96\times 10^{-8}\cdot f_{cl}\cdot\left[(t_{cl}+273)^4-(t_{mr}+273)^4\right]+f_{cl}\cdot h_{cl}\cdot(t_{cl}-t_a)\end{aligned} \quad (3.47)$$

measured in W/m^2 of human body, therefore in the definitive version:

$$\begin{aligned}&(M-W)-\left\{3.05\times 10^{-3}\cdot[5733-6.99\cdot(M-W)]-p_a\right\}\\&-\{0.42\cdot[(M-W)-58.15]\}-\left\{1.72\times 10^{-5}\cdot M\cdot(2587-RH\cdot p_a)\right\}\\&-\{0.0014\cdot M\cdot(34-t_a)\}\\&=3.96\times 10^{-8}\cdot f_{cl}\cdot\left[(t_{cl}+273)^4-(t_{mr}+273)^4\right]+f_{cl}\cdot h_{cl}\cdot(t_{cl}-t_a)\end{aligned} \quad (3.48)$$

where

$$
\begin{aligned}
t_{cl} = {} & 35.7 - 0.028 \cdot (M - W) - 0.155 \cdot I_{cl} \\
& \cdot \{(M - W) - 3.05 \times 10^{-3} \cdot [5733 - 6.99 \cdot (M - W) - p_a] \\
& - 0.42 \cdot [(M - W) \cdot 58.18]\} - 1.7 \times 10^{-5} \\
& \cdot M \cdot (5867 - p_a) - 0.0014 \cdot M \cdot (34 - t_a)
\end{aligned}
\tag{3.49}
$$

with the following values of h_{cl} and f_{cl}:

$$h_c = 2.38 \cdot (t_{cl} - t_a)^{0.25} \quad \text{for} \quad 2.38 \cdot (t_{cl} - t_a)^{0.25} \geq 12.1 \cdot \sqrt{v_{air}}$$

$$h_c = 12.1 \cdot \sqrt{v_{air}} \quad \text{for} \quad 2.38 \cdot (t_{cl} - t_a)^{0.25} < 12.1 \cdot \sqrt{v_{air}}$$

and

$$f_{cl} = 1.00 + 0.2 \cdot I_{cl} \quad \text{for} \quad I_{cl} \leq 0.5 \text{ clo}$$

$$f_{cl} = 1.05 + 0.1 \cdot I_{cl} \quad \text{for} \quad I_{cl} > 0.5 \text{ clo}$$

This is the final expression of the Heat Balance Equation or *Thermal Comfort Equation*, used to assess exchanges between Human Body and environmental, is missing in this equation the Resistance of clothing, described in the next section.

The formula (3.49) can be expressed based on work efficiency η

$$\eta = \frac{W}{M} \tag{3.50}$$

in the following formulation (Bartal et al. 2012; Butera 1998)

$$
\begin{aligned}
& \left[\left(\frac{M}{A_{Du}}\right) \cdot \left(\frac{1}{\eta}\right)\right] - \left\{3.05 \times 10^{-3} \cdot \left[5733 - 6.99 \cdot \left(\frac{M}{A_{Du}}\right) \cdot \left(\frac{1}{\eta}\right)\right] - p_a\right\} \\
& \quad - \left\{0.42 \cdot \left[\left(\frac{M}{A_{Du}}\right) \cdot \left(\frac{1}{\eta}\right) - 58.15\right]\right\} \\
& \quad - \left\{1.72 \times 10^{-5} \cdot \left(\frac{M}{A_{Du}}\right) \cdot (2587 - RH \cdot p_a)\right\} \\
& \quad - \{0.0014 \cdot M \cdot (34 - t_a)\} \\
& = 3.96 \times 10^{-8} \cdot f_{cl} \cdot \left[(t_{cl} + 273)^4 - (t_{mr} + 273)^4\right] + f_{cl} \cdot h_{cl} \cdot (t_{cl} - t_a)
\end{aligned}
\tag{3.51}
$$

3.4 Resistance of Clothing

In the equation of Body Heat Balance (formula 3.1) the Heat Conduction through the Clothing (K), expresses the heat transfer by conduction through clothes, between the temperature of the skin and the external environment.

The clothes are the main factor in climate adaptation of man to the environment, followed using fire to control the microclimate of indoor environments.

Moreover, the clothes are an element of recognition; they depend on climate, the availability of materials, and are part of the cultural tradition of the people. In the choice of different dress, we consider:

- Social habits and modesty, being naked is a condition of intimacy accepted only under certain conditions (sea, bathroom, intimacy) or in some populations (indigenous tribes, aboriginal, etc.);
- The characteristics of the indoor microclimate environment where you can be found or expect to reside in the day;
- The characteristics of outdoor climate, or the season, by the external temperature, the lower the temperature is the greater the amount of thermal insulation required in the clothing;
- In the clothing strategies and adaptation (human behavior), the choice of the combination of clothes, one or more sweaters, shoes or boots, or shirt and jacket instead of a single sweater, etc. depends on the choices of the individual in relation to the above considerations, by habit and by possibility to adapt their clothing in relation to climate and microclimate during the day;
- The rules dictated by fashion and aesthetics (psychologists and social aspects), the symbolic meaning attributed to the different clothes, gender (male/female), age, social context (dressing code, casual dressing, Casual Friday, uniform, livery, etc.) in addition to age and autonomy in the choice and opportunity to dress (disabled and children).

In the specific case of children, depending on age, autonomy varies in choosing clothes, usually for children up to 3 or 4 years of age, the clothes are chosen, and the child dressed directly by mothers and parents. From 5 years and older children are autonomous in dressing but not in choosing clothes, which are always prepared by mothers, who have a uniform sensitivity on Thermal Comfort with respect to metabolism and of the perception of children.

The fact of not being able to decide the type of clothing, for children, implies the need to take other adaptation strategies with respect to clothing.

Around age of 10 or 12 years old young people become autonomy to choose their dressing and clothes from their wardrobe, they are comparable than of an adult.

In the case of an adult, you can increase or decrease the isolation layer due to dress according to the context, for example, you can wear a jacket when you are outside and will take off when you are in a room, and if you are in a particularly warm environment, it is possible to eliminate the excess layers such as a jacket and remain only in a shirt, or in "shirtsleeves" if the dress-code allows.

This type of control may be missing in the case of vulnerable users, people with physical or mental disabilities.

Considering the human body, from the thermodynamic point of view, as an open system that exchanges energy through the outline (boundary). In this model, the boundary is the skin, and clothing constitutes a resistance to the passage of heat between the outline and indoor environment.

It should be considered that the Thermal Comfort assessment depends on factors and personal choices related to the activity and clothing, as mentioned by Havenith et al. (2002).

The clothing constitutes a separation between the human body with respect to thermal insulation and water vapor resistance, since even the water vapor of perspiration passes through the clothing, and the several types of clothing can promote or reduce the passage of the vapor through the clothes themselves.

Through the clothing there is a passage of heat by conduction, the clothing allows such passage to reduce thanks to the thermal resistance of the materials.

The Standard ISO 9920 *Ergonomic of the thermal environment—Estimation of thermal insulation and water vapor resistance of a clothing ensemble* defines the methods for estimating for a clothing ensemble (*resistance to dry heat loss and evaporative heat loss*) in steady state conditions for a clothing ensemble based on values for known garments, ensembles and textiles.

The clothing insulation is defined as:

> basic clothing insulation, that is the resistance of a uniform layer of insulation covering the entire body that has the same effect on sensible heat flow as the actual clothing under standardized (static, wind-still) conditions. (ISO 13731)

The general definition of the insulation of clothing (I) is given by ratio between temperature gradient and heat loss per unit of body surface area.

The evaluation of the characteristics of the clothing takes account of:

(a) *effective thermal insulation for individual garments* (I_{clu})

evaluated with following formula

$$I_{clu} = 0.0061 \cdot A_{COV} \quad \text{(clo)} \tag{3.52}$$

where

A_{COV} is the body surface area covered by clothing, which can be calculated from the *clothing area factor* (f_{cl})

$$f_{cl} = \frac{A_{cl}}{A_{Du}} \tag{3.53}$$

where

A_{cl} is a surface area of a clothed person (m^2)
A_{Du} is the surface area of a nude body (m^2) (Figs. 3.4, 3.5, 3.6 and 3.7).

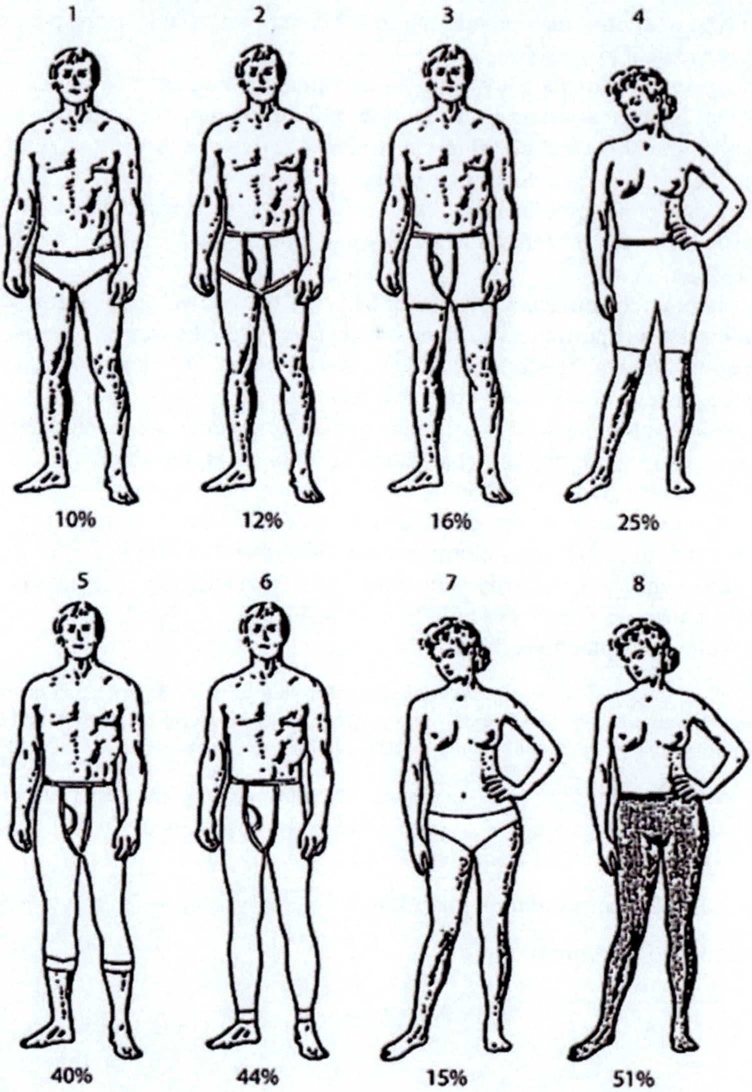

Fig. 3.4 Body coverage percentage. Example: underwear—pants (ISO 9920)

(b) *ensemble thermal insulation on individual garments* ($I_{cl,u}$)

all garments each have their own level of effective insulation ($I_{cl,u}$), which depends on the characteristics of single garment compared to the bare manikin insulation. Annex B of ISO 9920 contains a table with the isolation level for each standard garment. The calculation of ($I_{cl,u}$) depends on the Effective Clothing Insulation defined as

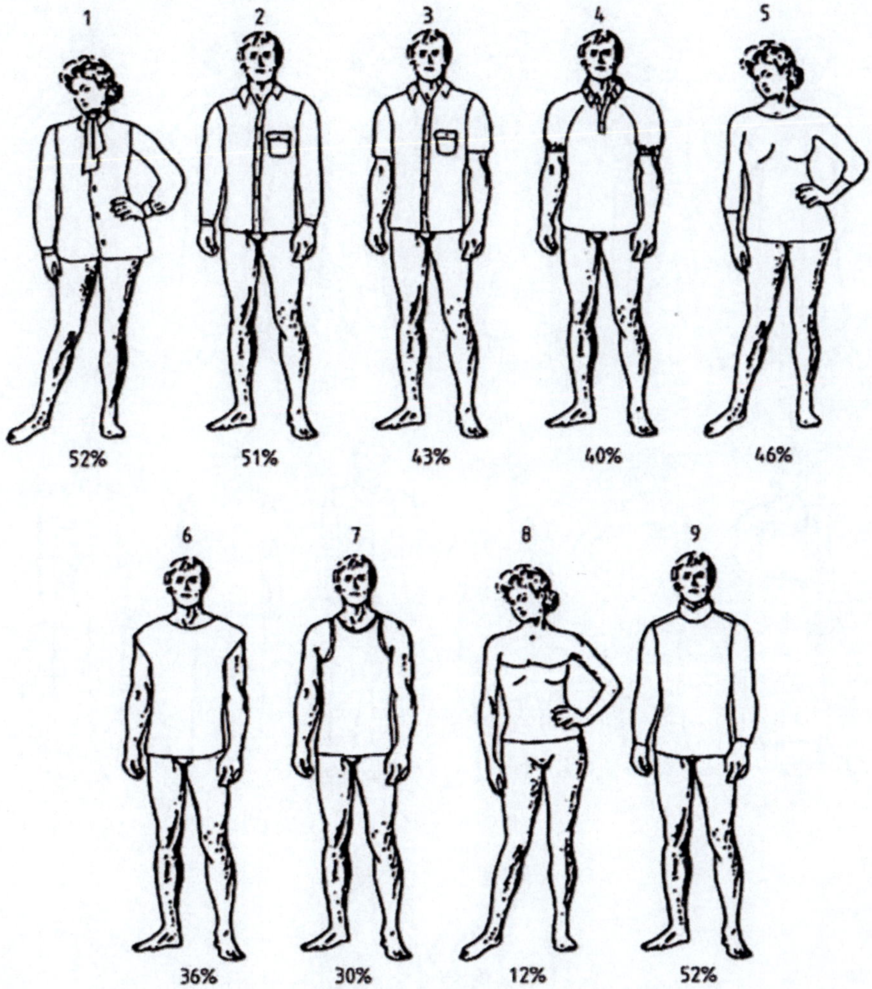

Fig. 3.5 Body coverage percentage. Example: underwear—shirts (ISO 9920)

> the increased body insulation due to clothing as compared to the nude state. It is the difference between the total insulation and the boundary layer insulation. (ISO 9920)

and is calculated with the la formula:

$$I_{clu} = I_T - I_a = \left[\frac{t_{sk} - t_o}{H}\right] - I_a \tag{3.54}$$

Fig. 3.6 Body coverage percentage. Example: sleepwear (ISO 9920)

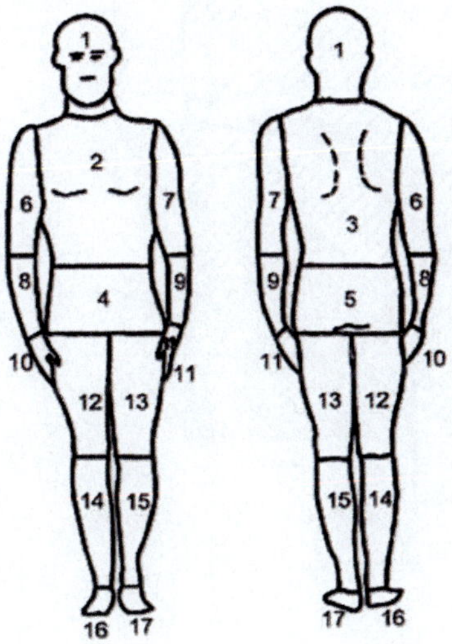

	Segment	% total area
1	Head + neck	8,7
2	Chest	10,2
3	Back	9,2
4	Abdomen	6,1
5	Buttocks	6,6
6	Right upper arm	5,0
7	Left upper arm	5,0
8	Right lower arm	3,1
9	Left lower arm	3,1
10	Right hand	2,5
11	Left hand	2,5
12	Right thigh	9,2
13	Left thigh	9,2
14	Right calf	6,1
15	Left calf	6,1
16	Right foot	3,7
17	Left foot	3,7
Total		100

Fig. 3.7 Template for estimation of body surface is covered by one or more layers of clothing and nude areas (ISO 13731)

where

I_a is air insulation, thermal insulation of the boundary surface air layer around the clothing that depend to convective heat transfer coefficient h_c, measured in m^2 K/W and radiative heat transfer coefficient h_r measured in m^2 K/W, and is calculated with the la formula:

$$I_a = \frac{1}{h_c + h_r} \tag{3.55}$$

I_T is the total thermal insulation of the garment measure in m^2 K/W or in clo;
t_{sk} is the skin temperature of the nude manikin insulation;
t_o is the operative temperature (°C) equal to air temperature (Fig. 3.8).

(c) *thermal insulation of clothing ensembles* (I_{cl})

which considers the type of clothing standards by type of activity, and typical clothing ensembles, calculated using the formula:

$$I_{cl} = 0.161 + 0.835 \cdot \sum I_{clu} \quad \text{(clo)} \tag{3.56}$$

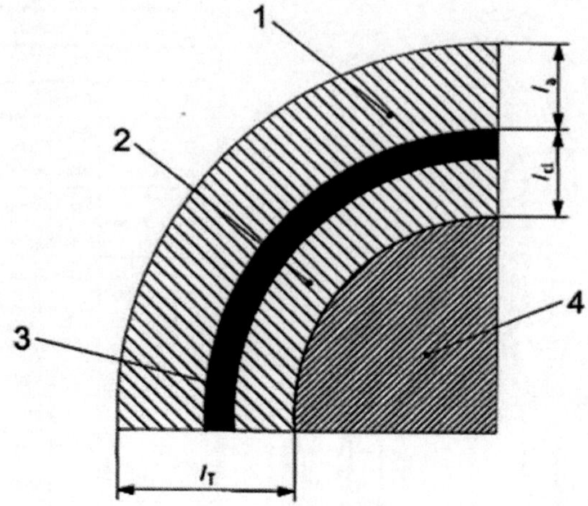

Key
1 surface (or boundary) air layer
2 enclosed air layer
3 clothing
4 body

Fig. 3.8 Schematic representation of total, basic and air insulations

or, with reduce accuracy

$$I_{cl} = 0.835 \cdot \sum I_{clu} \quad \text{(clo)} \tag{3.57}$$

$$I_{cl} = \sum I_{clu} \quad \text{(clo)} \tag{3.58}$$

(d) *estimation of water vapor resistance* ($R_{e,T}$)

the resistance to water vapor of the clothing is used to determine the flow of vapor between the skin and the environment and, therefore, allowing an evaluation of how clothing affects the Heat Loss by Skin Diffusion (Ed), and is given by ratio between vapor pressure gradient and evaporative heat loss per unit of body surface area. The total water vapor resistance ($R_{e,T}$), is measured in (m^2 kPa/W), can be estimated from the thermal insulation, with the equation

$$R_{e,t} = 0.16 \cdot \left[\frac{I_a}{f_{cl}} + I_{cl} \right] \quad (\text{m}^2\,\text{kPa/W}) \tag{3.59}$$

(e) *the influence of body movement and air movement*

compared to thermal insulation vapor resistance is not directly estimable or standardized. In actual behavior the effect of the motion of the body (pumping effect) and wind on the insulation should be considered. The Standard ISO 9920 provides a series of correction factors in relation to air speed.

(f) *other factors influencing clothing insulation are:*

- *body posture* (sitting, standup, etc.) that change the surface of exchange and the air layer insulation (Ia), in general reduce by 6–18%, if the person is sitting;
- effect of *seat*, in the case of people sitting the exchange of heat between the body—environment has a further element which is given by the isolation of the chair, the armchair (office chairs produces an increase of insulation from 0.04 to 0.17 clo) or seat (e.g. in particular in car);
- effect of *air pressure*;
- effect of *clothes wetting* when clothing gets wet it loses part of its insulation;
- effect of *clothes washing*, that washing, alters the fibers and reduces the insulation capacity of the clothing.

The measurement of the relative dimensions relating to the clothing (I_{cl} and $R_{e,T}$) is achieved using a *Thermal Manikin*, as described in Annex D of ISO 9920.

In the international standards for the evaluation of the isolation of garments and/ or clothing ensembles (I_{cl}) uses a manikin with the following characteristics:

- manikin height 170 ± 10 cm,
- manikin surface area 1.8 ± 0.3 m^2,
- position standing during test, with Mean Skin Temperature between 32 and 34 °C,
- Room Operative/Air Temperature greater than 12 °C respect to the manikin, Heat loss from thermal manikin more than 20 W/m^2,
- Relative Humidity 30–70% (preferably 50%) and
- Air Velocity less than 0.2 m/s,

for a Test Period greater than 20 min (Fig. 3.9).

Fig. 3.9 Thermal resistance of clothing (clo)

In summary the heat exchange through clothing depends on the ability of clothing to conduct heat, characteristics that are expressed with a grandeur called "**clo**", a contraction of clothing and parameterized according to regulatory standards. The clo expresses the resistance to the passage of heat and is equal to 0.155 m^2 °C/W, or for each square meter of surface of clothing, and for each degree of temperature difference between the skin and outer clothing, allows the passage through the clothes of 0.155 W.

The value of a clo corresponds to a winter clothing type slip, shirt, pants, socks and shoes and jacket, while it is equal to zero in the case of a naked human body in an environment, or 0.30 clo in case you wear only undergarments, socks and sandals, until reaching values of 2.5 clo in case of winter clothing that includes jackets with heavy padding, cap and gloves with extreme values of 3 or 4 clo for winter clothing for polar exploration.

The unit of measurement of thermal resistance (I_{cl}) is the clo defined as:

Clo: A unit to express the relative thermal insulation values of various clothing assemblies.

$$1\,\text{clo} = 0.18\left(^\circ\,\text{cm}^2\,\text{h/kcal}\right) = 0.155\left(\text{m}^2\,\text{K/W}\right) \tag{3.60}$$

The clo is a unit developed to express thermal insulation in practical terms and represents the insulation provided by the normal indoor clothing of a sedentary worker in comfortable indoor surroundings (Gagge et al. 1941). The term is used in heating and ventilation engineering in the determination of environmental conditions for human comfort. (Glossary of terms for thermal physiology 1987)

The values are defined in the Annex della ISO 9920 by means of tables (e.g. reported in Tables 3.5, 3.6, 3.7 and 3.8), of which are some examples.

3.5 Indoor Microclimate: The Building Parameters

In the exchange of energy between the human body and environment, those defined are the physiological parameters (Human Body Heat balance) and clothing, we must describe the parameters that affect the outline system, i.e. the parameters for indoor (or outdoor) Environment Microclimate.

The indoor parameters are:

- *Air Temperature* (t_a measured in °C), which characterizes the exchange of thermal energy between body (Skin Temperature Clothes or Temperature) and the environment;
- *Relative Humidity* (RH measure in %), that characterizes the thermal exchanges of mass and energy due to Heat Loss by Skin Diffusion (E_d) Respiration and Latent Heat Loss (E_{re}), in the case of high values of RH (e.g. > 75%) of the Indoor Environment, the amount of energy and mass that can be exchanged is less than the lower values of RH;

Table 3.5 Insulation values of typical clothing ensembles

Work clothing	I_{cl}		Daily wear clothing	I_{cl}	
	clo	m^2 K/W		clo	m^2 K/W
Underpants, boiler suit, socks, shoes	0.70	0.11	Panties, T-shirt, shorts, light socks, sandals	0.3	0.05
Underpants, shirt, trousers, socks, shoes	0.75	0.115	Panties, petticoat, stockings, light dress with sleeves, sandals	0.45	0.07
Underpants, shirt, boiler suit, socks, shoes	0.80	0.125	Underpants, shirt with short sleeves, light trousers, light socks, shoes	0.50	0.08
Underpants, shirt, trousers, jacket, socks, shoes	0.85	0.135	Panties, stockings, shirt with short sleeves, skirt, sandals	0.55	0.085
Underpants, shirt, trousers, smock, socks, shoes	0.90	0.140	Underpants, shirt, light-weight trousers, socks, shoes	0.60	0.095
Underwear with short sleeves and legs, shirt, trousers, jacket, socks, shoes	1.00	0.155	Panties, petticoat, stockings, dress, shoes	0.70	0.105
Underwear with short legs and sleeves, shirt, trousers, boiler suit, socks, shoes	1.10	0.170	Underwear, shirt, trousers, socks, shoes	0.70	0.11
Underwear with long legs and sleeves, thermojacket, trousers, socks, shoes	1.20	0.185	Underwear, track suit (sweater and trousers), long socks, runners	0.75	0.115
Underwear with short sleeves and legs, shirt, trousers, jacket, thermojacket, socks, shoes	1.25	0.190	Panties, petticoat, shirt, skirt, thick knee socks, shoes	0.80	0.12
Underwear with short sleeves and legs, boiler suit, thermojacket and trousers, socks, shoes	1.40	0.220	Panties, shirt, skirt, roundneck sweater, thick knee socks, shoes	0.90	0.14
Underwear with short sleeves and legs, shirt, trousers, jacket, thermojacket and trousers, socks, shoes	1.55		Underpants, singlet with short sleeves, shirt, trousers, V-neck sweater, socks, shoes	0.95	0.145

(continued)

Table 3.5 (continued)

Work clothing	I_{cl}		Daily wear clothing	I_{cl}	
	clo	m^2 K/W		clo	m^2 K/W
Underwear with short sleeves and legs, shirt, trousers, jacket, heavy quilted outer jacket and overalls, socks, shoes	1.85	0.285	Panties, shirt, trousers, jacket, socks, shoes	1.00	0.155
Underwear with short sleeves and legs, shirt, trousers, jacket, heavy quilted outer jacket and overalls, socks, shoes, cap, gloves	2.00	0.310	Pants, stockings, shirt, skirt, vest, jacket	1.00	0.155
Underwear with long sleeves and legs, thermojacket and trousers, thermojacket and trousers, socks, shoes	2.20	0.340	Panties, stockings, blouse, long skirt, jacket, shoes	1.10	0.170
Underwear with long sleeves and legs, thermojacket and trousers, parka with heavy quilting, overalls with heavy quilting, socks, shoes, cap, gloves	2.55	0.395	Underwear, singlet with short sleeves, shirt, trousers, jacket, socks, shoes	1.10	0.170
			Underwear, singlet with short sleeves, shirt, trousers, vest, jacket, socks, shoes	1.15	0.180
			Underwear with long sleeves and legs, shirt, trousers, V-neck sweater, jacket, socks, shoes	1.30	0.20
			Underwear with short sleeves and legs, shirt, trousers, vest, jacket, coat, socks, shoes	1.50	0.23

- *Air Velocity* (v_{air} measured in m/s), characterizing the thermal exchanges for Convection (Heat Loss by convection) between Human or Clothes Body Surface and air of the environment;
- *Mean Radiant Temperature* (t_{mr} measured in °C), that characterizes the heat exchange by radiation (Heat Loss by Radiation) between or Clothes Human Body Surface and the walls at a constant temperature, described in the following chapter.

The feeling of comfort and judgment is given by the relation between these parameters and those relating to Hand Body and Clothes.

Table 3.6 Additional insulation values of various clothing ensembles—daily wear clothing—example (by ISO 9920 and McCullough and Jones 1984)

Ensemble description	Mass	f_{cl}	I_{cl}		I_T	
	g		clo	m^2 K/W	clo	m^2 K/W
Briefs, T-shir, undershorts, calf-length socks, athletic shoes	318	1.10	0.33	0.051	0.95	0.151
Panties, tube to, short shorts, sandals	258	1.07	0.23	0.036	0.89	0.139
Brides, short-sleeve shirt, shorts, calf-length, socks, shoes	622	1.11	0.41	0.064	1.05	0.163
Briefs, 3/r-length-sleeve shirt, socks, athletic shoes	451	1.14	0.52	0.081	1.13	0.175
Panties, sleeveless blouse, fitted trousers sandals	523	1.14	0.44	0.068	1.06	0.165
Briefs, short-sleeve shirt, fitted trousers, calf length socks, shoes	725	1.14	0.50	0.078	1.12	0.174

The Indexes of Comfort are described in the following Chap. 4.

In this chapter, we only report the definition of "*Operating Temperature*" an empirical index to determine the comfort conditions and measure the air temperature considering heat exchange by convection and radiation and was introduced by Herringot and Gagge (1937).

3.5.1 Operating Temperature

The operating temperature (°C) or operator temperature, expresses the relation between the mean radiant temperature of the walls and the temperature of the environment, considering the phenomena due to the heat exchange by convection and radiation, and is given by the formula:

$$t_p = \frac{(h_r \cdot t_{mr}) + (h_c \cdot t_a)}{h_r + h_c} \quad (°C) \tag{3.61}$$

where

- t_o is Operative Temperature (°C)
- t_a is Air-Temperature (°C)
- h_c is the *convective heat transfer* coefficient (W/m^2 K), that depend on natural or forced convection
- h_r is the radiative heat transfer coefficient (W/m^2 K)

Table 3.7 Insulation values of various clothing ensembles—cold protective clothing—with dynamic insulation values for walking (no wind), body surface—example (by ISO 9920 and Kim and McCullough 2000)

Ensemble description	BSAC (%)	Mass g	f_{cl}	I_{cl} (clo)		I_T (clo)	
				Static	Dynamic	Static	Dynamic
Extreme cold weather expedition suit with hood (down-filled, one-piece suit), thermal long underwear top and bottoms, mittens with fleece liners, thick socks, insulated waterproof boots	98.9	2804	1.5	3.67	3.21	4.12	3.54
One-piece ski suit, thermal long underwear top and bottoms, knitted head/ear band, goggles, insulated ski gloves, thin knee-length ski socks, insulated waterproof boots	97.1	2357	1.28	1.6	1.13	2.13	1.51
One-piece fiberfill ski suit with hood, thermal long underwear top and bottoms, goggles, insulated ski gloves, thin knee-length ski socks, insulated waterproof boots	98.8	2494	1.27	1.97	1.53	2.51	1.92

(continued)

Table 3.7 (continued)

Ensemble description	BSAC (%)	Mass g	f_{cl}	I_{cl} (clo)		I_T (clo)	
				Static	Dynamic	Static	Dynamic
Ski jacket with detachable fiberfill liner, thermal long underwear bottoms, knitted turtleneck sweater, fiberfill ski pants, knitted hat, goggles, mitten shell with fleece glove inserts, thin knee length ski socks, insulated waterproof boots	95.8	2943	1.34	2.3	1.75	2.81	1.12
Refill jacket, jeans, T-shirt, thermal long underwear bottoms, long-sleeve flannel shirt, baseball cap, thick socks, low-cut leather work boots	91	2283	1.4	1.68	1.3	2.17	1.65
Fleece long-sleeve shirt, fleece pants, briefs, thermal long underwear top and bottoms, athletic socks, athletic shoes	88.1	1370	1.29	1.34	1.04	1.87	1.42
Heavy fiberfill hunting jacket with detachable liner, matching hunting bib overalls, briefs, knitted turtleneck sweater, jeans, hat with fleece liner and ear flaps, fiberfill gloves, thick socks, insulated waterproof boots	97.8	4386	1.5	2.98	2.32	3.43	2.65

Table 3.8 Insulation values of various typical garments

Garment	I_{cl} (clo)	Garment	I_{cl} (clo)
Underwear		Thin sweater	0.20
Panties	0.03	Sweater	0.28
Underpants with long legs	0.10	Thick sweater	0.35
Singlet	0.04	*Jackets*	
T-shirt	0.09	Light summer jacket	0.25
Shirt with long sleeves	0.12	Jacket	0.35
Panties and bra	0.03	Smock	0.30
Shirts, blouses		*Highly insulating, fiber-pelt*	
Short sleeves	0.15	Boiler suit	0.90
Lightweight, long sleeves	0.20	Trousers	0.35
Normal, long sleeves	0.25	Jacket	0.40
Flannel shirt, long sleeves	0.30	Vest	0.20
Lightweight blouse, long sleeves	0.15	*Outdoor clothing*	
Trousers		Coat	0.60
Shorts	0.06	Down jacket	0.55
Lightweight	0.20	Parka	0.70
Normal	0.25	Fiber-pelt overalls	0.55
Flannel	0.28	*Sundries*	
Dresses, skirts		Socks	0.02
Light skirt (summer)	0.15	Thick ankle socks	0.05
Heavy skirt (winter)	0.25	Thick long socks	0.10
Light dress, short sleeves	0.20	Nylon stockings	0.03
Winter dress, long sleeves	0.40	Shoes (thin-soled)	0.02
Boiler suit	0.55	Shoes (thick-soled)	0.04
Sweaters		Boots	0.10
Sleeveless vest	0.12	Gloves	0.05

t_{mr} is the Mean Radiant Temperature, measure following ISO 7726.

In moderate environments where the difference between the temperature of the walls and the air is less than di 4 °C ($t_{mr} - t_a < 4$ °C) the operating temperature can be obtained by the relation:

$$t_o = A \cdot t_a + (1 - A) \cdot t_{mr} \quad (^\circ\mathrm{C}) \tag{3.62}$$

where

A is an a-dimensional value that depend on air velocity:

$$A = 0.5 \text{ if } v_{air} > 0.2\,m/s,$$
$$A = 0.6 \text{ if } v_{air} \text{ is between } 0.2 \text{ and } 0.6\,m/s,$$
$$A = 0.7 \text{ if } v_{air} \text{ is between } 0.2 \text{ and } 0.6\,m/s.$$

3.6 Comfort and Health in Indoor Environment: Indoor Air Quality

The study of well-being and comfort in confined environments regards other aspects. Besides the Thermal Comfort concerning control of exchange in energy and thermoregulation, also ensures the well-being conditions for other Human Senses. Comfort also covers:

- the Visual and/or Lighting Comfort, regarding factors diseases and disorders that improper lighting, natural or artificial, can have on Human Eyes (e.g. Glare effect, illuminance, lux level, etc.);
- the Acoustic Comfort regarding the factors of diseases and disorders that affect the Human Ear (e.g. problems given by noisy environments, chronic noise, high-pitched and precise noises, Decibel Level, threshold of hearing and threshold of pain, etc.) or on the correct acoustic perception in special cases (e.g. Auditorium, Theatre, Classroom, etc.) speech or music;
- the Indoor Air Quality (IAQ), which covers both the effects on the Human Nose, as an instrument of reception and perception of odors, and the Human Respirator Tract, which part of the human body through which air enters and chemical elements, including pollutants contained in the air.

In this book we deal with aspects related to the Lighting and Acoustic Comfort, which require a specific treatment, while we treat aspects related to Indoor Air Quality (IAQ), since the room ventilation and air movement have direct effect even on Thermal Comfort.

The study of the Comfort in the Indoor Environment ensures factors, indoor and outdoor, which have a direct effect on the health, because of the presence of pollutants in the air which are introduced into the environment through breathing.

The study of IAQ concerns the architectural design of Natural Ventilation systems, design of Heating, Ventilating and Air-Conditioning (HVAC) systems, as well as sources of pollutants present in indoor environments or coming from outdoor the environment.

The factors that determine the IAQ pollutants, reported in Table 3.9, can be removed thanks to the ventilation, which in turn can be carried out automatically by the HVAC, or made according to the habit of regularly ventilating the environments by the occupants, or thanks to the presence of the receptors of bad odors present in the Human Nose.

Table 3.9 Factors and parameters for the quality of the indoor environment (Bluyssen 2009)

	Thermal comfort	Indoor air quality
Parameter	Air temperature, mean radiant temperature, relative humidity, air-velocity, human metabolism, clothes	Pollution sources and air concentration: allergic, irritational, carcinogenic, etc.; ventilation rate and efficiency
Control	Design of building (architecture, insulation, windows, etc.), design of HVAC, IAQ	HVAC, ventilation system and maintenance air cleaning, activity control
Issues	Thermal adaptation, energy use, HVAC systema (e.g. radiators, fan-coil etc.), human behaviors	Secondary pollution: indoor chemistry, micro-organism, dust; energy use

The guarantee of the indoor air quality (IAQ) is necessary because children are more susceptible to indoor pollutants, of biological (colds and exanthemata's diseases). Frontczak and Wargocki (2011) report a literature survey on the different factors that affect indoor comfort environments. Other studies relate IAQ and labor productivity (Kosonen and Tan 2004).

Some diseases are caused by the presence of pollutants in the workplace or home, and there are a number of symptoms generated by the poor condition of the building, such as Sick Building Syndrome (SBS). The SBS is a disease recognized by the World Health Organization (WHO) and affects approximately 20% of the occupants of the buildings. The symptomatology is manifested by:

- Respiratory system: feelings of irritation and dryness of nasal and pharynx, light forms of asthma;
- Ophthalmic manifestations, dryness and irritation of the eyes, unpleasant smells;
- Neuropsychological manifestations: numbness, weakness, drowsiness, difficulty concentrating (Fig. 3.10).

The air we breathe must be purified by room ventilation and dilution of pollutants. The sources of indoor pollution can have different origins:

- Outdoor: by atmospheric pollution, dust, pollen, Industrial and Traffic pollution as Benzene, Tetraethyllead, Sulfur dioxide (SO_2), Ozone (O_3) etc.;
- Indoor, that depend by:
  - people and human crowd: olf, CO_2, water vapor, bacteria, virus, etc.;
  - tobacco smoke, hydrocarbure;
  - dust;
  - VOCs (Volatile Organic Compounds);
  - the result of combustion due to manufacturing processes or for cooking food (Nitrous Oxide NO_x, CO or Carbon Monoxide Carbon Dioxide CO_2);
  - insulation panels, furniture that may contain formaldehyde (HCHO), freon, mites, dust;
  - detergent and maintenance products e.g. detergent, solvents, stain removers, anti-pesticides, deodorant, cosmetic;

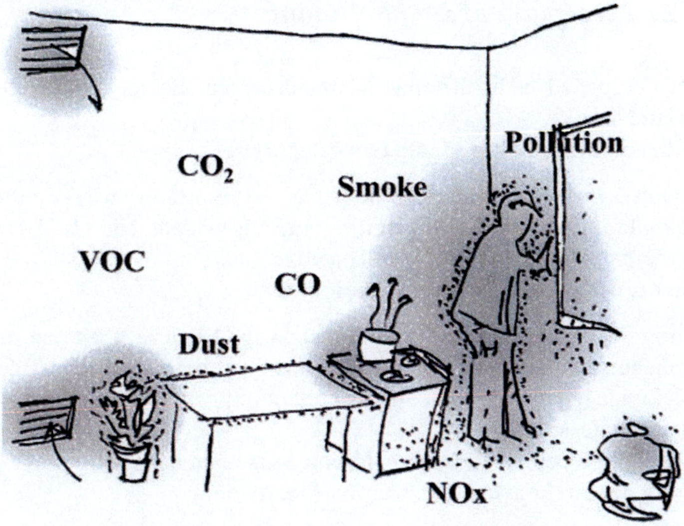

Fig. 3.10 Pollutant sources (CO_2 and other substances) in confined spaces

- Office Furniture: toner, printer and photocopier;
- Pollution by HVAC disfunction e.g. dust, fibers, bacteria, virus, mud, Legionellosis;
- by water technical systems: e.g. water vapor, VOCs, Radon, etc.;
- by building construction, e.g. radon by gypsum, clay, tuff, asbestos, mineral fiber, natural fiber, etc.

Table 3.10 report air pollutant by origin.

The IAQ and Ventilation Rate is discussed in Chap. 4, in this paragraph we simply highlight the aspects of perception and effects of IAQ.

Table 3.10 Air pollutants originating indoor and outdoor

Origin	Pollutants
Chemical	Nitrogen dioxide and sulfur (NO_2–SO_2); ozone (O_3); carbon monoxide (CO); carbon dioxide (CO_2); tobacco smoke (ETS); pesticides; volatile organic compounds (VOC)
Physical	Radon; asbestos; man-made mineral fibers (glass wool and rock); dusts
Microbiological	Viruses and bacteria; mold fungi; pollen; mites; microorganisms

3.6.1 The Perception of IAQ Pollutants

Pollution IAQ is perceived by the olfaction thanks at the olfactory epithelium present in the Human Nose.

The pollutants can be divided into two categories:

(a) pollutants *with odor*, which can be perceived in a direct way, by the Human Nose, and for which the feeling of discomfort is evident, bad smell, Dust, Mud, acids, detergents, perfumes, Benzene, etc.;
(b) pollutants *without odor*, which in their turn:
    (i) may be perceived in an indirect way, because they cause allergic phenomena, irritation of the mucous membranes, such as the Carbon Dioxide (CO_2) that alters the amount of oxygen and thus affect the feeling of tiredness;
    (ii) cannot be perceived by the Human Sense and cause chronic phenomena (e.g. radon) or acute (e.g. Carbon Monoxide).

The perception of the intensity of odors agrees with Stevens' law (Steven 1957) according to the level of concentration according to the law:

$$Perceived\ Odour\ Intensity = b \cdot C^{a} \tag{3.63}$$

where

C is pollution concentration (g/m^3)
a, b are constants.

The stimulation is given by the ratio of the number of molecules for each type of molecule.

Human Pollution: Olf and Decipol

The description of the pollutant load and techniques for the removal of polluted indoor help you understand and take action to ensure the respiratory-olfactory comfort of the indoor air.

The methodology of the study can be distinguished:

- Methodologies and descriptors to determine the level of air pollution and the definition of the pollutant load;
- Criteria and systems for the removal of pollutants by ventilation;
- Methods of measuring air quality and discomfort indexes (method PPD Predicted percentage of dissatisfied).

The pollution load due to the presence of people is described with two units of measure:

- the *Olf* (symbol olf, units [olf]), from Latin "olfactus", which measures the rate of air pollution produced by a normal person who carries out the activity of 1.2 met under conditions of thermal comfort standards and standard hygiene of 0.7 bathrooms/day. The unit of measurement was introduced by Fanger (1988), Fanger et al. (1988) to describe the pollution load produced by a person in relation to the activity carried out;
- the *Decipol* (symbol ζ, units [dp]), from Latin "pollutio", expresses the perceived air pollution due to the concentration of human effluent, and is defined as the pollution caused by a normal person (1 olf) subject to a ventilation of 10 (l/s liter per second) of unpolluted air:

$$\zeta = 0.1\,\text{olf} \quad (\text{dp}) \tag{3.64}$$

The level of perception of pollution limit (ζ_{lim}) due to odors for the presence of the person within the environment, is measured by the Predicted Percentage of Dissatisfied (PPD) desired, is calculated following Standard CEN-CR 1752 Ventilation for buildings—Design criteria for the indoor environment, by formula (Fig. 3.11):

$$\zeta_{lim} = 112 \cdot (PPD - 5.98)^{0.25} \quad (\%) \tag{3.65}$$

Fanger's methodology relates the air quality based on the indices of "perception" linked to the fulfillment of the occupants. The PPD (Predicted percentage of dissatisfied) is an index that expresses the percentage of people not satisfied in each condition microclimate. With respect to the pollutant load an environment with values less than 1 decipol the percentage of PPD is less than 15% (less than 15 out of 100 people are annoyed by the indoor air quality), for values of 10 decipol the PPD percentage rises to 60% of dissatisfied. Table 3.11 report pollution load by human in olf.

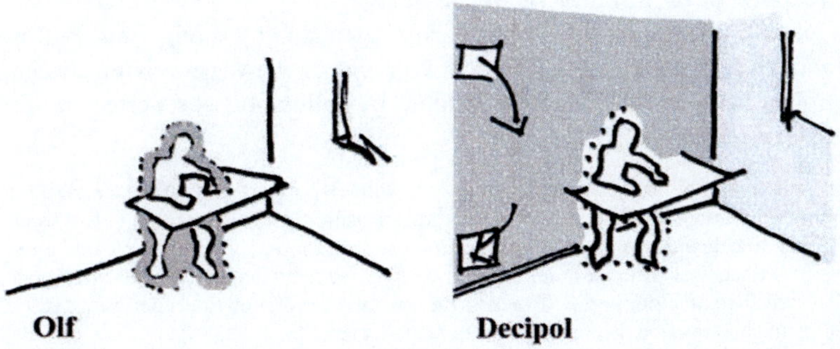

Fig. 3.11 Olf and decipol

Table 3.11 Pollution load in olf and functions of metabolic activity

Sources of pollution	Olf/m^2	Sources of pollution	Olf
		Non sedentary activities	
Crowding one person every 10 m^2		Light: 3 met	4
Biological emanations	0.1	Average: 6 met	10
+ with 20% of smokers	0.1	Heavy: 10 met	20
+ with 40% of smokers	0.2	Children 3–6 years: 2.7 met	1.2
+ with 60% of smokers	0.3	Young people 14–16 years: 1.2 met	1.3
Existing buildings with ventilation system	0.4	Person sitting con 1 met	1
Office buildings with 40% smokers	0.7	Person in light movement 4 met	5
No smokers	0.2	Person in movement 6 met	11
Office materials	0.05	*Sedentary activities (1.2 met)*	
		0% smokers	1
		20% smokers	2
		40% smokers	3
		100% smokers	6

Human Pollution: Carbon Dioxide (CO_2)

Carbon Dioxide (CO_2) is an indicator of the IAQ that helps determine whether the rate of pollution is due to the presence of people, given that CO_2 is mainly due to Human Respiration, and the quality of Ventilation rate in the environment. The excess CO_2 concentration affects the ability of concentration, can cause headaches or drowsiness effects that are excellent indicators on the quality of ventilation of an environment and that can also be perceived by children. The level of attention concentration in a game or activity at the school of the child are an indicator of air quality. The difficulty is to understand when these effects are due to physical fatigue and when they are the characteristics of the environment and its use, and what is the self-perception of the children on the air quality.

CO_2 is an indicator of the IAQ is easily detectable by the instrumental point of view, and, especially, from the perceptive point of view, perceiving the air of an environment such as "stale" and then loads of pollutants, as reported in the WHO guide for IAQ in School:

> "Methods for monitoring indoor air quality in schools" report that "Air exchange rate in a room is not constant. Therefore, longer-term measurements covering an entire week are necessary to characterize prevailing conditions in each room. Using CO_2 exhaled by occupants as a tracer gas offers a number of advantages, such as the ease of measurements and well established methodologies. To assess the dynamic ventilation rate, continuous CO_2 data from inside and outside the school have to be collected." […]

Table 3.12 Recommended CO_2 concentration above outdoor concentration

Category	Corresponding CO_2 above outdoor in ppm for energy calculations
I	350
II	500
III	800
IV	> 800

Table 3.13 Categories of air quality and levels of CO_2

Category	Description	CO_2 (ppm) level in rooms	
		Typical range	Default value
IDA 1	High indoor air quality	≤ 400	350
IDA 2	Medium indoor air quality	400–600	500
IDA 3	Moderate indoor air quality	600–1000	800
IDA 4	Low indoor air quality	> 1000	1200

> A suitable device should be capable of measuring CO_2 concentrations in a wide range from the ambient air level (300 to 400 ppm) up to 5,000 to 6,000 ppm level, that can be found in poorly ventilated classrooms. [...]
>
> The CO_2 method is applicable in schools with natural ventilation, mechanical exhaust ventilation and also for full mechanical ventilation systems if air is not re-circulated. [...]
>
> The raw CO_2, temperature and relative humidity monitoring data should be provided along with room occupancy and activity data for centralized data processing and analysis of air exchange rates. (WHO Regional Office for Europe 2011, pp. 10–12)

The use of CO_2 as an indicator of the Indoor Air Quality is provided in EN 15251 defines the criteria for the design of the internal environment and for the evaluation of the energy performance of buildings, in relation to indoor air quality, thermal environment, lighting and acoustics. CO_2 is the indicator that the EN 15251 identifies as people are the main cause of pollution (in buildings where people are the main pollution sources) as in the case of the kindergarten (nursery school) case study of this paper.

The standard EN 13779 defines the classification levels of indoor air (IDA indoor air quality) and CO_2 concentration that is a good indicator for the emission of human bio effluents. Tables 3.12 and 3.13 report IDA values following EN 13779.

References

Arens E, Zhang H, Huizenga C (2006) Partial and whole-body thermal sensation and comfort, part I: uniform environmental conditions. J Therm Biol 31:53–59

ASHRAE (1993) Physiological principles and thermal comfort. In: ASHRAE handbook-fundamentals. American Society of Heating, Refrigeration and Air-Conditioning Engineers, Atlanta, GA

ASHRAE 55 (2004) Thermal environmental conditions for human occupancy. American Society of Heating, Refrigerating and Air-Conditioning Engineers, Atlanta

Azer NZ, Hsu S (1977) The prediction of thermal sensation from a simple thermoregulatory model. ASHRAE Trans 83(Part 1)

Bartal I, Banhidi L, Garbai L (2012) Analysis of the static thermal comfort equation. Energy Build 49:188–191

Bellone E (2011) Qualcosa là fuori. Come il cervello crea la realtà (Something out there. How the brain creates reality). Codice Edizioni, Torino

Bluyssen PM (2009) The indoor environment handbook, how to make buildings healthy and comfortable. Earthscan, London

Butera FM (1998) Chapter 3—principles of thermal comfort. Renew Sustain Energy Rev 2:39–66

CEN prENV (1752) Ventilation for buildings design criteria for the indoor environment

Cheng Y, Niu J, Gao N (2012) Thermal comfort models: a review and numerical investigation. Build Environ 47:13–22

de Dear RJ, Ring JW (1990) Human subjective experience of ambient temperature step-changes: experimental results compared to the predictions of a numerical model. Indoor Air

de Dear R, Ring JW, Fanger PO (1993) Thermal sensation resulting from sudden ambient temperature changes. Indoor Air 3:181–192

EN 13779 Ventilation for non-residential buildings—performance requirements for ventilation and room-conditioning systems

EN 15251 Indoor environmental input parameters for design and assessment of energy performance of buildings addressing indoor air quality, thermal environment, lighting and acoustics

Fanger PO (1970) Thermal comfort. Analysis and application in environmental engineering. McGraw-Hill Book Company, New York

Fanger PO (1988) Introduction of the olf and the decipol units to quantify air pollution perceived by humans indoors and outdoors. Energy Build 12(1):1–6

Fanger PO, Lauridsen J, Bluyssen P, Clausen G (1988) Air pollution sources in offices and assembly halls, quantified by the olf unit. Energy Build 12(1):7–19

Fiala D, Kevin JL, Stohrer M (1999) A computer model of human thermoregulation for a wide range of environmental conditions: the passive system. J Appl Physiol 87:1957–1972

Fiala D, Psikuta A, Jendritzky G (2010) Physiological modeling for technical, clinical and research applications. Front Biosci S2:939–968

Frontczak M, Wargocki P (2011) Literature survey on how different factors influence human comfort in indoor environments. Build Environ 46:922–937

Gagge AP, Burton AC, Bazett HC (1941) Science 94:428

Gagge AP, Stolwijk JAJ, Nishi Y (1971) An effective temperature scale based on a simple model of human physiological regulatory response. ASHRAE Trans 77(1):247–262

Givoni B, Goldman R (1971) Predicting metabolic energy cost. J Appl Physiol 30:429–433

Glossary of terms for thermal physiology (1987) Pflugers Archiv 410:567–587

Guan Y, Hosni MH, Jones BW, Gielda TP (2003) Investigation of human thermal comfort under highly transient conditions for automobile applications—part 1: experimental design and human subject testing implementation. ASHRAE Trans 109(2):885–897

Havenith G, Holmèeer I, Parson K (2002) Personal factors in thermal comfort assessment: clothing proprieties and metabolic heat production. Energy Build 34:581–591

Herringot W, Gagge AP (1937) Physiological reactions to environmental temperature. Am J Physiol 120:1–22

Huizenga C, Zhang H, Arens E (2001) A model of human physiology and comfort for assessing complex thermal environments. Build Environ 36:691–699
International Ergonomics Association. www.iea.cc
ISO 11339:2001 Ergonomics of the thermal environment—principles and application of relevant international standards
ISO 11399 Ergonomics of the thermal environment—principles and application of relevant international standards
ISO 13731 Ergonomics of the thermal environment—vocabulary and symbols
ISO 26800 Ergonomics. General approach, principles and concepts
ISO 28802:2012 Ergonomics of the physical environment—assessment of environments by means of an environmental survey involving physical measurements of the environment and subjective responses of people
ISO 28803:2012 Ergonomics of the physical environment—application of international standards to people with special requirement
ISO 8996 Ergonomics of the thermal environment—determination of metabolic rate
ISO 9886 Ergonomics—evaluation of thermal strain by physiological measurements
ISO 9920 Ergonomics of the thermal environment—estimation of thermal insulation and water vapour resistance of a clothing ensemble
Jones B, Ogawa Y (1992) Transient interaction between the human body and the thermal environment. ASHRAE Trans 98:189–195
Kim CS, McCullough EA (2000) Static and dynamic insulation values for cold weather protective clothing. In: Nelson CN, Henry NW (eds) Performance of protective clothing: issues and priorities for the 21st century, vol 7. ASTM STP 1386. ASTM, Conshohocken, PA, pp 233–247
Kosonen R, Tan F (2004) The effect of perceived indoor air quality on productivity loss. Energy Build 36(10):981–986
McCullough EA, Jones BW (1984) A comprehensive data base for estimating clothing insulation. IER technical report 84-01. Institute for Environmental Research, p 1
Nilsson HO, Holmér I (2003) Comfort climate evaluation with thermal manikin methods and computer simulation models. Indoor Air 13:28–37
Poincarè H (1905) London and Newcastle-on-Tyne. The Walter Scott Publishing Co., LTD, New York
Schellen L, Loomans MGLC, Kingma BRM, de Wit MH, Frijns AJH, van Marken Lichtenbelt WD (2013) The use of a thermophysiological model in the built environment to predict thermal sensation. Coupling with the indoor environment and thermal sensation. Build Environ 59:10–22
Steven SS (1957) On the psycho-physical law. Psychol Rev 64:153–181
van Marken Lichtenbelt WD, Frijns AJH, van Ooijen MJ, Fiala D, Kester AM, van Steenhoven AA (2007) Validation of an individualised model of human thermoregulation for predicting responses to cold air. Int J Biometeorol 51:169–179
WHO Regional Office for Europe (2011) Methods for monitoring indoor air quality in schools. World Health Organization
World Health Organisation (WHO) (1999) Basic documents, 42nd edn. World Health Organisation, Geneva
Yao Y, Lian Z, Liu W, Jiang C, Liu Y, Lu H (2009) Heart rate variation and electroencephalograph—the potential physiological factors for thermal comfort study. Indoor Air 19:93–101

Chapter 4
The Indoor Thermal Comfort Indexes PMV and PPD

Abstract The study of the energy exchange between human body and environment (Human Body heat exchange) does not allow you to evaluate the view that people have with respect to comfort. Comfort can be expressed with indices and temperature sensation. This chapter describes the various indexes, in particular the Predicted Mean Vote and the related measurement tools as provided by the regulations. Added to this are considerations on the development of research in the field of comfort that includes the assessment of the social and cultural and tied to the Human Behavior, in particular the Adaptive Thermal Comfort, and the simultaneous evaluation of comfort parameters such as the Indoor Air Quality and the Indoor Environmental Quality.

4.1 The Research to Measure Comfort

The measurement of physical variables of the environment and physiological variables allow us to assess the Human or Thermal Energy Balance. The equations to evaluate the Human Thermal Energy balance allow us to measure the energy exchanges from the point of view of physical and physiological variables. These equations do not allow us to know the feeling and the perceived level of comfort of people.

The thermodynamic model proposed by Gagge (1946) excludes judgment on the subject: the human body is modelled as an open-system that exchanges energy and the level of well-being is related to the equilibrium conditions of the exchange of energy and mass, such as to ensure the optimal "physiological functioning" of the human body. In this approach with the Core-Skin, the feeling is absent, as well as all that concerns the cognitive aspect, namely understanding of its feeling on the part of the human body.

Gagge et al. (1986) proposed a "Two-node model" (core temperature and skin temperature) to analyze energy exchanges between human body and surrounding environment. In the 50s Houghten and Yagloglou wrote "Determining lines of equal comfort" (Houghten and Yagloglou 1953), introducing the notion of "Effective Temperature" index, which empirically resumed the concept of thermal comfort.

K. Fabbri, *Thermal Comfort Perception*, https://doi.org/10.1007/978-3-031-52610-7_4

"Effective Temperature" is the temperature of a fictitious room with 100% relative humidity.

In 1970, Fanger (1970), following a series of experiments, the indices of feeling Predicted Mean Vote (PMV) and Percentage People Dissatisfied (PPD) by adopting a statistical approach to define and measure, in relation to the metabolic characteristics, the clothing and the thermo-physical characteristics of the environment and the subjective judgment to feelings about the comfort conditions. Fanger's experiments confirmed and expanded the scientific literature allowing us to embed the individual cognitive process, about comfort, and express it as a judgment standard.

The approach of Fanger and the feelings indexes PMV and PPD included, also according to the statistical approach, also the cognitive aspect, the understanding of the question by the person. The PMV and PPD presuppose that the entity to which it raised the question is a healthy person and that has very clear meaning of the thermal heat, cold, etc.

The scientific literature has continued according to the two approaches, reaching an accuracy and robustness of the results that can process the International Standards, for parameterization and evaluation of the Feeling of Thermal Comfort.

The studies of Gagge, Fanger and others on the measurement of well-being has led to the formulation of two standards with international standards: ISO 7730 "*Ergonomics of the thermal environment—Analytical determination and interpretation of thermal comfort using calculation of the PMV and PPD*", and the ASHRAE 55 "*Thermal environmental conditions for human occupancy.*" Both standards have been updated over time as a result of studies and experiments and identify two approaches, one European and one American, more analytical to ensure the conditions of comfort.

Ten years ago, Olesen and Parson (2002) drew up a history of comfort:

> The main thermal comfort standard is ISO 7730, which is based upon the predicted mean vote (PMV) and predicted percentage of dissatisfied (PPD) thermal comfort indices.

In '*The indoor environment handbook*', Bluyssen (2009) describes a short history of comfort and indoor environment. Then Frontczak and Wargocki (2011) reported a literature survey about factors that can change comfort perception.

In this chapter we provide a summary of the contents of such standards and indices:

(a) on the one hand research and the standardization bodies (ISO International Standard Organization and ASHRAE);
(b) on the other hand, the research and other scholars have raised questions and sought to address other aspects of the relationship between man, human behavior and Environment, between feelings of comfort and physical parameters, physiological and psychological.

In particular Givoni (1969), contemporaneously to studies of Povl Ole Fanger, sought to study the relationship between the feeling of comfort and the cultural and social aspects.

Some areas of research, related to the study of biometeorology, conducted by a research team composed of de Dear, Brager and others (de Dear and Brager 1998a, b, 2011; de Dear et al. 1997), proposed the study of comfort according to the Model Adaptive Thermal Comfort (Nicol et al. 2012).

The New Guideline in the Netherlands (Auliciemes 1983) and by van der Linden et al. (2006) gives a definition of Adaptive Thermal Comfort:

> When a change occurs causing thermal discomfort, people react in such a way that their thermal comfort is reestablished.

Other areas of research focus on the study of the relationship between comfort and cognitive aspects, gender or for particular users (patients). Schellen et al. (2012, 2013) and Yao et al. (2008) studied physiological response of thermal comfort, on the other hand other kind of studies concern medicine approach like Choi and Loftness (2012) and Choi et al. (2012), or thermal sensation like Jin et al. (2012). Another interesting area of psychological research is the relationship between the thermal sensation and the understanding of heat (Gaoua et al. 2012).

In this way, more fascinating studies are about the relation between physiology and clothes, Ying et al. (2004), and application on the main indices in Ferrari and Zanotto (2012).

Figure 4.1 report different typologies of indoor temperature.

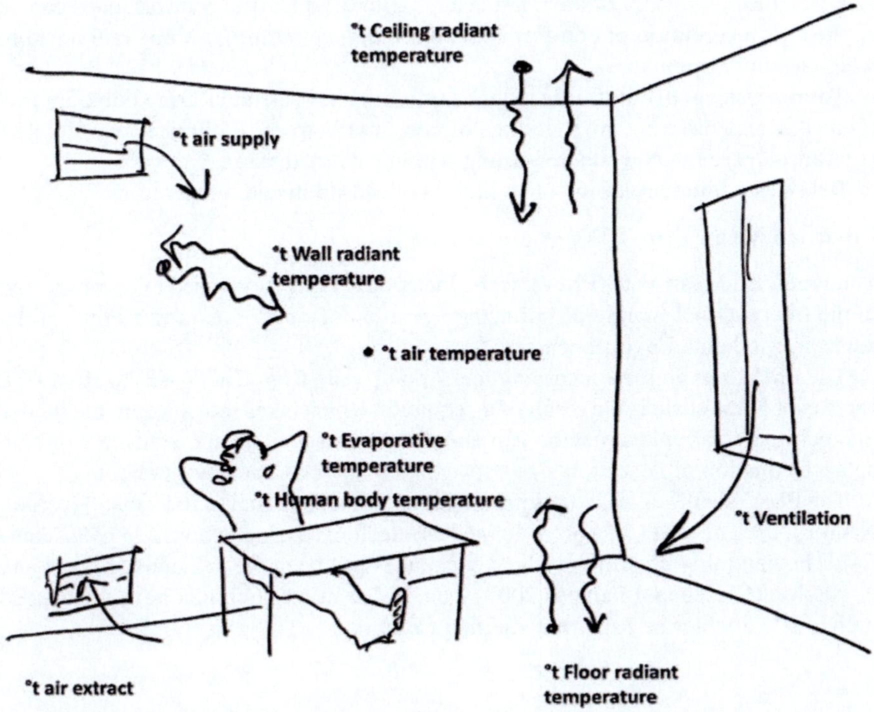

Fig. 4.1 Different typologies of indoor temperatures

4.2 The Comfort Indexes

The comfort study requires indicators, specifically the scientific literature uses the *indexes of comfort*, temperature or feeling, that allow you to express an opinion that people will go to express about the comfort in the environment, indoor or outdoor.

The indices of comfort are useful in summarizing the complexity of the reactions that occur between physiological human body, activity, clothing, habits, individual variables and the variability of the physical quantities relating to the environment.

Air-temperature, mean-radiant-temperature, air-velocity, relative humidity, absolute humidity, plan radiant-temperature, etc.

The index is thus a tool for *predictive* character, since it allows to speculate what will be the sensation of thermal comfort in relation to the choices of design, or evaluative to confirm the characteristics of an environment, indoor or outdoor, with respect to the choices or to guide the design choices.

The comfort index is a value that expresses the relationship between Human-and-Environment.

For this reason, it is necessary to define the characteristics of the subject (age, clothing, activities, habits, etc.) and the Environment (Moderate, Hot Environment, Cold Environment, Outdoor, etc.).

Given the complexity of the relationships and variables, the comfort index can be applied once condition of comfort defined (*boundary setting*), i.e. the range within which it can be applied.

The assessment of comfort in indoor environments is different from outdoor environments, also different are the range of comforts in tropical climates or cold. The comfort indexes may be adopted acting within a defined range.

Below is a summary of the main indices of sensation and temperature.

Predicted Mean Vote (PMV)

The Predicted Mean Vote (PMV), is the index of *Feeling* more used in literature and in the International Standards, allowing valuation of opinion expressed by healthy adults in Moderate Environmental.

The PMV is expressed according to a 7 point scale from Cold (– 3) to Hot (+ 3), each level of the scale is the result of an equation which takes into account the indoor microclimatic variables, metabolism and clothing of the subject, starting from the balance equation of heat exchange between human body and environment.

The PMV identifies as an optimal situation judgment equal to the value 0 (zero = Neutral), the Thermal Comfort is defined as a feeling in which you are in no discomfort. The neutrality condition of heat exchanges has been the subject of review and discussion (Olesen and Parsons 2002) from different research that have introduced individual variables or Adaptive Thermal Comfort.

Predicted Percentage of Dissatisfied Index (PPD)

The Predicted Percentage of Dissatisfied (PPD), is an *Statistical* index correlated to the PMV, but can also be expressed with respect to other indicators such as CO_2, the level of illumination, or local situations of discomfort.

It is an index that expresses the percentage expected of people who complain, which express a state of malaise with respect to the environment: the worse the condition of comfort, the greater the amount of people who have a negative opinion.

The Physiological Equivalent Temperature Index (PET)

The index Physiological Equivalent Temperature (PET) is an index of temperature, introduced in 1987 by Höppe and Mayer (1987) and Hoppe (1999), which expresses the real Air-temperature of an Indoor or Outdoor Environment, like the Air-Temperature Equivalent which would be in a Virtual Environment, in which are realized the same heat exchange of the human body of a real situation.

It is therefore a comparison, in which the heat exchange of the human body is kept constant, while varying the conditions of temperature to the boundary, on the one hand those defined for Virtual Environment, on the other the real conditions.

The human body heat exchange balance is fixed with a metabolism of 80 W/m^2 and the resistance of clothing equal to 0.9 clo.

The Virtual Environment reference is assumed with:

- a Mean Radiant Temperature equal to Air-Temperature ($t_r = t_a$);
- an Air-Velocity 0.1 m/s;
- a Relative Humidity 50% with 20 °C of air temperature.

The PET index is used both to evaluate indoor and outdoor environments and is an common index in biometeorological evaluations.

Relative Strain Index (RSI)

The Relative Strain Index (RSI) (Kyle 1992) is an index of feeling and statistics evaluating the response of individuals with regard to a condition of heat stress and is used to assess the conditions of heat stress due to the excess of heat exchange between the human body and hot environment, or stress that the body in trying to dispose of for energy to ensure the metabolic equilibrium.

The index RSI aims to describe three variables:

- The effects of heat on health (Heat Stroke or Hyperthermia);
- The effect on the Human Body, metabolism, age, sex;
- The criteria of cause and effect: hot sensation, tolerance, thirst sensation.

The index is calculated by taking an average man of 25 years old, healthy, sitting, dressed in full work clothes and not acclimatized to the heat.

The calculation is given by the following formula:

$$RSI = \frac{[10.7 + 0.74 \cdot (t_a - 35)]}{(44 - p_a)} \tag{4.1}$$

where

The meteorological data considered are the air temperature (°C) and the vapour pressure of the air (hPa) and the equation is:

where

t_a is the air temperature, in °C Expressed derived from meteorological data;
p_a is the pressure of water vapour, Expressed in hPa.

The measurement of the water vapor pressure is obtained by measuring the air temperature (t_a) and Relative Humidity (%), with the formula:

$$p_a = 0.006 \cdot HR \cdot 10^{(0.03 \cdot t_a)} \tag{4.1.a}$$

The range of applicability of this index is between 26 and 35 °C, and below 26 °C, considering the human body in a condition of homeothermy without the risk of heat stroke.

The classification of the index RSI is shown in Table 4.1.

In the table you can find the threshold values with the IHR and alongside the corresponding description of the conditions of discomfort.

In the case of people who are sick or elderly or at risk of Heat Stroke Hyperthermia occurs with RSI = 0.20.

The index is used in the field of meteorological and bioclimatological proposals, in particular, to warn the Health Authority and the population about risk of Heat Wave.

Actual Sensation Vote (ASV)

The Actual Sensation Vote index (ASV) (Nikolopoulou et al. 2001; Nikolopoulou 2004; Thorsson et al. 2004) is a bioclimatic index, empirical, mainly used to evaluate the outdoor amenities, and takes account of Heat exchange by convection, calculated using the formula

$$ASV = 0.034 \cdot t_a + 0.0001 \cdot SR - 0.086 \cdot WS - 0.001 \cdot RH - 0.412 \tag{4.2}$$

$$ASV_{Europe} = 0.049 \cdot t_a + 0.0001 \cdot SR - 0.051 \cdot WS - 0.014 \cdot RH - 2.079 \tag{4.3}$$

Table 4.1 Classification index RSI

RSI value	Sensation
≤ 0.15	Thermal comfort
0.15 ≤ RSI < 0.20	Less than 25% of the total population feels some discomfort
0.20 ≤ RSI < 0.35	All 100% of the total population feels some discomfort
0.35 ≤ RSI < 0.45	Less than 75% of the total population feels high discomfort
RSI ≥ 0.45	All 100% of the total population feels high discomfort

where

t_a is the air-temperature (°C)
RH is the Relative Humidity
WS is wind speed (m/s)
SR is Downwelling solar radiation.

ASV is the index is based on a 7-point scale as in P. O. Fanger for the index PMV, where zero represents "as usual".

Discomfort Index (DI) or Thom Index

The Discomfort Index (DI) or Thom Index (Thom 1959; Clarke and Bach 1971; Giles et al. 1990), is an index used in bioclimatology to evaluate, in an empirical way the feeling of discomfort in open spaces, in particular is designed to prevent heat waves and the Heat Stroke.

The Thom Index considers the evaporative heat loss required for thermal equilibrium, lack of balance or excess heat loss, involving the collapse of the human body and Heat Stroke. The index is calculated using the formulas

$$DI = t_a - 0.55 \cdot (1 - 0.01 \cdot RH) \cdot (t_a - 14.5) \tag{4.4}$$

$$DI_{index} = 8.3 + 0.4 \cdot t_a + 0.4 \cdot t_w \tag{4.5}$$

where

t_a is the dry-bulb air temperature (°C)
t_w is the wet-bulb air temperature (°C)
RH is the Relative Humidity (%).

The scale of sensation/feelings is shown in Table 4.2.

Effective Temperature (ET), Temperature Equivalent Perception (PET) and Equivalent Temperature (T_{eq})

The Effective Temperature Indexes (ET) (Fountain and Huizenga 1995; Houghton and Yaglou 1923) was developed by Gagge according to the 2-Node Model and

Table 4.2 Classification of human comfort-discomfort sensation

DI index (°C)	Comfort sensation
DI < 21	No discomfort feeling
21 ≤ DI < 24	Less than 50% of the total population feels discomfort
24 ≤ DI < 27	More than 50% of the total population feels discomfort
27 ≤ DI < 29	Most of the population feels discomfort
29 ≤ DI < 32	The discomfort is very strong and dangerous
DI ≥ 32	State of medical emergency

considers heat exchanges that occur between the center of the body and the skin (core-Skin). It considers the human body as a system which exchanges mass and energy in order to ensure a balance between environment and core (metabolic activity) that occur through exchanges in its contour (skin).

The model allows to evaluate the heat exchange between body and environment and as such correlate the human body heat exchange due to Air-Temperature and the real Relative Humidity with respect to that you would with a fictional environment, where the unknown factor of equation is the temperature of the fictitious environment, called Effective Temperature (ET).

The same principle, but with other formulas and parameters, is adopted in the Temperature Perception Equivalent Index (TEP) (Monteiro and Alucci 2008) or Standard Effective Temperature (SET) (Gagge et al. 1971 and 1986), and Equivalent Temperature (T_{eq}) (Bründl and Hoppe 1984).

The calculation is given by the formulas:

$$ET = t_a - 0.4 \cdot (t_a - 10) \cdot \left(1 - \frac{RH}{100}\right) \quad (4.6)$$

$$TEP = -3.77 + 0.4828 \cdot t_a + 0.5172 \cdot t_{mrt} + 0.0802 \cdot RH + 2.322 \cdot WS \quad (4.7)$$

$$T_{eq} = t_a + m \cdot \frac{(r - 2.326 \cdot t_a)}{(c_p + m \cdot c_w)} \quad (4.8)$$

where

t_a is air temperature (°C)
t_{mr} is mean radiant temperature (°C)
RH is Relative Humidity (%)
WS is wind speed (m/s)
w is specific humidity (g_v/kg_a)
m is the mixing ratio of water vapor (kg_v/kg_a)
c_p is the specific heat of air at constant pressure
c_w is the specific heat of water
r is the latent heat of vaporization of water.

As you can see both formulas reflect only air-temperature and Relative Humidity, and radiative and convective heat transfer are not included.

The classification of the index T_{eq} is shown in Table 4.3.

Operative Temperature (OT)

The Operative Temperature (OT) (ISO 7933; ASHRAE 1989) is an index that takes into account the air-temperatures (convective), the mean radiant temperature (radiation) and of the phenomena of heat exchange by radiation and convection-which entail a change of the indoor temperature over time; for this is used to detect the set-point temperature of the installations.

Table 4.3 Classification of Equivalent Temperature (Teq) by Bründl and Hoppe (1984)

T_{eq} (°C)	Comfort sensation
$51 < T_{eq}$	Sultry
$47 < T_{eq} \leq 51$	Slightly sultry
$34 < T_{eq} \leq 47$	Neutral (comfortable)
$27 < T_{eq} \leq 34$	Moderately cool
$T_{eq} \leq 27$	Cool

The calculation is given by the formula:

$$OT = \frac{(h_r \cdot t_{mr} + h_c \cdot t_a)}{(h_r + h_c)} \tag{4.9}$$

where

t_a is air temperature (°C)
t_{mr} is mean radiant temperature (°C)
h_r is Radiation coefficient
h_c is Convective coefficient.

The operative temperature of a room in the warm period, to define the characteristics of a room in order to avoid overheating in summer at the design stage, or to define whether the installation of a cooling system is necessary.

In the ISO 13792 OT defined as:

> *operative temperature*: uniform temperature of an enclosure in which an occupant would exchange the same amount of heat by radiation plus convection as in the actual non-uniform enclosure.

Universal Thermal Climate Index (UTCI)

The Universal Thermal Climate Index was introduced in the scientific literature with the research program COST ACTION 730 (COST UTCI 2004; Jendritzky et al. 2001, 2012) in order to make an assessment of the outdoor thermal environment, and is used within bioclimatological, urban design (Park et al. 2014), epidemiology, etc. to evaluate the heat exchange conditions of the human being.

The calculation of UTCI (Bröde et al. 2012) is given by 6th order polynomial Calculated by t_a (air temperatures in °C), t_r (mean radiant temperatures in °C) and WS10m (wind speed at 10 m above ground floor in m/s).

4.3 The Predicted Mean Vote (PMV) and the Predicted Percentage Dissatisfied (PPD) According to the Standard ISO 7730

Among the various indicators of comfort, the most common is the one developed by P. O. Fanger the Predicted Mean Vote. To confirm the validity of this index and the diffusion of index feeling is the fact that it was developed as an international standard for the calculation of the PMV and PPD from data indoor microclimate. In this section we describe the main contents of the standard.

The ISO 7730 has the purpose of predicting the general thermal sensation of a moderate environment. It also provides the criteria for calculating the Local Discomfort.

The standard applies to men and women, adults in good health who are in environments with microclimatic conditions that may have moderate deviations from the neutral condition of thermal comfort; in other words, apply to Moderate Thermal Environments.

In the case of Thermal Environments in conditions of stress such as Hot Environments (workplaces with more severe climatic conditions, kitchens, pizzerias, etc.), Very Hot Environments (working environments such as blast furnaces, or similar) or Cold Environments (work environments such as cold storage, or the like) should be done in reference to other standards such as ISO 7243 and the index WBGT (Wet Bulb Globe Temperature), ISO 11079, ISO 15743 for Cold Environment, and ISO 12894 for the risks due to extreme hot or cold environments.

The criterion for calculating the ISO 7730 considers the conditions of the environment in steady-state condition, no changes to the outline of the environment due to special situations (Local Discomfort). The same standard defines the conditions local discomfort always considering the steady-state condition.

The PMV index is defined as:

> The PMV is and index that predicts the mean value of the cote of a large group of person on the 7-point thermal sensation scale, based on the heat balance of the human body. Thermal balance is obtained when the internal heat production in the body is equal of the loss of heat to the environment.

The temperature control of the human comfort automatically controlling the temperature of the skin (skin temperature) and secretions of sweat (sweat secretion) (Table 4.4).

The PMV is calculated with the following equation by ISO 7730:

$$PMV = \left(0.303 \cdot e^{(-0.036 \cdot M)} + 0.028\right) \cdot \{(M - W)$$

Table 4.4 The 7-point thermal sensation scale

PMV	+ 3	+ 2	+ 1	0	− 1	− 2	− 3
Sensation	Hot	Warm	Slightly warm	Neutral	Slightly cool	Cool	Cold

$$
\begin{aligned}
& - 3.05 \times 10^{-3} \cdot [5733 - 6.99 \cdot (M - W) - p_a] \\
& - 0.42 \cdot [(M - W) - 58.15] - 1.7 \times 10^{-5} \cdot M \cdot (5867 - p_a) \\
& - 0.0014 \cdot M \cdot (34 - t_a) - 3.96 \times 10^{-8} \\
& \cdot f_{cl} \cdot \left[(t_{cl} \cdot 273)^4 - (t_r \cdot 273)^4\right] - f_{cl} \cdot h_c \cdot (t_c - t_a)\} \qquad (4.10)
\end{aligned}
$$

where

t_{cl} is the *clothing surface temperature* (°C), between 0 and 0.310 m^2 K/W (0–2 clo), calculated with the formula

$$
\begin{aligned}
t_{cl} = {} & 35.7 - 0.028 \cdot (M - W) - I_{cl} \\
& \cdot \left\{3.96 - 10^{-8} \cdot \left[(t_{cl} \cdot 273)^4 - (t_r \cdot 273)^4\right] - f_{cl} \cdot h_c \cdot (t_c - t_a)\right\} \qquad (4.11)
\end{aligned}
$$

h_c is the convective heat transfer coefficient (W/m^2 K), calculated with the two formulas

$$
\begin{aligned}
h_c &= 2.38 \cdot (t_{cl} - t_a)^{0.25} && \text{if } 2.38 \cdot (t_{cl} - t_a)^{0.25} > 12.1\sqrt{v_{ar}} \\
h_c &= 12.1\sqrt{v_{ar}} && \text{if } 2.38 \cdot (t_{cl} - t_a)^{0.25} < 12.1\sqrt{v_{ar}}
\end{aligned} \qquad (4.12)
$$

f_{cl} is the clothing surface area factor, calculated with the one of the due formulas

$$
\begin{aligned}
f_{cl} &= 1.00 + 1.29 \cdot I_{cl} \ \text{if } I_{cl} \le 0.078 \left(\frac{\mathrm{m^2\,K}}{\mathrm{W}}\right) \\
f_{cl} &= 1.05 + 0.645 \cdot I_{cl} \ \text{if } I_{cl} > 0.078 \left(\frac{\mathrm{m^2\,K}}{\mathrm{W}}\right)
\end{aligned} \qquad (4.13)
$$

the other terms in the equation are

- M is the metabolic rate (W/m^2), calculated according to ISO 8996 which depends on the subject and is calculated in relation to the area of Dubois, given that the equation relates to an adult to use in the case of children it is necessary identify the metabolic rate equivalent of the child rather than an adult. The equation is valid in the range of 46–232 W/m^2 (0.8–4 met);
- L is the effective mechanical power (W/m^2), usually equal to 0 if it is conducted moderate physical activities;
- t_a is the air temperature (°C) of indoor environment (range between 10 and 40 °C);
- t_r is the mean radiant temperature (°C) of indoor environment, measured with a globe thermometer (range between 10 and 40 °C);
- v_{ar} is the relative air velocity (m/s), affecting convective thermal exchange of skin/clothing and air, with a range from 0 and 1 m/s;
- p_a is the water vapour partial pressure (Pa), that affects thermal exchanges evapotranspiration due to breathing and sweat (range between 0 and 2700 Pa).

The indoor microclimatic data environment is calculated as required by ISO 7726.

The PMV index is added to the PPD index that allows us to quantify how many people have a negative opinion or dissatisfied with respect to the environment. The

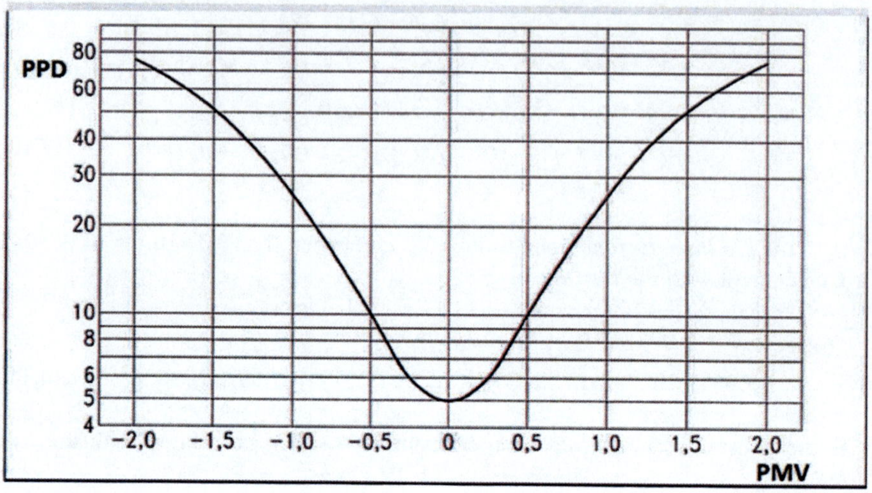

Fig. 4.2 Relationship between PPD and PMV (ISO 7730)

PPD is calculated with the formula

$$PPD = 100 - 95 \cdot e^{(-0.3353 \cdot PMV^4 - 0.2179 \cdot PMV^2)} \quad (4.14)$$

It follows that the condition of neutrality PMV index corresponds to 5% of people, however, have a negative opinion on Thermal Comfort.

The relationship between PMV and PPD is shown in Fig. 4.2.

4.4 Local Discomfort

Thermal comfort indices PMV and PPD refer to average conditions of the environment. In addition to these factors must be considered Local Thermal Comfort that can be caused by:

- Specific microclimatic situations localized in a part of the space, for example close to the windows or in the vicinity of the terminals of the heating system;
- Individual variables, floor temperature, vertical asymmetry, drafts, etc.

The Local Discomfort has a negative effect on the perception of the quality of indoor environments because it directly affects individuals. In environments for collective and continuous use, such as offices, schools and the like, in which people remain, often, in the same places, this can generate situations of chronic malaise, and diffuse physiological discomfort, there are also reports that a social microcosm is created between the occupants.

For example, students who occupy desks near the window can be in a condition of discomfort due to air-leakage (drafts) or Radiant Asymmetry. This factor can influence negatively on the perception of comfort by the student who will perceive their situation as disadvantaged with respect to the more fortunate classmates and this factor may acquire their negative perception on Thermal Comfort, or, on the contrary lead to situations of complaint chronic.

The factors of Local Discomfort, showed in Figs. 4.3 and 4.4, are:

- Drafts;
- Vertical Air Temperature Difference;
- Warm and Cool Floor;
- Radiant Asymmetry.

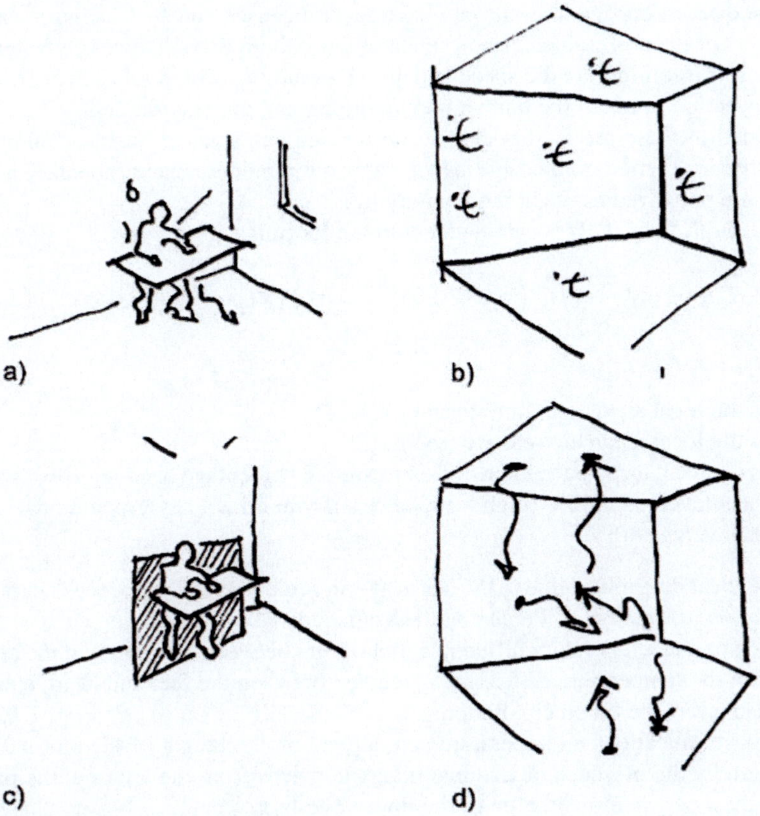

Fig. 4.3 Local discomfort factors: **a** the energy exchange of the body with the environment, **b** temperature and surface, **c** asymmetry of the radiant temperature and **d** the presence of localized air currents

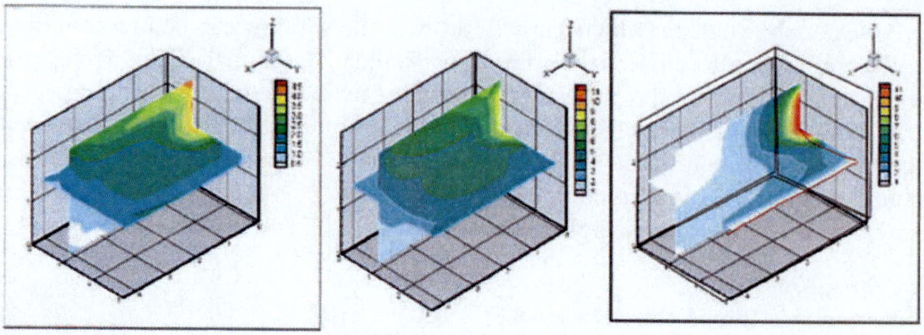

Fig. 4.4 Study of the vertical and horizontal distribution of temperature in an environment (Saro et al. 2006)

The discomfort due to drafts can have several causes, the air leakage of exterior windows or internal frames, the presence of fan-coil or air-vent, etc. The factors that affect the discomfort are the speed and air temperatures, which affect heat exchange by convection between the human body/clothing and the environment.

Added to these are factors depend on the subject, state of health, clothing and the activities carried out because if you carry out a moderate and sedentary activity drafts are perceived as much more annoying.

The Draft Rate (DR) is calculated with the formula:

$$DR = \left(34 - t_{a,1}\right) \cdot \left(v_{a,1} - 0.05\right)^{0.62} \cdot \left(0.37 \cdot v_{a,1} \cdot T_u + 3.14\right) \tag{4.15}$$

where

- $t_{a,l}$ is the local air-temperature locale (°C)
- $v_{a,l}$ is the local mean air velocity (m/s)
- T_u is the local turbulence intensity, expressed in percentage between 10% (without turbulence) and 60% (high turbulence), if you do not have specific data using the average 40%.

The local discomfort due to the Vertical Air Temperature Difference is expressed directly as a function of Percentage Dissatisfied (PD). The feeling of discomfort depends on the temperature difference that occurs between the floor and the ceiling, and then the temperature difference perceived between the feet and head, elements that also affect the blood circulation.

This stratification of temperature can depend on the height of the internal environment, by the presence of terminal plants in places from the top or at the base of the space, or by poor insulation of the floor or ceiling.

The formula to calculate the Percentage Dissatisfied (PD) is

$$PD = \frac{100}{1 + \exp\left(5.76 - 0.856 \cdot \Delta t_{a,v}\right)} \tag{4.16}$$

where

$\Delta t_{a,v}$ is the difference in vertical temperature between head and feet (°C).

Figure 4.5 show a graphic of local discomfort by vertical air temperature difference.

The temperature of the floor can cause a Local Thermal Discomfort, due to the different thermal sensation in the foot area, both for people who wear shoes or are barefoot, in relation to the activity carried out. The floor temperature may depend on the floor coating materials (marble, wood, ceramic, etc.); or by the presence or absence of thermal insulation between the layers of the floor and from the local to the adjacent beneath the floor (heated environment, non-heated, ground) that affects the surface temperature, or by the presence of heating systems with floor radiant floor panels; or other factors that may affect the surface temperature or near the floor.

The feeling of Thermal Discomfort also affects clothing proposals, especially the use of shoes and footwear, the type of shoes and socks; the size, height and age of the subject and the work done: standing, static or moving, or sitting, all of which affect the body's circulation close to the full.

To determine the conditions Percentage Dissatisfied (PD in %) using the formula:

$$PD = 100 - 94 \cdot \exp(-1.387 + 0.118 \cdot t_f - 0.0025 \cdot t_f^2) \tag{4.17}$$

where

t_f is the surface temperature of the floor.

Figure 4.6 show a graphic of local discomfort by pavement temperature.

In the event that the environment that is occupied by people with bare feet, the situation in cases of children in kindergartens, who can also play barefoot, you must follow the directions provided by the ISO/TS 13732-2.

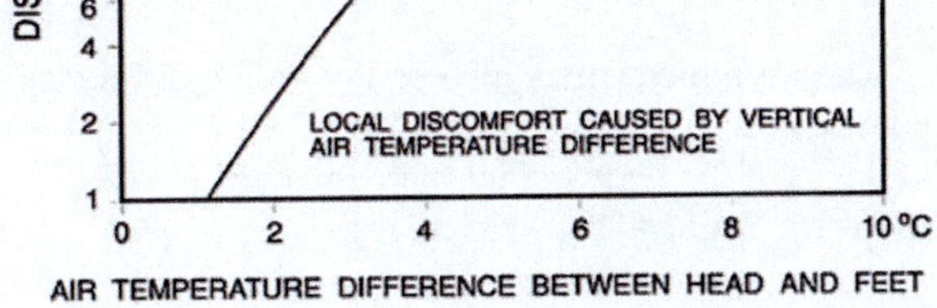

Fig. 4.5 Local discomfort caused by vertical air temperature difference (ASHRAE 55)

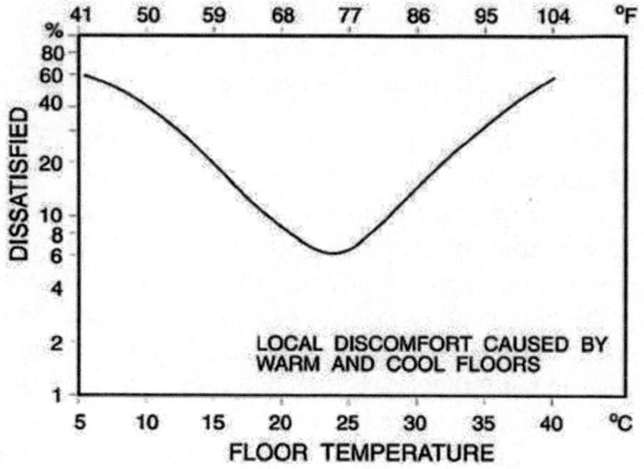

Fig. 4.6 Local discomfort caused by pavement temperature (ASHRAE 55)

The discomfort due to Radiant Asymmetry depends on the different distribution of surface temperature of the walls or ceiling, a factor that determines an asymmetry heat exchange by radiation between the human body and environment. In these environments will have the sensation of having one side of the body colder (or hotter) than the other.

The discomfort may be due in Warm Ceiling, in the case of radiant ceiling panels or roofs overheated, a phenomenon that can give the feeling of warmth in the head (or circle in the head), or Cool Ceiling, when the ceiling is cold and involves a greater feeling of cold in the vicinity of the head. The formulas to calculate the Percentage Dissatisfied (PD) are, respectively:

Warm Ceiling (if $\Delta t_{pr} < 23$ °C)

$$PD = \frac{100}{1 + \exp\cdot\left(2.84 - 0.174 \cdot \Delta t_{pr}\right)} - 5.5 \tag{4.18}$$

Cool Ceiling (if $\Delta t_{pr} < 15$ °C)

$$PD = \frac{100}{1 + \exp\cdot\left(9.93 - 0.5 \cdot \Delta t_{pr}\right)} \tag{4.19}$$

where

Δt_{pr} is radiant temperature asymmetry (°C).

The asymmetry may also be due to the difference of mean radiant temperatures between vertical walls (Wall Warm, Cool Wall) in this case there will be the sensation of a part of the body colder than another:

Warm Wall (if $\Delta t_{pr} < 35$ °C)

$$PD = \frac{100}{1 + \exp \cdot (3.72 - 0.052 \cdot \Delta t_{pr})} - 3.5 \quad (4.20)$$

Cool Wall (if $\Delta t_{pr} < 15$ °C)

$$PD = \frac{100}{1 + \exp \cdot (6.61 - 0.345 \cdot \Delta t_{pr})} \quad (4.21)$$

Figure 4.7 show a graphic of local discomfort by vertical wall/ceiling temperature asymmetry.

ISO 7730 identifies, in relation to values of PMV and PPD and local discomfort PD 3 categories of comfort, from Category A best to Category C, worst but still tolerable (Table 4.5).

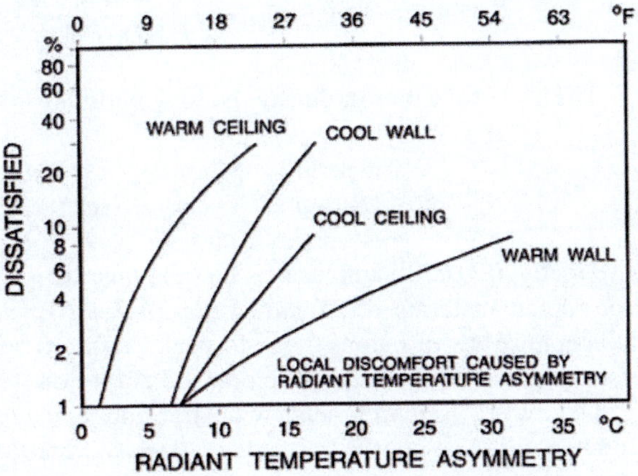

Fig. 4.7 Local discomfort caused by temperature asymmetry (ASHRAE 55)

Table 4.5 Thermal comfort category (ISO 7730)

Category	Thermal state of the body as a whole		Local discomfort			
	PPD (%)	PMV	DR (%)	PD (%) caused by		
				Vertical air temperature difference	Warm or cool floor	Radiant asymmetry
A	< 6	− 0.2 < PMV < + 0.2	< 10	< 3	< 10	< 5
B	< 10	− 0.5 < PMV < + 0.5	< 20	< 5	< 10	< 5
C	< 15	− 0.7 < PMV < + 0.7	< 30	< 10	< 15	< 10

The Standard 15251 extends this approach by introducing the categories relating to "*Indoor Environmental Quality*", which includes the Thermal Comfort, the Indoor Air Quality (and Ventilation Rates), Humidity, Lighting and Noise comfort levels.

In assessing the Thermal Comfort in the long period (Long-term evaluation) two approaches can be adopted (the ISO 7730 shows 5 approaches):

(a) calculate the number of hours (or percentage) during which the building is occupied and the value of PMV in this specific range;
(b) the time during which the value of PMV is outside of comfort Boundary (see previous Table) is weighed (weighted) with a factor based on the index PPD, and therefore

if PMV = PMV limit the weight factor $w_f = 1$
if |PMV| > |PMV limit| the weight factor w_f is given that

$$w_f = \frac{PPD_{actual,PMV}}{PPD_{PMV,limit}}$$

The Thermal Comfort during the long period (t = time) is given that:

$$\begin{array}{lll} \text{Warm period} & \sum w_f \cdot t & \text{for PMV} > \text{PMV}_{\text{limit}} \\ \text{Cold period} & \sum w_f \cdot t & \text{for PMV} < \text{PMV}_{\text{limit}} \end{array}$$

Finally, ISO 7730 emphasizes that the method is applicable to the environments considered in terms of "*Steady-State Thermal Environments*", or if the internal temperature has variations (peak-to-peak variation) below 1 K, and if there are rapid rises or descents of temperature below 2 K for now.

Otherwise, in environments with large and rapid changes in temperature the environment must be considered as "*Non-steady-state thermal environment*", and you can not apply the conditions of the long term.

4.4.1 Comfort Diagrams

The study of the thermal comfort of the indoor environment has affected several authors in the field of environmental (Peter Ole Fanger, Baruch Givoni), in the field of plant design (as ASHRAE or ISO standards) and in the approach to bioclimatic architecture (Victor Olgyay, 1963). These studies have led to the drafting of maps or diagrams of well-being, the graphical representations in which are the parameters that influence the microclimate of the indoor environment and the range of values within which you are in situations of comfort.

The diagrams are taken from the psychrometric chart for the study of the transformations of the moist air, fixed metabolic activity, the air speed and the type of clothing, the parameters which can vary are the temperature and the humidity; you

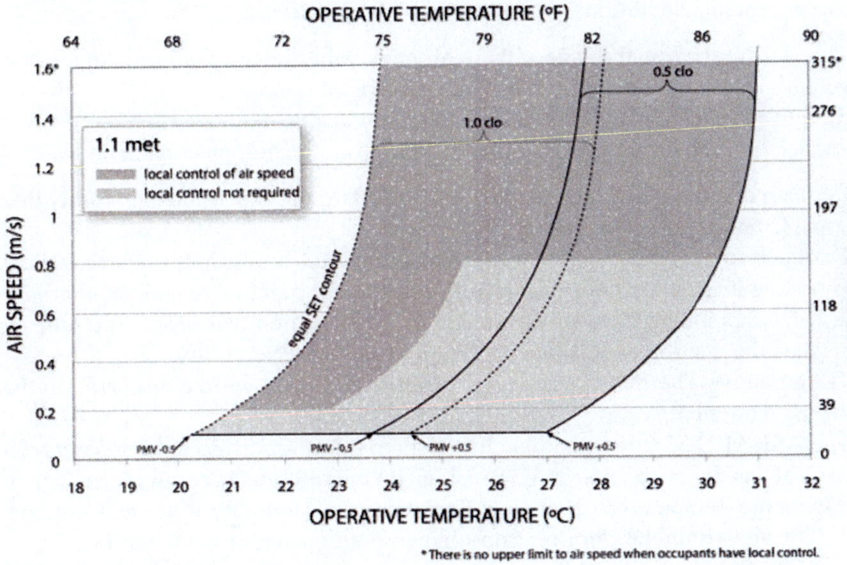

Fig. 4.8 Acceptable ranges of operative temperature to and average air speed Va for the 1.0 and 0.5 clo comfort zones at humidity ratio 0.010 (ASHRAE 55)

can then draw on a psychrometric chart to place the representative points of the well-being conditions. Figure 4.8 report and example of are well being in psychrometric chart.

4.5 Thermal Environmental Condition Form Human Occupancy Following ASHRAE Standard 55-2014

The ASHRAE Standard 55-2014 is the other international standard of reference, besides the ISO 7730, about the feeling of Thermal Comfort, and is a standard issued directly by ASHRAE (American Society of Heating, Refrigerating and Air-Conditioning Engineers) and constitutes the American standard (USA).

The standard defines the acceptable Thermal Environmental condition for healthy adults and uses the same PMV and PPD to assess the condition of indoor comfort.

The *Thermal Comfort* is defined as

> That condition of min that expresses satisfaction with the thermal environment and is assessed by subjective evaluation;

and the *Thermal Environmental* is defined as

> the characteristics of the environment that affect a person's heat loss

and the *Acceptable Thermal Environmental* is defined as

> an environmental that substantial majority of the occupants would find thermally acceptable.

The acceptable conditions are expressed with two criteria:

- *Thermal Neutrality*, the indoor thermal index value corresponding with a mean vote of neutral on the thermal sensation scale (PMV index);
- *Adaptive Model*, a model that relates to indoor design temperature or acceptable temperature ranges to outdoor meteorological or climatological parameters.

Compared to the ISO 7730 the ASHRAE Standard introduces the possibility that Thermal Comfort varies in time.

Consider in detail the thermal comfort compared to variable meteorological conditions outside, therefore considering only the variability of indoor microclimate parameters, regardless of the processes of adaptation that enacts the subject, even when the environment is in a condition of Thermal Neutrality.

The Adaptive Thermal Comfort is a specific mode that will be discussed in the following paragraph.

The ASHARE 55-2014 traditionally refers to the Comfort Zone Method, which consists of an area on the ASHRAE Diagram: all micro-climatic conditions in which the Operating Temperature, Wet-Bulb Temperatures, Humidity Ratio and Relative Humidity, are within the Comfort Zone and ensure the condition of comfort.

The Comfort Zone (Fig. 4.9) are defined according to the metabolic activity (1.0 and 1.3 met) and to clothing (0.5 and 1.0 clo), consider an air speed of 0.20 m/s.

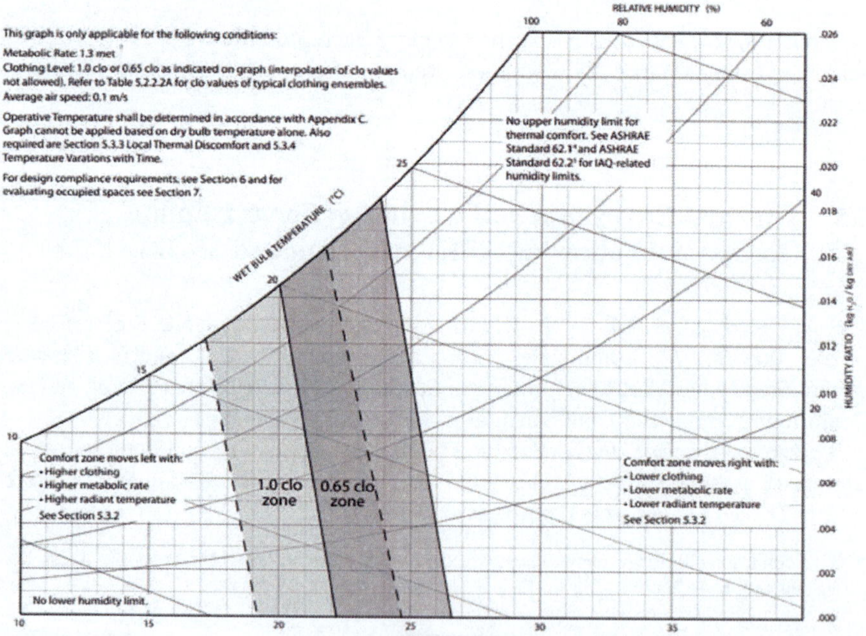

Fig. 4.9 Graphic Comfort Zone Method. Acceptable range of operative temperature and humidity for space (1.1 met: 0.5 and 1.00 clo) (ASHRAE 55)

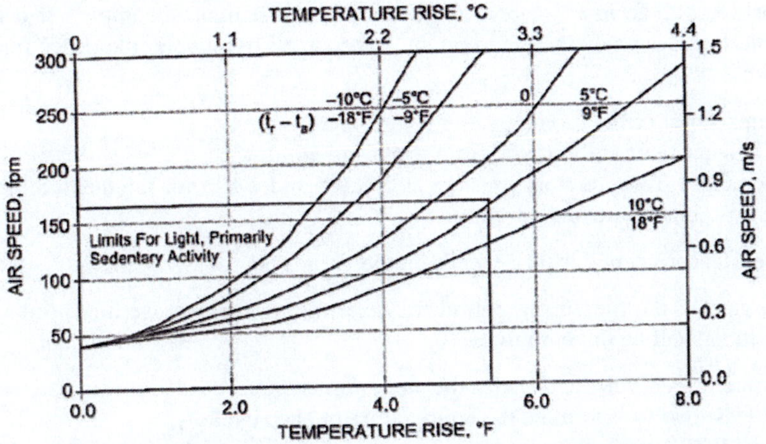

Fig. 4.10 Air speed required to offset increased air and radiant temperature (ASHRAE 55)

The Operative temperature is the uniform temperature of an imaginary black enclosure in which an occupant would exchange the same amount of heat by radiation plus convection as in the actual non-uniform environment.

In the case in an Elevated Air Speed, it is necessary to increase the values of the temperature in relation to the air-speed (Fig. 4.10), or to adopt SET Method (Standard Effective Temperature model).

The SET model uses a thermophysiological simulation of the human body to reduce any combination of real environmental and personal variables.

The method identifies the ASHRAE Standard Percentage Dissatisfied (PD) of the conditions of Local Thermal Discomfort, compared to:

- Radiant Temperature Asymmetry;
- Draft;
- Vertical Air Temperature Difference;
- Floor Surface Temperature.

The Standard considers the conditions vary over time, in particular the temperature, the Air-Temperature and/or Mean Radiant Temperatures; and shows that fluctuations under the direct control of the occupants do not have negative impact on Thermal Comfort, while fluctuations not under the control of the users can have a negative effect on the occupants.

In the case of Cyclic Variations, these are perceived as conditions of discomfort in the case where there is a Drift or a Ramps variation within 15 min.

The operations to carry out Evaluation of the Thermal Environment allow simple manual calculation or by computer models.

In the case of measurement of comfort with the use of instrumental equipment (probes, datalogger, etc.) are provided with the criteria for each variable. The Measurement Device must be located in an area of the building occupied by the people, so as to form a "*representative sample of occupant locations*". If it is not possible to assess how the environment is occupied by people, the probe must be located:

- In the center of the room;
- At a distance of at least 1.00 from walls and windows;
- The height from the floor must be at 0.1, 0.6 and 1.1 m for the measurement of the air-temperature and air-speed.

The Measurement Period depends on the variables:

- Air Speed, the measuring period for determining the average air speed at any location shall be three minutes.

Temperature Cycles and Drifts, the measurement must be made every five minutes and at least two to determine the temperature of each cycle.

The ASHRAE 55 gives an informative Appendix E to perform the Thermal Environment Survey, similar to ISO 10551. The questionnaire also includes questions to assess the position of people (Tables 4.6 and 4.7).

Table 4.6 Thermal Environment point-in-time survey

1. Record the approximate outside-air temperature ____ and seasonal condition	5. Are you near an exterior wall (within 15 ft)
☐ Winter ☐ Spring ☐ Summer ☐ Fall	☐ Yes ☐ No
2. What is your general thermal sensation?	6. Are you near a window (within 15 ft)?
☐ Hot ☐ Warm ☐ Slightly warm ☐ Neutral ☐ Slightly cool ☐ Cool ☐ Cold	☐ Yes ☐ No 7. Using the list below, please check each item of clothing that you are wearing right now ☐ Short-sleeve shirt ☐ Dress ☐ Nylons ☐ Long-sleeve shirt ☐ Shorts ☐ Socks ☐ T-shirt ☐ Athletic sweatpants ☐ Boots ☐ Long-sleeve sweatshirt ☐ Trousers ☐ Shoes ☐ Sweater ☐ Undershirt ☐ Sandals ☐ Vest ☐ Long underwear bottom ☐ Jacket ☐ Knee-length skirt ☐ Overalls ☐ Ankle length skirt ☐ Slip
3. Either a place on a X in the appropriate place where you are located now (Draw of room)	8. What is your activity level right now?
4. On which floor of the building are you located now? ☐ 1st ☐ 2nd ☐ 3rd ☐ Other (provide the floor number)	☐ Reclining ☐ Seated ☐ Standing relaxed ☐ Light activity standing ☐ Medium activity standing ☐ High activity

Table 4.7 Thermal Environment Satisfaction Survey (adapted from the CBE occupant IEQ survey)

1. Either a place on a X in the appropriate place where you are located now (Draw of room)	Please respond to the following questions base on your overall average experience in the past (six) months		
2. On which floor of the building are you located now?	6. How satisfied are you with the temperature in your space?		
☐ 1st ☐ 2nd ☐ 3rd ☐ Other (provide the floor number)	Very Satisfied	☐☐☐☐☐☐☐☐☐☐☐☐☐	Very Dissatisfied
3. Are you near an exterior wall (within 15 ft)	7. If you are dissatisfied with the temperature in your space, which of the following contribute to your dissatisfaction:		
☐ Yes ☐ No	*a. In warm/hot weather temperature in my space is:* ☐ Always too hot ☐ Often too hot ☐ Occasionally too cold ☐ Often too cold ☐ Always too cold		
4. Are you near a window (within 15 ft)? ☐ Yes ☐ No	*b. In cool/cold weather temperature in my space is:* ☐ Always too hot ☐ Often too hot ☐ Occasionally too cold ☐ Often too cold ☐ Always too cold		

(continued)

Table 4.7 (continued)

5. Which of the following do you personally adjust or control in you space?	*c. When is this most often problem?*
☐ Window blind or shades ☐ Room air-conditioning unit ☐ Portable heater ☐ Permanent heater ☐ Door in interior space ☐ Door in exterior space ☐ Adjustable air vent in wall or ceiling ☐ Ceiling fan ☐ Adjustable air vent (diffuser) ☐ Portable fan ☐ Thermostat ☐ Operable window ☐ None of these ☐ Other: ___________	☐ Morning (before 11 am) ☐ Midday (11 am–2 pm) ☐ Afternoon (2–5 pm) ☐ Evening (after 5 pm) ☐ Weekend/holidays ☐ Monday morning ☐ No particular time ☐ Always ☐ Other
	d. How do you best describe the source of this discomfort?
	☐ Humidity too high (damp) ☐ Humidity too low (dry) ☐ Air movement too high ☐ Air movement too low ☐ Incoming sun ☐ Heat from office equipment ☐ Draft from windows ☐ Draft form vents ☐ My area is hotter/cooler than other areas ☐ Thermostat is inaccessible ☐ Thermostat is adjusted by other people ☐ Clothing policy is not flexible ☐ Heating/cooling system does not respond quickly enough to the thermostat ☐ Hot/cold surrounding surfaces (floor, ceiling, walls, or windows) ☐ Deficient window (not operable) ☐ Other: ___________

4.6 Ergonomics of Thermal Environment and People with Special Requirements

The goal of ergonomics is to know the response of the human body with respect to the indoor microclimate.

The definition of Ergonomics of thermal environment includes a set of parameters and standards that take into account the parameters related to the physical environment, the metabolic characteristics and clothing of the subjects (people) and working conditions or user space.

The Standard ISO 11399 Ergonomics of the thermal environment—Principles and application of relevant international Standards, summarizes the framework in which to study and measure the comfort according to the criteria of international standards. Then there are the studies and research for specific cases (as are those reported in the second part of the book) that take into account the non-standard data.

The study of the Standard allows the use of definitions and calculation criteria consolidated under the Scientific Literature proposals, especially for the Comfort valuation sector in which it is possible to have different interpretations and application contexts.

To make a distinction between the different study Environments:

- *Moderate Environments* (ISO 7730) or which he refers to the index PMV and the standards ISO 7726 (*Thermal environments—Instruments and methods for measuring physical quantities*) with regard to measuring instruments, ISO 9886 (*Evaluation of thermal strain by physiological measurements*) to characterize the parameters of physiologic human body (body core temperature, skin temperature, heart rate, body mass loss), ISO 8996 (*Ergonomics—Determination of metabolic heat production*) that allows you to define the metabolism of the subjects, useful if calibration of the metabolism of children than that of adults and the standard ISO 9920 (*Ergonomics of the thermal environment—Estimation of the thermal insulation and evaporative resistance of a clothing ensemble*) that allow us to define and calculate the thermal resistance of clothing;
- In the working environment *Hot Environmental* (ISO 7243—*Hot environments—estimation of the heat stress on working man, based on the WBGT-index* (*wet bulb globe temperature*)) for which we calculate the index WBGT (Wet Bulb Globe Temperature), which assesses the risk of Thermal Stress of the human body subjected to high energy exchange and mass (sweat and evapotranspiration);
- The working environments *Cold Environments* (ISO/TR 11079) that allows you to calculate the index $IREQ_{neutral}$ and $IREQ_{min}$ and based on the that the type of appropriate clothing, calculated according to ISO 9920;
- These final two environments do not affect the objectives of this book.

The standards exclude the relative aspects:

- To the psychological variables or individual context, including cognitive aspects related to age of the subjects (in other words standards always refer to adults in good health);

- To the individual physiological variables, due to physical and/or mental illness;
- To the variable of cultural, social, income etc.;
- To the Adaptive Thermal Comfort;
- To the effects of other factors in the environment perceptual or synthetic that affect the comfort with Humidity, Lighting, Indoor Air Quality or Noise.

All these aspects relate primarily to research, to Scientific Literature and to the Design Environment.

The "specific users" also known as people with special requirements, are the subject of the Standard ISO/TS 11415, which identifies the need to define criteria and range of Thermal Comfort, by virtue of the uniqueness of health of the subject.

The diversity of the subject, in particular a subject's health and hospital care in buildings, implies the need to take into account some special factors to consider in assessing the Thermal Environment. The special factors are:

- *Diversity or physical disability* and the method of treatment and care, which affects the thermal sensation, some people have their thermal needs in relation to their condition, such as pregnant women, the elderly, children and the sick in general;
- The *body shape*, the Thermal Comfort allows for heat exchange between the human body and environment, considering the human body as an open system and the skin surface contour through which energy exchanges take place. ISO 7730 in the human body and its exchange surface is calculated according to the formula of DuBois. The formula to calculate the body surface area, results in errors in applying generalized to the elderly, children, pregnant women, etc. whose body proportions are different than adults, and this has an influence on heat exchange, and impact of thermal radiation, convection and evaporation;
- The impact of heat exchanges due to *perspiration* (sweat secretion), linked to the metabolic characteristics of the subjects, and that varies in relation to physical activity and to particular disease conditions, the ISO/TS 14415 gives the example of quadriplegic (high-level spinal-cord-injured persons) and some other paralytic disease;
- The impact of the vasomotor system (*vasomotor control*) that regulates the circulation in the peripheral areas of the body (hands, feet, head, or Peripheral vasomotor disorders) and which can be altered in relation to conditions caused by disease or specific treatments that alter the metabolism or mobility;
- The different metabolism (*metabolism rate*), parameter standardized by the rules, but that varies appreciably from person to person in order to have an impact on comfort. The metabolic activity depends on the activity (sitting, standing, walking, etc.) and specific conditions of health, some individuals may have high levels of metabolism, ISO/TS 14415 gives the example of the high metabolism of sufferers of athetosis cerebral palsy, and low metabolism from adults, and the elderly;
- The influence of situations of *Thermal Stress on physiological function*, such as, for example in cases of *Cerebral Apoplexy* or *Cardiovascular attacks* increases the feeling of cold.

People with particular health conditions, are more sensitive compared to the microclimatic conditions of an environment, both from the physiological point of view and, consequently, from the psychological point of view.

The ISO/TS 14415 Annex has placed some disabilities that require specific attention with respect to *Indoor Thermal Comfort* (accompanying thermo-physiological disabilities). In particular responses that:

- In the case of *Spinal cord injury*, you do not have thermal sensations due to the sensations on the skin of the affected part, so the body does not use this type of temperature control, while you have disorder of micturition, which can be aggravated in the case of Thermal Stress;
- *Cerebral palsy*, usually they want a cooler environment because of a higher metabolic rate;
- *Diseases of circulatory organs* (cerebral infarction, myocardial infarction, nephritis, hypertension, arteriosclerosis, etc.) in these cases the parties are no longer able to properly regulate vasodilation, so their body does not respond and adapt to the environment in case of hot or cold (e.g. $-1 > \text{PMV} > +1$).

This book focuses on the perception of the comfort of children, it is nevertheless useful to have a reference about specific situations in the standard that deviate from the ISO 7730 and are not related to work situations. It can be a useful reference on how to share and analyze specific situations related to children: healthy, with physical or mental disorders.

Children are an important part of the population, as they are in school buildings and, in their case, the study of Thermal Comfort includes related aspects:

- To the body, or to the metabolism and physiological factors (*Thermal Children Body Area and Metabolism*);
- Age and corresponding cognitive level (***Thermal Cognitive Knowledge***);
- The state of health, physical and psychological, of the child (e.g. what can be the impact on the comfort of children with Autism, Asperger Syndrome ADHD (Attention-Deficit/Hyperactivity Disorder));
- Legal issues related to the "*Designing for Education*" (OECD 2011), i.e. the building projects and HVAC, where there are children, and in particular school buildings;
- Design buildings proposals, especially those schools in which there are children.

All aspects are not (yet) standardized.

In the design of school buildings research and literature in the field of architecture, technical and scientific, since the early pedagogical models, there has been interest in the relationship between the space and the behavior of children, including for education. In pedagogy refers to the expression "*space that educates*" intending to report the relationship that the spatial arrangement has on the child's behavior in space and educational process.

Nevertheless, the school buildings are designed according to the standards and comfort levels of adults (teachers).

It remains to be studied the relationship between the Indoor Environment, Thermal Comfort and Children in relation to the pedagogical aspect: how they perceive and understand the microclimate of children? that can have influence on their cognitive level? or school learning? how it can affect the understanding of environmental sustainability, the scientific method and art, in their construction and comprehension of the world?

4.7 The Instruments of Measurement of the Indoor Microclimate

The measurement of physical sizes required to define the indoor microclimate must be conducted in accordance with the ISO 7726 Standard.

The physical quantities (Table 4.8) that characterized the environment are:

Table 4.8 Physics quantities, measure units, definition and instrument

Physics quantities	Measure units	Definition	Instrument
Air temperature	Kelvin K (T_a) Celsius °C (t_a)	Is the temperature of air around human body	Thermometer
Mean radiant temperature	Kelvin K (T_r) Celsius °C (t_r)	Is the uniform temperature of an imaginary enclosure in which radiant heat transfer from the human body is equal to radiant heat transfer in the actual non-uniform enclosure	Globe-thermometer
Plane radiant temperature	Kelvin K (T_{pr}) Celsius °C (t_{pr})	Is the uniform temperature of an enclosure where radiance on one side of a small plane elements is the same in the non-uniform environment	Net-radiometer
Absolute humidity of the air	Partial vapour pressure (pa) in kilopascals (kPa)	Is the actual amount of water vapour contained in the air	(Directly) dew-point instrument (Indirectly) by psychrometer
Relative humidity of the air	(%)	Is which gives the amount of water vapour in the air in relation to the maximum that it can contain at a given temperature and pressure	Psychrometer
Air velocity (va), expressed	Meter per second (m/s)	Speed of the air, average of the velocity	Hot-wire anemometer
Surface temperature	Kelvin K (T_r) Celsius °C (t_r)	The temperature of a given surface	Contact thermometer Infrared sensor

- *Air Temperature*, expressed in Kelvin K (T_a) or in Celsius °C (t_a), is the temperature of air around human body;
- *Mean radiant temperature*, expressed in Kelvin K (T_r) or in Celsius °C (t_r);
- *Absolute Humidity* of the air, expressed by partial vapour pressure (p_a) in kilopascals (kPa);
- *Air Velocity* (v_a), expressed in meter per second (m/s);
- *Surface Temperature* expressed in Kelvin K (T_r) or in Celsius °C (t_r).

From the measurement of these quantities, you can get the magnitudes and related PMV or PPD indexes. These quantities can be related with the variables related to the Human Body to evaluate the Thermal Balance between human body and environment (Table 4.9). The measuring instruments must ensure accuracy and 90% response times of the sensors for the basic quantities.

An environment can be considered as "homogeneous" from the point of view bi-climate if, at the same time, air temperature, radiation, humidity and air velocity, can be considered uniform around the subject. If the environment is heterogeneous, local variations must be identified as factors of discomfort.

The physical quantities are functions of time, change in the environment and in time in relation to the operating temperature of the HVAC, the outside temperature, the presence and activities of people, etc. considering the environment stationary conditions over time if the fluctuation does not exceed the measurement error of accuracy required by ISO 7723.

If the environment can not be considered stationary in relation to the subject it should be taken into consideration the physical variables in function of time (Table 4.10).

The ISO 7726, attached, reports the "*Principle for measuring*" for each variable reporting the methods of calculation and precautions in the use of probes, and types of sensors.

Figures 4.11 and 4.12 report an example of sensors: a Globe thermometer and hot-wire anemometer.

In the case of the measurement of the Air-Temperature (t_a) attention must be paid to the effect of radiation on the probe as direct radiation may alter the measurement of air temperature. In addition, the sensor should have elapsed periods equal to at least 1.5 times the response time (90%) of the probe.

Table 4.9 Relation between physics quantities and thermal balance between man and the thermal environment

Elements in the thermal balance	Quantities							
	t_a	t_r	v_a	p_a	I_{cl}	R_{cl}	M	W
	Air temperature	Mean radiant temperature	Air velocity	Absolut humidity of the air	Insulation of clothing	Evaporative resistance of clothing	Metabolism	External work
Internal heat production, M + W							X	X
Heat transfer by radiation, R		X						
Heat transfer by convection, C	X		X		X			
Heat losses through evaporation: • From the skin, E			X	X		X		
• By respiration E_{res}				X			X	
Convection by respiration, C_{res}	X						X	

Table 4.10 Characteristic of measuring instrument in cases of measurement of comfort (Class C)

Physics quantities	Measuring range	Accuracy	Response time (90%)
Air temperature (t_a)	10–40 °C	Required: ± 0.5 °C Desirable: ± 0.2 °C Deviation $\|t_r - t_a\|$ equal to 10 °C	The shortest possible. Value to be specified as characteristic of the measuring instrument
Mean radiant temperature (t_r)	10–40 °C	Required: ± 2 °C Desirable: ± 0.2 °C	The shortest possible
Plane radiant temperature (t_{pr})	0–50 °C	Required: ± 0.5 °C Desirable: ± 0.2 °C Deviation $\|t_{pr} - t_a\|$ z 10 °C	The shortest possible
Absolute humidity of the air (p_a)	0.5–3.0 kPa	± 0.5 kPa Deviation $\|t_r - t_a\|$ at least 10 °C	The shortest possible
Air velocity (v_a), expressed	0.05–1 m/s	Required: ± (0.05 + 0.05v_a) m/s Desirable: (0.02 + 0.07v_a) m/s	Required: 0.5 s Desirable: 0.2 s
Surface temperature (t_s)	0–50 °C	Required: ± 1 °C Desirable: ± 0.5 °C	The shortest possible

The Temperature Sensor can be:

- expansion thermometers: (a) liquid expansion thermometer (mercury), (b) solid expansion thermometer;
- electrical thermometer: (a) variable resistance thermometer (platinum resistor or thermistor), (b) thermometer base on the generation of an electromotive force (thermocouple);
- thermo-manometers, with variation in the pressure of a liquid as a function of temperature.

The measurement of Mean Radiant Temperature (tr) can be made with different instruments, the most widely used is the black-globe thermometer. The black-globe

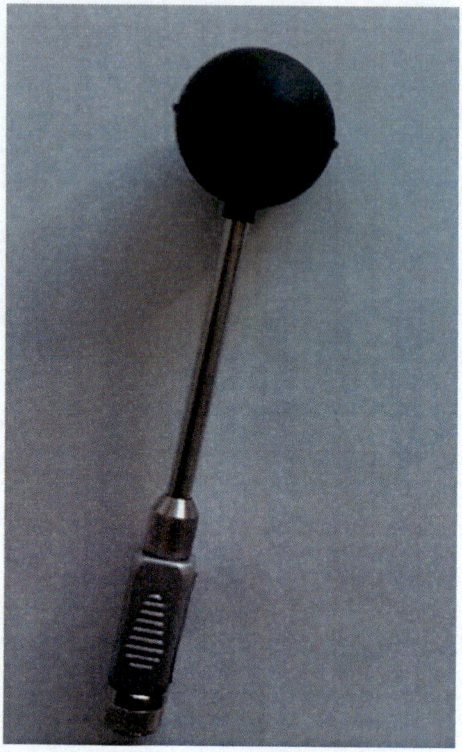

Fig. 4.11
Globe-thermometer

thermometer consists of a black globe (from 0.05 to 0.15 m of diameter) in the center of which is placed in temperature sensor: such as the bulb of a mercury thermometer, or a thermocouple (more common), or resistance to probe.

In the use of a black-globe thermometer remembering that the response time is about 20–30 min, that people are not equivalent to a sphere of globe thermometer, and that the measurement is strongly influenced by direct solar radiation (the sun for example) and the color of the globe that should remain black, without powder or other alterations.

The other measuring instruments are:

- *Two-sphere radiometer*, consisting of two spheres at the same temperature, with different emissivity in which one is black is the other is polished. The emittance of the black sphere is higher than the polished, the measurement of radiant temperature occurs due to the difference of the heat supply between the two spheres;
- *Constant air temperature sensor*, in this method a sensor is controlled at the same temperature as the surrounding air temperature, there being no convection heat loss and the necessary heat supply (cooling supply) to the sensor being equal to the radiant heat loss.

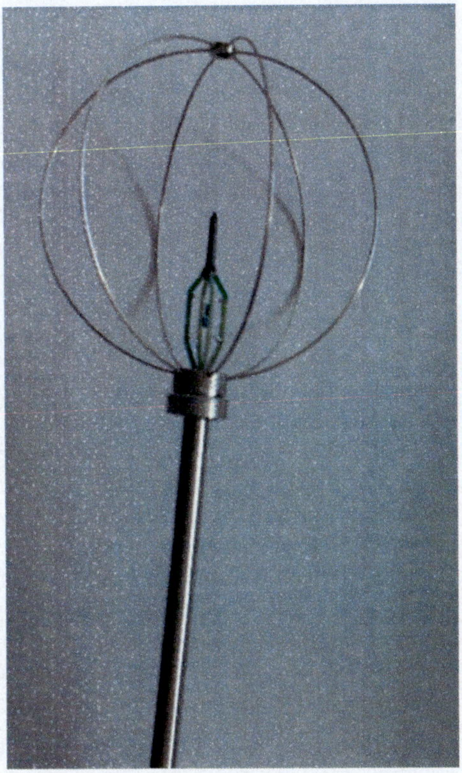

Fig. 4.12 Hot-wire anemometer

The measurement of the *Mean Radiant Temperature* can also be obtained:

- from the plane radiant temperature (t_{pr});
- from the temperature of the surrounding surface: in this case the mean radiant temperature is obtained in function of the form factor (function of the shape) and the emissivity of the walls.

To measure the *Plane Radiant Temperature* (t_{pr}) can be used:

- A *Heated sensor* consisting of a reflective disc (gold-platter) and an absorbing disc (matt black painted). The two disks have a different emissivity, the difference in heat supply between the two discs to keep them at the same temperature allows for the measurement of the plane radiant temperatures;
- with *Net-radiometer* (more common), that consists of a small black plane element with heat flow meter (thermopile) between the two side of the element. The net heat flow between the two sides is equal to the difference between the radiant heat transfer at the level of the side of the element.

The *Absolute Humidity of the air*, can be calculated with a *Psychrometer* (more common), or a Lithium Chloride Hygrometer, or using the psychometric chart from Dry-bulb Temperature Thermometer and Wet-bulb temperature thermometer.

The Air Velocity is measured with an anemometer, which can be bi-directional or use sensors in three directions. Anemometers can be of different types: Vane and cup anemometers; Hot-wire anemometers (more common); Pulsed wire anemometer; Hot-sphere and thermistor anemometer (more accuracy); Ultrasonic anemometer or Laser-Doppler anemometer.

In the evaluation of the comfort in moderate environments, given the limited air velocity (< 1.5 m/s) using Hot-wire anemometers, constituted by a pair of heating elements placed very close to each other that, due to the difference in measurement of current intensity, are able to determine the air velocity.

Finally for the measurement of Surface Temperature can be with used:

(a) *contact thermometers* (resistance, or thermocouples), which consist of sensors placed in contact directly with the surface of the wall; the temperature stability is obtained by measuring the heat exchange between surface and environment and then the measured surface temperature.
(b) *infrared sensor* (or remote temperature sensor) does not provide the contact with the wall, the measurement is made with Infrared radiometers; a radiometer only measures the energy level of the radiation detector incident on it, and its incident radiation includes radiation emitted by the object and radiation reflected from the surface of the object.

The ASHRAE Standards 55 and EN ISO 7730 are recognized as the correct standardized procedure to evaluate comfort, in contexts outside of the scope of the research, such as Building Design, HVAC Design, Healthy Evaluation, etc. or in professional contexts in which it is necessary to ensure a standardized assessment procedure. The ISO Standards should be valid, reliable and useable, this is concerned with whether the users of a standard can use it correctly.

The standards are constantly being revised in particular Olesen (Olesen and Parson 2002) raised the need for an update. Olesen reviews the *Validity* of ISO 7730,

> Laboratory studies have often supported the validity of ISO 7703, whereas field study has not.

and this is a critical element and weakness from the theory, that dynamic model of human thermoregulation, and wonders

> Why continue to improve the heat balance equation or develop thermal models when the complexity of 'reality' will undermine any improvement? The adaptive model of thermal comfort ha questioned the validity of ISO 7730 and this is also related to scope.

The proposed revision of ISO 7730 takes into account dynamically varying variables; in particular of how varied the temperature-rise is with increasing air speed, but especially the incidence of "*Adaptation*" by the subjects.

The Adaptation Thermal Comfort is the behavior implemented by the subject in an Indoor or Outdoor Microclimate Environment, to dynamically vary their thermoregulation in time. This adaptation is not about the clothing or the Company principal activities, but the micro-actions that, in time, are put in place for each subject even in neutral thermal comfort (PMV-based).

The micro-actions are continuous and are referred to as "*preferable*", which is the currency of the subject and which will change their preference of the microclimate in the future and adapt to it.

The other two observations concern "*Non-steady state thermal environment*" and "*Long-term evaluation of the general thermal comfort condition*", which is the time during which the PMV exceeds the comfort boundary, which must be weighed by means of a weighting factor (wf) given by:

$$\text{weighting factor (wf)} = \left(\text{PPD}_{\text{actual,PMV}}\right) / \left(\text{PPD}_{\text{PMV,limit}}\right)$$

In order to predict the combination of factors that influence the Thermal Environment on People, p must take into account the variables and inter-individual variance in response to thermal environment, therefore

> A key objective of standards is to transfer the latest scientific knowledge into practice.

The observations of Olesen are the subject of research under the Adaptive Thermal Comfort in the next section.

This book starting from these objections, nevertheless, considered that the very fact that the heat balance equation is defined by the Standards, confirms its reliability as a means of comparison.

In the studies reported in the second part of the book, the goal is to assess the gap between standard and real response (with Questionnaire survey). The ISO 7730 standard is considered still valid in order to study the feeling of comfort measuring microclimatic parameters in situ.

It is believed that the equation of heat balance is not sufficient, not only for the complexity of the dynamics of the thermoregulation (observation of Olesen), but also because the comfort includes a variable that is not expressed by the equations of thermodynamics, psychological or cognitive variables.

4.8 Adaptive Thermal Comfort

The study consists of Thermal Comfort in the synthesis of the complex system of interactions between man and environment, indoor and outdoor environment.

The response of the individual microclimatic conditions present in the interior of the body takes place in a *continuum* over time, a continuous process of exchanges, behavior and metabolic adaptations, related to clothing and psychological (Brager and De Dear 1998).

At the same time these responses may be related to all the individuals and population. In his "*A map of the world*" of 1939 by Markham (1947) shows a map of the world divided by climate zone and years in which the "*Climate Controls tool*" was introduced. The map shows a relationship between climate zone and earliness of adoption of systems for climate control.

Climate Control strategies are part of the adaptive strategies that man has taken to cope with the weather conditions. The current standards are referring to the possibility of controlling the Indoor Microclimate by building and mechanical conditioning. The term Adaptation allows the expression of a series of actions and processes that occur in continuous to ensure the condition of comfort on the part of the individual subject.

The *Adaptive Thermal Comfort* has as its object the study of this process that is also the case in which the environment is in terms of *Thermal Comfort: Neutral*, a condition partially introduced by ASHRAE Standard 55 and EN 15251.

The Adaptive Principle can be defined (Nicol et al. 2002; Humphreys and Nicol 1998; Humphreys 1973):

> If a change occurs such as to produce discomfort, people react in ways which tend to restore their comfort. (Nicol et al. 2002, p. 8)

Referring only to the relationship between comfort and indoor temperatures, compared to condition of thermal neutrality of the Operating Temperature, the condition of comfort varies with respect to the climatic context: in very hot environments the comfort temperature is slightly lower than the operating temperature; vice versa in colder environments where people prefer an environment "slightly warmer", with a temperature slightly higher than neutral.

The actions to adapt to their own comfort temperature affect the regulation of the internal heat generation (metabolism), reduce/increase body heat loss, as regards the Human Body, while for Indoor Environment it is possible to adjust the temperature, or other thermal environmental feelings or act by modifying the body's physiological condition comfort e.g. walking, rubbing hands together, increase the clothing when it is cold, fanning or taking off your clothes in hot weather.

The general principle of "Adaptive Thermal Comfort" takes into account the fact that even in a thermally neutral environment; people react in such a way as to meet their condition of comfort. Thus, there is a difference between the condition of comfort perceived and desired, the well-being is therefore not tied to the condition of neutrality, which allows to express the PMV index, but the difference between what is perceived and what is desired, the greater this difference that cannot be filled through adjustments of clothing, activity or position, or a psychological level, the more the conditions of discomfort.

According to individual characteristics an environment considered neutral, depending on the activity and characteristics of the clothing, cannot be "desired" for those subjects that in performing certain activities prefer, however, to "keep cool" and therefore have a PMV index slightly lower than the neutral value.

For example, a subject even though in a "neutral" environment can accept a slightly lower temperature if he carries out an activity for which concentration is required (reading, work) because "the cool keeps him awake." Conversely often older people tend to prefer a slightly warmer environment, or to get close to the heat source (radiator), not because it is a requirement of the body, but because they stay close to the heat source (radiator, stove, fireplace) it is "psychologically" pleasant.

The model of the Adaptive Thermal Comfort summarizes the actions that people take through three different mechanisms:

- behavioural adaptation: clothing, activities, open window, fan etc.;
- physiological adaptation: acclimatization after a period of a few hours, days or weeks following the persistence of the factors of thermal stress (the body gets used to the different temperature, as in the case of climatic areas such as the tropics or the polar areas);
- psychological adaptation: related to the conditions of indoor comfort expected by the occupiers and tolerance compared to the change in temperature.

The adaptive thermal comfort is linked with the choices of architectural design, in the broadest sense as it involves both the microclimatic parameters, dependent on environmental conditions (orientation) and the technology choices, both psychological and perceptual parameters as influenced by the above factors such as the choice of the sequence of environments, size and proportions, colors, available equipment plant, internal and external relationships etc.

The above aspects seem bound to "extreme" conditions, severe thermal environments of work (hot or cold) or "special" activities such as schools, kindergartens, in-patient clinics/hospitals.

To be sure the comfort conditions determine directly how to use the environments and particularly of heating systems, there where the passive temperature control systems (masonry insulation, orientation, openings, etc.) and natural ventilation systems do not allow compliance with the conditions of well-being in a position "neutral" or "adaptive".

The direct consequence is the different management systems of regulation of the heating systems, or the thermoregulation or all of the indirect systems that allow to "adjust" an environment or individually the condition of comfort. The temperature control in order to ensure the comfort conditions directly affects energy consumption for winter heating or summer air conditioning, in buildings where it is not present at the thermoregulation (TABS—Thermally Activated Building System), for example in centralized systems, or where there is a collective use of the same environment (offices, schools, etc.) there is adopted the solution of adaptation for thermoregulation called "a window" indoor temperature cannot be controlled with the heating controls, so the windows are opened, even if the heating is turned on, to reduce the internal temperature.

This mode of adaptation is present in all those buildings where you can open the windows, even if there are systems of controlled mechanical ventilation, precisely because of what is specified with respect to the conditions of adaptive comfort, although the indoor conditions are "neutral" the adaptive thermal comfort condition, cannot be "locked" to the respect of these parameters.

Users tend to want (and have the power) to change the distance between the conditions of perceived comfort and that desired.

In cases of Social Housing add to that the cultural habits and climate adaptation of the occupants from different regions of the world.

The studies of Baker and Standeved emphasize that the "adaptive opportunity" (possibility of adaptation) is interpreted as the ability to open the windows, obscuring the openings, turn on or off the wind convectors or radiators, and can also include the option of not using a clothing standard (e.g. a uniform or suit jacket + tie) for the conduct of activities.

The ability to independently manage these aspects makes it easier to adapt and therefore the perception of a comfort condition. The same studies propose to take into account, in the comfort standard UNI EN 7730 ASHARE 55, the component "*adaptation time*", as in Humphreys and Nicol (2002).

The scientific literature has dealt with several studies of the field Adaptive Thermal Comfort, with literature, applications, and critics (Halawa and van Hoof 2012; Sanderter et al. 2011; van der Linden et al. 2006; Ferrari and Zanotto 2012; Nicol and Humphreys 2010; Sourbron and Helsen 2011; Buratti and Ricciardi 2009). The description of the calculation algorithms to evaluate the Adaptive Thermal Comfort by Building dynamic performance simulation was developed by Nicol and Humphreys (2004) and Haldi and Robinson (2010).

In the last decade, several studies have proposed different approaches to deal with adaptive comfort, like de Dear and Brager (1997), who used "indoor comfort operative temperature". In Europe, Ferrari and Zanotto (2012), McCartney and Nicol (2002), Nicol and Humphreys (2010), Halawa and van Hoof (2012) and Goto et al. (2007) who showed potential future developments.

In this field Surveys to assess the Post Occupancy Evaluation of buildings (POE) are useful. POE is used to evaluate the building and not the occupants and considers:

- The use of questionnaires administered directly to the occupants to know information such as "how often do feel too hot in the summer," or ask for a judgment with respect to a feeling dependent on the building
- The geometrical and physical conditions of the building and the type of control that occupants have compared to indoor microclimatic conditions, such as ventilation of rooms, control of the heating systems, etc.
- The energy audit of the building by means of a building simulation.

The *Adaptive Hypothesis* takes into account the contextual factors and past history of the occupants from the expectations of Thermal Comfort (de Dear et al. 1997), so:

> People in warm climate zones would prefer higher indoor temperatures than people living in cold climate zones, which is in contrast to the assumptions underlying comfort standards based on the PMV/PPD-model (quoted in abstract of de Dear et al. 1997).

In this sense the Indoor Comfort temperature as a function of outdoor temperature, following formula

$$T_{comf} = A \cdot T_{a,out} + B \tag{4.22}$$

where

T_{comf} is comfort temperature (°C)
$T_{a,out}$ is monthly mean outdoor air temperature (°C)
A, B are constants.

Humphreys and Nicol (2000) propose as constants A and B of the formula, the following values

$$T_{comf} = 13.5 + 0.54 \cdot T_0 \tag{4.23}$$

where

T_0 is the monthly mean of the outdoor air temperature.

Morgan and de Dear (2003) set the condition for the adaptive approach to work:

> It is essential that building occupants are free to adapt themselves primarily through clothing adjustment, to the variable indoor climate regime prevailing inside such buildings. (Morgan and de Dear 2003, p. 267)

Adaptation actions (Adaptive Actions) require time to adapt to the comfort temperature, time that varies continuously with respect to time fluctuation outdoor, the season and to fluctuations in the environment minute bye minute due to the variation in the activity, the presence of people and of the operation of the HVAC. Figure 4.13 report the adaptive comfort region following de Dear and Brager (1998a, b).

4.9 Indoor Air Quality Assessment

The Thermal Comfort is the main parameter to evaluate the quality of an Indoor Environment, from the physical point of view. The other aspect to consider is the approach "chemical" or linked to interaction of chemicals that are created between the human body and the indoor air, in particular through breathing.

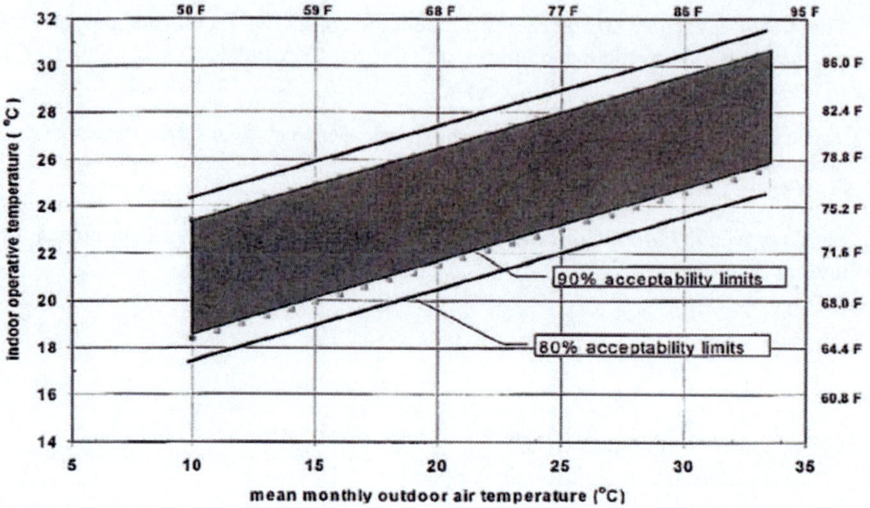

Fig. 4.13 The original adaptive comfort chart as presented by de Dear and Brager (1998a, b) prior to its inclusion (after amendments) in the ANSI/ASHRAE Standard 55-2004 (ASHRAE 55)

In this case no mention of Indoor Comfort but of Indoor Air Quality: the assessment is made as of the concentration of pollutants in indoor environments, compared to the concentration of pollutants that can have negative influences on human health.

Unlike studies on Thermal Comfort, whose extreme health heat stroke and freezing studies on Indoor Air Quality (IAQ) are more focused on the measurement of pollutants rather than the feeling of the people.

The feeling of people (Subjective Evaluation) regarding the IAQ are a negative type, and data from the negative feedback due to the reaction between the human body and chemicals pollutants, such as burning eyes, headache, allergic reactions, diseases, etc. are all factors that cannot be expressed as a range of values of feeling, but, possibly as reactions to IAQ conditions.

The perceptive approach, prevalent in the evaluation of Thermal Comfort, as regards the IAQ is less used: ***the subject (human) is used as the "first sensor"*** of situations that can lead to conditions of damage to health, chronic, acute or pathological. The study of IAQ focuses on ventilation strategies to reduce the concentration of pollutants. A summary of the contents of the main regulations on the ventilation and perception of IAQ by the occupants.

The ASHRAE Standard 62.1-2013 defines the requirements for the ventilation of rooms, as well as air-cleaning-system use, installation, and maintenance commission. These aspects do not interest us.

The Standard defines "*Acceptable Indoor Air Quality*": air in which there are no known contaminants at harmful Concentrations as determined by cognizant Authorities and with which a substantial Majority (80% or more) of the people exposed do not express dissatisfaction.

The parameter that characterizes the *System* and *Equipment for Ventilating systems*, is the Ventilation Rate, which depends on the Occupancy Category (Dayroom, Classroom, Dwelling, Bedroom, Corridors, Office, etc.) for each of which is defined a Minimum Ventilation Rates in Breathing Zone.

The procedure for assessing Indoor Air Quality (IAQ) is to define:

- The Contaminant Sources or mixtures of concern, given that each it is necessary to ensure a maximum level of concentration. ANNEX B,
- The Contaminant Concentration, limit exposure period, and finally the Perceived Indoor Air Quality, expressed by the percentage of occupants or visitors who express satisfaction on IAQ.

The IAQ may also be assessed using the Subjective Evaluation, since often the Indoor Air contains a complex mixture of pollution from various sources, tobacco smoke, infectious and allergenic biological aerosols, and of chemical emissions from commercial and consumer products.

The subjective perception depends on the interaction between the human body and those pollutants (by human nose), due to irritation of the mucosal tissues and eyes, these irritations are the answer of "irritant receptor" of individuals (allergies) and populations and depends also upon age and state of health.

Not always is the subject able to assess pollutants, for example, the Carbon Monoxide and radon are two pollutants without odor but with high Health Risk.

The levels of pollutants are defined in the table (an example show in Table 4.11) depending on the source of the pollutants: industrial environmental, outdoor environments, general pollutants (by World Health Organization WHO) and Residential Indoor Environment.

4.9.1 The CO_2 Concentration as an Indicator of Indoor Pollution Due to the Occupants

The concentration of CO_2 (Carbon Dioxide) constitutes the main indicator for the Indoor Air Quality, to measure the effect of indoor pollution due to the presence of people.

The human body needs for food, water (and water vapor) and Oxygen (respiration), to ensure their metabolic activity. The contribution of oxygen is measured by Respiratory Quotient (RQ) which expresses the amount of Volumetric Ratio of CO_2 produced to oxygen consumed. Figure 4.14 report relation between oxygen consumption, CO_2 concentration and physical activity.

Table 4.11 Levels of maximum concentration of pollutants (WHO)

Indoor pollution	WHO/Europe levels [time exposure]	Sources
Carbon dioxide	3500 ppm [Long-term]	
Carbon monoxide (MO)	90 ppm [15 min]	Leaking vented combustion appliances
	50 ppm [30 min]	Unvented combustion appliances
	25 ppm [1 h]	Parking garages Outdoor air
	10 ppm [8 h]	The concentration considers acceptable for non industrial
Formaldehyde (HCHO)	0.1 mg/m^3 (0.081 ppm) [30 min]	Press-wood products Furniture and furnishings Epidemiological studies suggest a casual relationship between exposure to formaldehyde and nasopharyngeal cancer
Lead (Pb)	0.05 μg/m^3 [1 year]	Paint dust, outdoor air
Nitrogen dioxide (NO_2)	0.1 ppm [1 h]	Leaking vented combustion appliances
	0.02 ppm [1 year]	Unvented combustion appliance Outdoor air Parking garages
Ozone (O_3)	0.064 ppm [8 h]	Electrostatic appliances, office machine, ozone generators, outdoor air
Odors	Predicted or measure acceptability	*CO_2 concentration can be used as a surrogate for occupant odor (odorous bio effluents)* Occupants, VOC sources, cooking, food processing, sewage, biowaste facilities
Particles ($PM_{2.5}$)	15 μg/m^3	Combustion products, cooking, candles, incense, resuspension, outdoor air, diesel exhaust, parking
Particles (PM_{10})	50 μg/m^3	Dust, smoke, deteriorating materials, outdoor air
Radon	4 pCi/L (0.024 μg/m^3)	Soil gas
TVOCs		Total volatile organic compounds. New building materials and furnishing, consumable products, maintenance materials, outdoor air
VOCs	Must be determined for each individual compound	Volatile organic compounds. New building materials and furnishing, consumable products, maintenance materials, outdoor air
Sulfur dioxide	0.048 ppm [24 h]	
	0.012 ppm [1 year]	

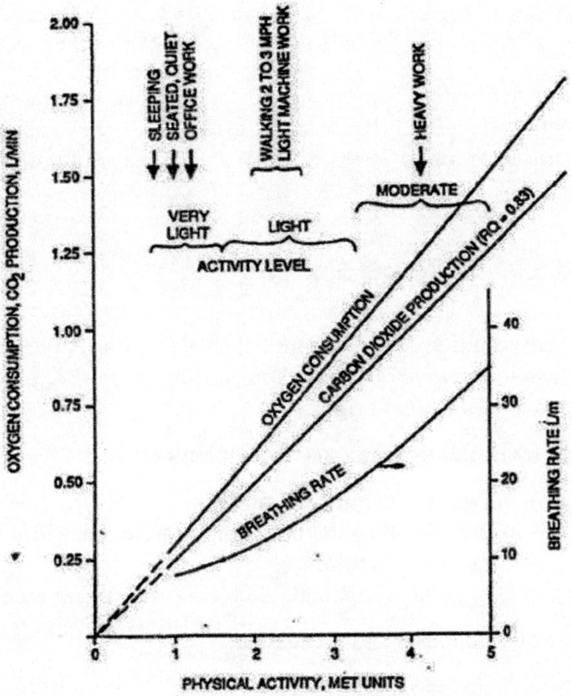

Fig. 4.14 Metabolic data: activity and concentration of CO_2 (ASHRAE 62)

The rate of oxygen consumed, and CO_2 generation depend on physical. The volume of outdoor airflow rate person is calculated using the formula:

$$V_0 = \frac{N}{C_s - C_0} \tag{4.24}$$

where

V_0 outdoor airflow rate per person
V_e breathing rate
N CO_2 generator rate per person
C_s CO_2 concentration in the space
C_0 CO_2 concentration in outdoor air.

(*by Appendix C ASHRAE 62.1-2013*) For example at an activity level of 1.2 met units corresponding to sedentary person, the CO_2 generation rate is 0.31 l/min (liters/minutes).

$$C_s - C_0 = \frac{N}{V_0} = 700\,\text{ppm} = 0.00689\ liters\ of\ CO_2\ per\ liters\ of\ air$$

then

$$C_s - C_0 = \frac{N}{V_0} \tag{4.25}$$

(by Appendix C ASHRAE 62.1-2013) For ventilation at a rate of 429 l/m and an activity level of 1.2 met units, the room oxygen level will be reduced from an outdoor concentration of 20.95–20.85% a percentage of 0.48% ([20.95/20.85]/20.95).

4.9.2 Pollutant Sources

The pollution sources, other than CO_2, are expressed as an "*emission rate*" (or "*emission factor*") or based on specific chemical pollution in $\mu g\ s^{-1}$ or $\mu g\ m^{-2}\ s^{-1}$ can have different origins:

- by Building, materials and equipment:
  - building structure and materials,
  - furniture and soft furnishing building services equipment (filters, dust, odor sources in ducts, etc.),
  - computer equipment, photocopying machines, etc.;
- by Occupants and processes:
  - body odor, skin cells, cosmetics,
  - food, drink, smoking, toilets,
  - printing, paper storage, equipment, dust,
  - biological contaminants, virus bacteria, fungi, muds, spores, etc.,
  - other processes;
- by External Environment:
  - particles,
  - gas e.g. VOC.

The Standards ISO 16000-6, ISO 16000-9, ISO 16000-10 and ISO 16000-11 define Labelling and certification scheme for low emission products, developed by the European Collaborative Action (ECA) started in 1989 with the ECA-COST project 613. The ISO 16814 Annex D shows the "*European Labelling schemes for low-emission flooring products*" and the list of European certification schemes, such as: EU Ecolabel scheme, Danish voluntary labelling scheme, GuT Environmental Quality Mark for Carpets, GEV EMICODE Labelling System for adhesive, primers and smoothing compounds.

The control of these pollutants can be done through the Elimination, Substitution (e.g. equipment or furniture), or the Modification of source pollution. The level of concentration of pollutants is assessed using a mass balance in two conditions:

(a) Steady-state calculation of mass balance, by equation

$$q_v = \frac{G_h}{C_{h,i} - C_{h,0}} \cdot \left(\frac{1}{\varepsilon_v}\right) \tag{4.26}$$

where

G_h is the emission rate of chemical, *h*, in milligram per second
$C_{h,i}$ is the guideline concentration value of a chemical, *h*, in a milligram per cubic meter
$C_{h,0}$ is the outdoor concentration value of a chemical, *h*, in a milligram per cubic metre
ε_v is the ventilation effectiveness.

(b) Non-Steady-state condition, by equation

$$C_h(z) - C_{h,o} = C_h(0) + \left(\frac{G_h}{q_v}\right) \cdot \left[1 - e^{\left(-\left(\frac{q_v}{V_r}\right) \cdot t\right)}\right] \tag{4.27}$$

where

(0), (z) are the start time (0) and final time (z)
V_r is the volume of air in the room, in cubic meters
t is the time, expressed in second
q_v is the air flow, expressed in cubic meters per second.

The Ventilation effectiveness as given by equation

$$\varepsilon_v = \frac{C_e - C_s}{C_i - C_s} \tag{4.28}$$

where

C_e is the pollution concentration in the exhaust air
C_s is the pollution concentration in the supply air
C_i is the pollution concentration in the breathing zone.

4.9.3 Methods of Expressing the IAQ for Human Occupancy

The evaluation of IAQ by persons covers two aspects:

- On the one hand the risk that the concentration of pollutants has effects on the health of people, as has been described above;
- On the other the fact that the quality of the air and smells influence the perception of indoor comfort;
- In the alternative it is possible to evaluate the IAQ, in an indirect manner, based on the Ventilation Rate.

The criteria for designing the IAQ take account of health (risk factors for disease and its symptoms, by WHO), or the comfort (or perceived IAQ Acceptability), or of Human Performance (for learning, work, sports, etc.).

The perception of comfort as regards the IAQ depends on the sense organs from: nose, throat, eyes and skin, all parts of the body are sensitive to temperature, humidity and to a wide range of pollutants present in the air. These are ways that determine the judgment of the people regarding IAQ comfort the "olfactory sensory receptor" (nasal cavity) and general chemical sensor receptors situated all over the mucous membranes in the nose and the eyes; these senses determine the feeling of "acceptability" of comfort.

The Indoor Environmental where people have more negative feelings, or negative reactions from those senses are judged as "Not-Acceptability", "Tolerable" up to "Not Tolerable".

The evaluation of the comfort IAQ can also be made from the level of efficiency and staff productivity. Environments with situations of discomfort, with greater concentration of bacteria, viruses (and therefore diseases), bad odors, in particular the excessive concentration of CO_2 and other pollutants affect the ability of concentration of people, in Office Buildings, and therefore on their productivity (e.g. speed and accuracy).

In this regard, the ISO 16814 emphasizes that, in practice, IAQ is not only a problem with regard to productivity, yet poor IAQ can promote or exacerbate conditions of SBS (Sick Building Syndrome) that depends on ventilation rates, surface pollution, as well as temperature, humidity, noise, lighting and personal control over the environment.

Finally, the control of IAQ, given that affects Ventilation Rate has consequences on the energy performance of the building.

4.10 Indoor Environmental Quality (IEQ) Assessment

Finally comfort also depends on other "physical" factors that characterize the environment, in addition to the Thermal Comfort, the IAQ, such as Lighting Comfort and Acoustic Comfort.

The subject is vast and here we only report a summary of the Standard EN 15251, which defines the criteria for evaluating all the parameters that affect the perception of indoor comfort.

The Standard is part of the set of rules that allow us to calculation the Building Energy Performance, defines the indoor environment parameters for energy calculation and conditions standard (Designed or Measured) when they are not defined by national legislation or other rules.

The Comfort Parameters concern:

- *Thermal Comfort*, described in the previous paragraph;
- *Indoor Air Quality*, described in the previous paragraph;
- *Humidity*, factors affecting the thermal perception, odors and the formation of phenomena of irritation of mucosal and spread of micro-biological factors;
- *Lighting Comfort*, factors that affect the visual perception of the environment, and in the execution of the visual task environments, working and not working;
- *Noise (Acoustic Comfort)*, due to the presence of noise sources in the internal environment, or external.

The Standard defines the criteria and classification certification of the indoor environment, based on the Indoor Environmental Comfort (IEQ) and defines categories of comfort for each of Comfort Parameters. Table 4.12 show IEQ categories.

The recommended criteria for *Lighting* do not refer to the different categories IEQ, following the Standard EN 12464-1 (*Light and lighting—Lighting of work places—Part 1: Indoor work places*), for every type of building (office, Education Dwelling, Hospital, etc.) are defined:

- Maintained illuminance (E_m) measured in lux (lx), (Luminous flux by square meters);
- UGR, Unified Glare Rating;
- Ra or CRI, Color Rendering Index.

Similarly, the Acoustic Comfort (absence of Noise) evaluates the indoor system noise criteria of some spaces and building, each room contains values of sound pressure level in decibels (dB): Typical Range and Default design value; e.g. in case of residential Living room typical range is 25–40 dB, and the default value is 32 dB design. In case of bedroom the values are 20–35 dB and 36 dB default.

Table 4.12 IEQ categories—general requirements

Category		Thermal Comfort		IAQ and ventilation rates		
		PPD (%)	PMV	PD (%)	q_v (l/s/m^2)	CO_2 (ppm)
I	High level of expectation and is recommended for spaces occupied by very sensitive and fragile persons with special requirements	< 6	− 0.2 < PMV < + 0.2	15	0.50	350
II	Normal level of expectation and should be used for new buildings and renovations	< 10	− 0.5 < PMV < + 0.5	20	0.35	500
III	An acceptable, moderate level of expectation and may be used for existing buildings	< 15	− 0.7 < PMV < + 0.7	30	0.30	800
IV	Values outside the criteria for the above categories	> 15	PMV < − 0.7 PMV > + 0.7	> 30	> 0.30	> 800

4.10.1 Methodologies for Subjective Evaluations

Finally, also the EN 15251 proposes a method for the evaluation of the parametric subjective thermal sensation, through questionnaires.

Questions are expressed according to the levels of judgment (e.g. Tables 4.13 and 4.14).

Table 4.13 Thermal sensation questionnaire by EN 15251

How do you rate your thermal sensation?	How do you perceive the temperature?	Do you want the room temperature?
☐ Hot	☐ Clearly acceptable	☐ (a) Higher
☐ Warm	☐	☐ (b) No change
☐ Slightly warm	☐ Just acceptable	☐ (c) Lower
☐ Neutral	☐	
☐ Slightly cool	☐ Just unacceptable	
☐ Cool	☐	
☐ Cold	☐ Clearly unacceptable	

Table 4.14 Air quality sensation questionnaire by EN 15251

How do you perceive the air quality?	How do you perceive the odor intensity?
☐ Clearly acceptable	☐ No odor
☐	☐ Weak odor
☐ Just acceptable	☐ Moderate odor
☐ Just unacceptable	☐ Strong odor
☐	☐ Very strong odor
☐ Clearly unacceptable	☐ Overpowering odor

4.11 Conclusions

In this chapter we have seen how the Standards define how to learn, plan and measure the indoor environmental comfort. The fact that a parameter so uncertain as the subjective judgment about the sensation of heat or air quality, has found its own methodology of investigation, and that this is standardized by the rules, you can start from a methodological basis, scientifically sound and recognized. Compared with the data reported in the scientific literature doubts are raised, studies and research with the aim of improving the accuracy in the evaluation of comfort inside buildings, in order to guide designers and correct any errors (d'Ambrosio Alfano et al. 2014; Toftum et al. 2009).

Other studies have addressed the issue looking for differences in gender, age, etc. bringing the various factors that influence the indoor comfort as in Frontczak and Wargocki (2011).

Among the various aspects not explored in the scientific literature there is the perception of comfort by children, which is the object of this book.

The results of experimental experience described in the second half, show how these studies can provide useful tools for the project architectural-plant engineer, and for the pedagogical-didactic aspects.

The following paragraphs describe the methods of subjective evaluation, and how to use these methods for children, taking into account their particular cognitive characteristics.

References

ASHRAE (1989) ASHRAE handbook fundamentals American Society of Heating, Refrigerating, and Air-Conditioning Engineers, Atlanta, GA

ASHRAE Standard 55-2014. Thermal environmental conditions for human occupancy. American Society of Heating, Refrigerating and Air-Conditioning Engineers, Atlanta

ASHRAE 62.1-2013. Ventilation for acceptable indoor air quality

Auliciemes A (1983) Psychophysical criteria for global thermal zones of building design. Int J Biometerorol 69–86

Bluyssen PM (2009) The indoor environment handbook. Earthscan, London

Brager GS, de Dear RJ (1998) Thermal adaptation in the built environment: a literature review. Energy Build 27(1):83–96

Bröde P, Fiala D, Błażejczyk K, Holmér I, Jendritzky G, Kampmann B, Tinz B, Havenith G (2012) Deriving the operational procedure for the Universal Thermal Climate Index (UTCI). Int J Biometeorol 56:481–494

Bründl W, Hoppe P (1984) Advantage and disadvantage of the urban heat island. An evaluation according to the hygro-thermic effects. Arch Met Geoph Biocl 35:55–66

Buratti C, Ricciardi P (2009) Adaptive analysis of thermal comfort in university classrooms: correlation between experimental data and mathematical models. Build Environ 44:674–687

Choi J-H, Loftness V (2012) Investigation of human body skin temperatures as a bio-signal to indicate overall thermal sensations. Build Environ 58:258–269

Choi J-H, Loftness V, Lee D-W (2012) Investigation of the possibility of the use of heart rate as a human factor for thermal sensation models. Build Environ 50:165–175

Clarke JF, Bach W (1971) Comparison of the comfort conditions in different urban and suburban microenvironments. Int J Biometeorol 15:41–54

COST UTCI (2004) Towards a Universal Thermal Climate Index UTCI for assessing the thermal environment of the human being. MoU of COST Action 730. www.utci.org

d'Ambrosio Alfano FR, Olesen BW, Palella BI, Riccio G (2014) Thermal comfort: design and assessment for energy saving. Energy Build 81:326–336

de Dear RJ, Brager GS (1997) Final report: developing an adaptive model of thermal comfort and preference. ASHRAE RP-884 1997. American Society of Heating, Refrigerating and Air-Conditioning Engineers, Atlanta

de Dear RJ, Brager GS (1998a) Towards an adaptive model of thermal comfort and preference. ASHRAE Trans 104(1):145–167

de Dear R, Brager GS (1998b) Developing an adaptive model of thermal comfort and preference. ASHRAE Trans 104(1a):145–167

de Dear R, Brager GS (2011) The adaptive model of thermal comfort and energy conservation in the built environment. Int J Biometeorol 45:100–108

de Dear RJ, Brager GS, Cooper (1997) Developing an adaptive model of thermal comfort and preference. Final report ASHRAE RP-884. ASHRAE Inc., Atlanta

EN 15251 Indoor environmental input parameters for design and assessment of energy performance of buildings addressing indoor air quality, thermal environment, lighting and acoustics

Fanger PO (1970) Thermal comfort-analysis and applications in environmental engineering. Danish Technical Press, Copenhagen

Ferrari S, Zanotto V (2012) Adaptive comfort: analysis and application of the main indices. Build Environ 49:25–32

Fountain M, Huizenga C (1995) A thermal sensation model for use by the engineering profession: results of cooperative research between the American Society of Heating, Refrigeration, and Air-Conditioning Engineers, Inc. and Environmental Analytics. Final report

Frontczak M, Wargocki P (2011) Literature survey on how different factors influence human comfort in indoor environments. Build Environ 46:922–937

Gagge AP (1946) The linearity criterion as applied to partitional calorimetry. Am J Physiol 116:656–668

Gagge AP, Stolwijk JAJ, Nishi Y (1971) An effective temperature scale based on a single model of human physiological temperature response. ASHRAE Trans 77:247–262

Gagge AP, Fobelets AP, Berglund LG (1986) A standard predictive index of human response to the thermal environment. ASHRAE Trans 92(2):709–731

Gaoua N, Granham J, Racinais S, El Masioui F (2012) Sensory displeasure reduces complex cognitive performance in the heat. J Environ Psychol 32:158–163

Giles BD, Balafoutis CH, Maheras P (1990) Too hot for comfort: the heatwaves in Greece in 1987 and 1988. Int J Biometeorol 34:98–104

Givoni B (1969) Man, climate and architecture. Elsevier Science Ltd, Amsterdam

Goto T, Mitamura T, Yoshino H, Tamura A, Inomata E (2007) Long-term field survey on thermal adaptation in office buildings in Japan. Build Environ 42:3944–3954

Halawa E, van Hoof J (2012) The adaptive approach to thermal comfort: a critical overview (review). Energy Build 51:101–110

Haldi F, Robinson D (2010) On the unification of thermal perception and adaptive actions. Build Environ 45(11):2440–2475

Hoppe P (1999) The physiological equivalent temperature PET—an universal index for the biometeorological assessment of the thermal environment. Int J Biometeorol 43:271–275

Hoppe P, Mayer H (1987) Planungsrelevante Bewertung der thermischen Komponente des Stadtklimas. Landsch Stadt 19:229

Houghten FC, Yagloglou CP (1953) Determining lines of equal comfort. Trans Am Soc Heat Vent Eng 29:165–176

Houghton FC, Yaglou CP (1923) Determining equal comfort lines. J Am Soc Heat Vent Eng 29:165–176

Humphreys MA (1973) Classroom temperature, clothing and thermal comfort. A study of secondary school children in summertime. J Inst Heat Vent Eng 41:191–202

Humphreys MA, Nicol JF (1998) Understanding the adaptive approach to thermal comfort. ASHRAE Trans 104(1):991–1004

Humphreys MA, Nicol JF (2000) Outdoor temperature and indoor thermal comfort: raising the precision of the relationship for the 1998 ASHRAE database of field studies. ASHRAE Trans 206(2):485–492

Humphreys MA, Nicol JF (2002) The validity of ISO-PMV for predicting comfort votes in every day thermal environments. Energy Build 34:667–684

ISO 10551 Ergonomics of the thermal environment. Assessment of the influence of the thermal environment using subjective judgement scales

ISO 11079 Ergonomics of the thermal environment—determination and interpretation of cold stress when using required clothing insulation (IREQ) and local cooling effects

ISO 11399:1995 Ergonomics of the thermal environment—principles and application of relevant international standards

ISO 12894 Ergonomics of the thermal environment—medical supervision of individuals exposed to extreme hot or cold environments

ISO 13731:2001 Ergonomics of the thermal environment—vocabulary and symbols

ISO 13792 Thermal performance of buildings—calculation of internal temperatures of a room in summer without mechanical cooling—simplified method

ISO 15743 Ergonomics of the thermal environment—cold workplaces—risk assessment and management

ISO 16000-9 Indoor air—part 6: determination of volatile organic compounds in indoor and test chamber air by active sampling on Tenax TA sorbent, thermal desorption and gas chromatography using MS/FID

ISO 16000-10 Indoor air—part 9: determination of the emission of volatile organic compounds from building products and furnishing—emission test chamber method

ISO 16000-11 Indoor air—part: 11 determination of the emission of volatile organic compounds from building products and furnishing—sampling, storage of samples and preparation of test specimens

ISO 16813 Building environment design—indoor environment—general principles

ISO 16814 Building environment design—indoor air quality—methods of expressing the quality of indoor air for human occupancy

ISO 7243 Hot environments—estimation of the heat stress on working man, based on the WBGT-index (wet bulb globe temperature)

ISO 7726 Thermal environments—instruments and methods for measuring physical quantities

ISO 7730 Ergonomics of the thermal environment—analytical determination and interpretation of thermal comfort using calculation of the PMV and PPD indices and local thermal comfort criteria

ISO 7933 Ergonomics of the thermal environment—analytical determination and interpretation of heat stress using calculation of the predicted heat strain.
ISO 8996 Ergonomics—determination of metabolic heat production
ISO 9886 Evaluation of thermal strain by physiological measurements
ISO 9920 Ergonomics of the thermal environment—estimation of the thermal insulation and evaporative resistance of a clothing ensemble
ISO/TR 11079 Evaluation of cold environments—determination of required clothing insulation (IREQ)
ISO/TS 11415 Ergonomics of the thermal environment—application of international standards to people with special requirements (technical specification)
ISO/TS 13732-2 Ergonomics of the thermal environment—methods for the assessment of human responses to contact with surfaces—part 2: human contact with surfaces at moderate temperature
Jendritzky G, Maarouf A, Staiger H (2001) Looking for a Universal Thermal Climate Index UTCI for outdoor applications. In: Windsor-conference on thermal standards, Windsor, UK, 5–8 Apr 2001
Jendritzky G, de Dear R, Havenith G (2012) UTCI—why another thermal index? Int J Biometeorol 56:421–428
Jin Q, Li X, Duanmu L, Shu H, Sun Y, Ding Q (2012) Predictive model of local and overall thermal sensations for non-uniform environments. Build Environ 51:330–344
Kyle WJ (1992) Summer and winter patterns of human thermal stress in Hong Kong. In: Kyle WJ, Chang CP (eds) Proceedings of the 2nd international conference on East Asia and Western Pacific meteorology and climate. World Scientific, Singapore-New Jersey-London-Hong Kong, Hong Kong, pp 575–583
Markham SF (1947) Climate and the energy of nations, Reprint edn. Ams Pr Inc. ISBN-13: 978-0404162146
McCartney KJ, Nicol JF (2002) Developing an adaptive algorithm for Europe. Energy Build 34:623–635
Monteiro LM, Alucci MP (2008) Outdoor thermal comfort modeling in Sao Paulo, Brazil. In: Proceedings of conference on passive and low energy architecture 2008
Morgan C, de Dear R (2003) Weather, clothing and thermal adaptation to indoor climate. Clim Res 24(3):267–284
Nicol JF, Humphreys MA (2004) A stochastic approach to thermal comfort, occupant behavior and energy use in buildings. AHRAE Trans 110(2):554–568
Nicol F, Humphreys N (2010) Derivation of the adaptive equations for thermal comfort in free-running buildings in European standard EN 15251. Build Environ 45:11–17
Nicol JF, Humphreys M, Roaf S (2002) Adaptive thermal comfort. Principles and practice. Routledge. ISBN: 978-0-415-69159-8
Nicol F, Humphreys M, Roaf S (2012) Adaptive thermal comfort. Principles and practice. Routledge Taylor & Francis Group, New York
Nikolopoulou M (2004) Designing open spaces in the urban environment: a bioclimatic approach. Centre Renewable Energy Sources (CRES)
Nikolopoulou M, Baker N, Steemers K (2001) Thermal comfort in outdoor urban space: understanding the human parameter. Sol Energy 70(3):227–235
OECD (2011) Designing for education. Compendium of exemplary educational facilities. Centre for Effective Learning Environments. ISBN: 9789264112292
Olesen BW, Parson KC (2002) Introduction of thermal comfort standards and to the proposed new version of EN ISO 7730. Energy Build 34:537–548
Olgyay V (1963) Design with climate: bioclimatic approach to architectural regionalism. Princeton University Press, Princeton, NJ
Park P, Tuller SE, Jo M (2014) Application of Universal Thermal Climate Index (UTCI) for microclimatic analysis in urban thermal environments. Landsc Urban Plan 125:146–155

Sanderter M, Hensen JLM, Loomans MGLC, Boerstra AC (2011) Adaptive thermal comfort in primary school classrooms: creating and validating PMV-based comfort charts. Build Environ 46:2454–2461

Saro O, De Angelis A, Feruglio B (2006) Numerical evaluation of local radiant asymmetry. In: 8th international conference and exhibition on healthy buildings—HB 2006, pp 161–166

Schellen L, Loomans MGLC, de Wit MH, Olesen BW, van Marken Lichtenbelt WD (2012) The influence of local effects on thermal sensation under non-uniform environmental conditions—gender differences in thermophysiology, thermal comfort and productivity during convective and radiant cooling. Physiol Behav 107(2):252–261

Schellen L, Loomans MGLC, Kingma BRM, de Wit MH, Frijns AJH, van Marken Lichtenbelt WD (2013) The use of a thermophysiological model in the built environment to predict thermal sensation: coupling with the indoor environment and thermal sensation. Build Environ 59:10–22

Sourbron M, Helsen L (2011) Evaluation of adaptive thermal comfort models in moderate climates and their impact on energy use in office buildings. Energy Build 43:423–432

Thom EC (1959) The discomfort index. Weatherwise 12:57–60

Thorsson S, Lindqvist M, Lindqvist S (2004) Thermal bioclimatic conditions and patterns of behavior in an urban park in Goteborg, Sweden. Int J Biometeorol 48:149–156

Toftum J, Andersen RV, Jensen KL (2009) Occupant performance and building energy consumption with different philosophies of determining acceptable thermal conditions. Build Environ 44:2009–2016

van der Linden AC, Boerstra AC, Raue AK, Kurvers SR, de Dear R (2006) Adaptive temperature limits: a new guideline in The Netherlands: a new approach for the assessment of building performance with respect to thermal indoor climate. Energy Build 38:8–17

Yao Y, Lian Z, Liu W, Shen Q (2008) Experimental study on physiological responses and thermal comfort under various ambient temperatures. Physiol Behav 93(1–2):310–321

Ying BA, Kwok YL, Li Y, Yeung CY, Li FZ, Li S (2004) Mathematical modeling of thermal physiological responses of clothed infants. J Therm Biol 29(7–8):559–565. In: International thermal physiology symposium: physiology and pharmacology of temperature regulation

Chapter 5
Assessment of the Influence of the Thermal Environment Using Subjective Judgement Scales

Abstract The evaluation of the feeling of Thermal Comfort can also be achieved by directly asking users of indoor environments. After all, this is the same approach taken by Fanger. In these cases, taking the questionnaire by which you can determine the indexes of judgment (PMV and PPD) and indexes of preference on the desirable conditions of Thermal Comfort. This chapter describes the contents of the survey questions, and criteria for administration through surveys and/or interviews.

5.1 The Use of Questionnaires in Scientific Literature

The evaluation of the comfort of the indoor or outdoor environments is also possible without tools for environmental monitoring and/or modelling using dynamic modelling software.

Since the studies of Fanger PO the study of comfort has used questionnaires to assess the user's sensation corresponding to a series of physical and physiological parameters. The user's feeling does not match the exchange of energy body-environment, because the user is affected by subjective judgment, the psychological aspect, social and cultural rights, as well as the state of mind of those who respond.

Questionnaires can be used to assess the relationship between feeling and psychological, cultural and social aspects.

The use of questionnaires requires expertise and experience in their (i) development, (ii) formulation, (iii) administration and (iv) interpretation of the results.

The phases of a research carried out by the use of questionnaires should include:

(i) The stage of development that includes:

   a. the definition of the objectives of the questionnaire,
   b. the identification of relationships between the variables physical, cultural and psychological, object of study and questions, with their relative margins of uncertainty that the answers may have;

(ii) the formulation of the questions, which must take into account:

K. Fabbri, *Thermal Comfort Perception*, https://doi.org/10.1007/978-3-031-52610-7_5

   a. the cognitive level and amplitude (quantity and availability) of subjects to which the questionnaire is to be submitted, content syntax and vocabulary used in the formulation of the questions, and the responses of applications, in relation to the previous point;

(iii) the mode of administration of the questionnaires, which should take into account:

   a. the type; if realized by direct interview with a face to face interviewer, remote (telephone), auto-filled, etc.,
   b. the context in which the research is conducted, therefore data and impressions of the environment should be collected during the administration of the questionnaires;

(iv) the interpretation of the results, which provides:

   a. data collection,
   b. statistical processing of results, graphs, tables, etc.,
   c. *the interpretation of results, taking into account the ability of the researcher to read the results in relation to content and environmental context*,
   d. the preparation of the final report of the research, which should clearly show the above steps with the relevant conclusions.

The questionnaire cannot be improvised, the risk is to carry out research that does not have a sound statistical basis (representativeness of the sample) or with too many questions that have margins of uncertainty in order to draw a clear conclusion and reliable. Questionnaires can be used:

- In the assessment "of laboratory", as done by Fanger, an approach typically used to identify the correlation between physical parameters, physiological and feeling, or to "find a formula";
- or directly "on site"; in confined spaces (Moderate Environments) such as offices, schools, theaters, housing, etc., or in open spaces (squares, parks, etc. for assessment of outdoor comfort). In their turn they can be related to individual categories of persons or population, etc. in user defined spaces (workers, etc.) or with mixed users (occasional visitors, etc.), and finally, may be refer to a single data collection or extended over time.

In the latter cases, the approach of mathematics and science is represented as the study of the statistical results.

The scientific literature contains many examples where questionnaires are used to evaluate indoor comfort.

The first study shows that uses questionnaires to evaluate Thermal Comfort dates back to 1936, created by Bedford (1936) which measured the Thermal Environment of about 2500 workers in a UK industry. The research involved the measurement of physical parameters (temperature of the air, mean radiant temperature, speed of the air movement, humidity) and the administration of questionnaires to people, in which they were asked for a judgment on Thermal Comfort using a scale of seven categories from Warm to Cool. Thomas Bedford identified the pattern of the questions to be used for the evaluation of Thermal Comfort.

In the seventies Humphreys (1975) collected data from the research carried out on Thermal Comfort, collecting more than 200,000 observations on all continents, Europe, America, Africa, Asia. From these observations he prepared a Subjective ranking to evaluate the heat with correspondence to the physical parameters microclimatic of the thermal environment. The questionnaire data was collected in a database. A upgrade of Humphrey research described in Humprey et al, 2007. The de Dear Database is an important contribution to the study of adaptive thermal comfort and the use of subjective Judgement. At the end of the 90s de Dear (de Dear et al. 1997; de Dear 1998; de Dear and Brager 1998) and colleagues collected more than 20,000 sets of data from various database researches from around the world, this collection is called the "de *Dear Database*", and is one of most extensive and solid references for the calculation of the comfort PMV scores. The database uses the scale to score with 7 category degrees provided by ASHRAE (American Society of Heating, Refrigeration and Air-Conditioning) that constitute the reference for the ASHRAE Thermal Comfort Standard (ASHRAE 55).

Another database is the SCAT (Smart Control and Thermal Comfort) Database, related to European studies and has as its object questionnaires administered in the office environment, by different research groups for a total of 4500 "desk visits". The study included the automatic collection of data on the Thermal Environment, the level of illumination (horizontal illuminance) and the concentration of CO_2 in the workplace.

The study of human behavior can involve two levels of inquiry: (a) the analysis of the medium or long-term on a limited sample, by means of continuous monitoring and measurement, as with the studies of Andersen et al. (2012) and the studies of Fabi et al. (2012), which suggest the literature review related to the study on ventilation of indoor environments, or (b) the *method of the social sciences*: the questionnaire by direct interview or survey which needs a large representative sample (confidence interval).

Questionnaires are also used by the head of epidemic studies and in those health (hygiene) studies like that of Hutter et al. (2006) relating to 65 subjects divided into three areas of questions: *perceptual environment room*, perceived symptoms and suggestions for improving on working environments. In other research the use of the questionnaire, allows the study of the "Sick Building Syndrome" as described by Raw et al. (1996). The questionnaire covers the interview of 556 occupants in an office building.

Questionnaires can be used to study the relationship between the architectural shape and culture, as described in the search for Dili et al. (2010) in India, where he studied the perceived comfort of those living in vernacular buildings. The research provided a comparison of 200 people living in traditional and modern buildings.

In the study of comfort through questionnaires it is worth quoting the use of "*the Swedish questionnaire MM*" applied in studies of Andersson et al. (1993) and Andersson (1998) similar to that used in epidemiological studies.

Some studies have dealt with defining the visible energy (keeping energy visible), including studies related to Indoor Environmental Quality (IEQ) use of quality assessment model, an example is the text of Bluyssen and Cox (2002) and the search for Heinzerling et al. (2013) with a thorough literature review on IEQ.

The questionnaire may be administered following a precise survey of research, as is the case of smart metering at the end of a monitoring activity, or by way of large samples. Some research shows that making users aware of consumption (keep energy visible) by monitoring tools or metering, affects their habits and reduces fuel consumption. Particularly if users can act directly on consumption, as often happens for housing. Hargreaves et al. (2010, 2013) uses Smart Energy Monitoring (SEM) for making energy visible allowing the user to have the energy feedback.

5.2 The Use of Questionnaires in Social Research Methods

Tests involving humans and their feelings require putting questions to them to verify the perception of a phenomenon. In medical studies, as part of physiology and pharmacology, it is necessary to know what could be answer, and the perception by the patient compared to a therapy (Bryman, 2004).

Several phenomena related to the human-environment can be evaluated only by directly asking the subjects. The response of the subjects is affected by their personal construction of the world; from the purely perceptual aspects to psychological ones, from cultural to social ones.

The relationship question/answer has in it a number of uncertainties due to the contents of the application, the mode of administration, the cognitive level and the availability of those who respond. Interviews with individuals can be used to express social considerations, only interpreting the results in relation to the type of questions asked.

In this sense social studies are in the driving seat, including studies related to marketing, which also use the instruments of the interview to know the opinion of groups of individuals. Questionnaires are investigative tools that allow you to measure the perception of a phenomenon in three stages:

(a) at the time of administration of the questions, in relation to the understanding of the questions and the availability of the users;
(b) at the time of compilation of results from the statistical point of view, including the correlation between the results of the questionnaires and those measured (or measurable effects of the phenomena);
(c) the interpretation of the results according to the objectives of the research.

Comfort in the environment is an area of research closely related with the opinion of the people, and only the statistical interpretation of the answers allows us to assume average responses to the conditions of the microclimate measured.

The opinion has the characteristic of being volatile, nevertheless the evaluation of comfort indexes has an amount of literature and case studies that allow the issue of technical rules, as described in the previous paragraph. The questionnaire, despite the margin of uncertainty of the answer, it is a useful tool to assess specific cases not provided for by the terms of the rules. It is an empirical assessment tool used to study reality as the perceived reality of the senses is acceptable.

The investigations of social character are distinguished in *quantitative* research, using a deductive method to test a theory, and *qualitative*, using a deductive method which allows the generalization of the theory and requires interpretation of the data. *The case studies described fall in this area.*

Research must be designed by defining the framework of the research and the limitations of collecting and analyzing data, checking: the reliability of the questionnaire and research that needs to be replicated in other contexts within the same framework of research, and finally the internal validity of methods, such as control of causal relationships internal to the question, and externally in other contexts.

The "classical" research experiment is that which includes two sets of people, one defined experimental group to which is administered a questionnaire or evaluated the variable, and the other control group. Another type of experiment is the laboratory experiment where the subjects and variables are controlled in the entrance and in the output. The Fanger experiments or questionnaires in offices in a single building fall in this area.

The Case Study design refers to an individual case, which can take on the characteristics of generality or replicability in other similar contexts, can be related to individual communities (Whyte 1955), such as school classes (Ball 1981), to individual families, organizations, to individuals, or to individual events (Allison 1971). The case studies can be: Critical houses, unique homes, revelatory houses, or exemplifying cases, as reported in this text.

Quantitative research: theory, hypothesis, research design, process data, data analysis, conclusion. Quantitative research involves the definition of a concept, of a theory to evaluate, or doubt to refute, defined by the *indicators* defined by data that can be measured (*measures*).

The structure of the interviews (quantitative research) the interview must reduce the variability of error, such as the understanding of the response. Common errors in the questionnaires are the poor order of questions, the sense of the questions not understood by the interviewers, incomprehension of the problems of memory in reading the questionnaire or sequence of responses. The variables of the interviews each have their own range of error.

The interviews (administration of questionnaires) can be:

- structured,
- standardized,
- *non-structured*, where the questionnaire is informal, based on a dialogue of interviewer and interviewee,
- *qualitative*, when they are in part responses to specific questions and answers and in part free answers or dialogue with the interviewer (who must be an expert),
- *focus interview*, with many open questions or multiple related to a specific question,

- *focus group*, and *group interview* when referring to a specific,
- *oral history interview* typical social studies historical, ethno-anthropological, where they collected stories and interpretations about a phenomenon,
- *life-history interview*, refers to the collection of biographical data of each respondent.

The Context of the Interviews

In the traditional version of the standard interviews the interviewer is standing or sitting in front of the interviewee who responds in a controlled environment. In some cases there may be multiple interviewers (Bechhofer et al. 1984).

The interview may be done *in person or remotely*, for example by means of telephone (with an operator), or by *online devices*—interviews online, without an operator, impersonal, in this case the risk of non-acceptance as spam.

In the case of interviews with children (the same is true with the sick) you may have the following cases:

- interviewers directly in the classroom, which poses the question of interviewing individual children;
- interviewers in attendance who administer the questions by explaining the purpose of the questionnaire, without describing the content (possibly clarifying aspects not clear, this is also an element of assessment);
- interviewers are not present in the classroom and the questions are submitted or delivered directly by the classroom teacher or other teachers;
- questionnaires are delivered and distributed independently by children.

All these modes may require preparation before the questionnaire.

The order of the questions has a role, as well as the number of questions. Answers may be:

- *A single response*, when compared to a series of options it is possible to respond with a single answer, in these cases it is good to evaluate the number of options-response, which can be from two (yes/no) up to a maximum of 5 or 6 answers, a greater number poses assessment problems for the interviewee, or the risk of forgetting the previous options, the number may increase if the options are expressed in terms of a scale (e.g. from very cold to cold);
- *Multiple choice*, when compared to a number of options it is possible to respond with than one answer, such as a list of devices or the possibility they are not mutually exclusive;
- "*Nested*" when clarifying a previous question, such as "if the previous question you answered" No, "which of the following options …", in children, especially the younger ones, this kind of question is likely to cause confusion because they tend to answer the question instinctively, without structuring the answers in a logical sequence;
- *Free*, when the respondents can write their opinion, are typically used to screen situations that can not be assessed with other methods of investigation, or to know

the opinion of the interviewee. In the case of children these questions allow us to know their opinion of the school.

The questions can include the "other" option in which the respondent may add, independently, their response. In the case of questions on comfort it is preferable to minimize or not to provide this possibility, in particular if the application refers to the measurement of a variable measured at the same location.

The interviewee must be clear to which category of question and answer, each belongs.

The questions must be easy to understand, questions of a personal nature, age, sex, weight, clothing, education etc., a structural range of values is preferable, placed at the beginning or end of the questionnaire, so as not to interrupt the discussion.

The order of the questions should be structured by subject (unless you want to check the consistency of the answers with the "*check questions*") in increasing order of detail and complexity.

In the case of use of a questionnaire to compare with data monitoring, as is the case with well-being, each question must be correlated with at least one physical variable measure, directly or indirectly, by the probes.

Questionnaires can be administered directly by an interviewer or self-compiled (*self-completion questionnaires*).

In the case of self-completed questionnaires, they must make it clear who is the proponent (institutions, universities) the purpose of the questionnaire (research, market research, etc.) and the processing of data, preferably anonymous. The number of questions must take into account:

- *The availability of the person*, if the person has little availability, they tend not to respond or tick the answers randomly; in the case of children, their availability can be given by the teacher as an obligation or a game or other activity seen positively by the student, the manner of presentation of the questionnaire may have influence on the answers;
- *The ease of understanding of the question*, the respondent cannot ask for clarification so the phrases and words used must match the level of cognition and culture of the person, avoiding technical terms that cannot be understood, or words or expressions that can have double meaning; in the case of children;
- *The number of questions, and pages*, keeping in mind that in general after 10/15 questions there is the risk of fatigue in reading, or reduced availability of the answers;
- *The number of pages*, not everyone who responds independently notices the back of the page;
- *Whether printed or online*, bearing in mind that this second mode, if conveyed through email increases the risk of spam; while in the case of paper applications you have to be careful in defining who distributes the questionnaires and those who collects them.

The responses with more options or multiple options, can be arranged vertically or horizontally, the layout must be such as to make clear the reading and the sequence of questions and answers.

In the case of children it is useful to add an explanatory *image*, for the questions and/or response for each option. Images can be photos, drawings, symbols, patterns, colours, in any case they must be agreed with the children and be part of the universe of children (cartoons, games, references to nature, etc.) and it *must* be agreed with the children, in relation to the procedures required by the pedagogical model or the present environmental situation at the time of administration of the questionnaire. Children associate different meanings than adults to the images, so the use of apparently unique images in a game (such as the sun for the heat, the snow for the cold) may have different meanings in the imagination of children, the sun can remind them of summer, and the snow of Christmas.

The research should include the collection of the whole process of administration of the questionnaire, including the difficulties encountered, both in the distribution and the collection of the questionnaire, opinions, comments and clarifications. Particularly in the case of children, because the collection phase of the questionnaires or requests for clarifications can help you understand different aspects with respect to the questions. For example, the distinction between the perception of the physical environment in which the children live, with respect to the business carried out or to social dynamics that are created between children within the same class, with other children or with adults.

The return of the results of the responses can be made using diagrams, pie charts or bar charts, and with data analysis, as required by the regulations. The analysis of the results can measure the Central Tendency, by calculating the *Arithmetic mean*, the *Medium* or the *mode* of the responses; generally in the case of the questions on comfort it is sufficient that one must take into account the dispersion of responses (*Standard Deviation*). A type of analysis, also used in Social Science, is the "*Bivariate analysis*" that compares two variables, one on the x-axis, the other placed on the y-axis, a solution that allows you to build a *Scatter diagram* and the trend that relates the two variables (what happens to a variable if the other increases or decreases).

The *Multivariate Analysis* consists of comparing the relationship between multiple variables, the analysis may be of the *spurious* type, when it appears to be a direct correlation between two variables, which are not related to each other, such as age, income, and the intention to vote, or for thermal comfort the feeling of comfort, the time of day, and/or the season and/or the age of the person and/or the state of their health; or *non-spurious* if the relationship between the two variables is direct, as between the feeling of comfort, indoor temperature, age.

The questionnaires for the evaluation of perceived comfort, are a type of *Quantitative Research*, since they allow the relationship of the data received with that measured. In the case of particular users, as in the case of children, this type of research may also include aspects related to *Qualitative Research*, or, according to the approach of the Social Sciences, a type of search in which the data are not collected, but are result of observations on site and interviews with the participation of the subjects. The result is therefore not only the numerical data to be compared with the measured data, but a story of the individual child or group, and constitutes a specific case study which in turn may be representative. In the case of children that occurs in three stages:

- During the preparation of the questionnaire and questions, where you can learn the cognitive universe;
- During the administration of the questionnaire, the collection and the eventual request for clarification on the questions from the children;
- In the open-ended responses, if the questionnaire provides.

The data thus collected should be interpreted to build the conceptual framework, and possibly theoretical, on the evaluation of comfort in that context on those aspects that make it subjective, not replicable, or generalizable.

5.3 Assessment Using Subjective Judgement Scale: Standard Approach

The construction of questionnaires to assess the subjective judgment on the thermal environment is defined by Standard ISO 10551 "*Ergonomics of the thermal environment. Assessment of the influence of the thermal environment using subjective judgement scale*". The fact that it was enacted as legislation confirms the validity and replicability of the method of investigation and provides the basic criteria in order to achieve results comparable to those provided by the Standards for measuring and monitoring Thermal Comfort in situ. The ISO 10551 can be applied in a work environment and to evaluate Thermal Stress. There are references about other comfort parameters such as Indoor Air Quality, Light Comfort Light or Sound.

The subjective nature of the questions can be distinguished in two phases:

- The first concerns the response to the objectivity of the method of investigation, not considered the norm, but that may be accompanied by measurement on site, and from the statistical interpretation of the results, as well as from the *purpose of the questionnaire*, which is expected to drive or look for better solutions or architectural and plant design;
- The second concerns the intrinsic values of the data and results obtained through the questionnaire, in this case the purpose of an International Standard is to specify the reliability.

The purpose of the questionnaire is an issue not addressed by the standard, in the specific case of this book, and in general the questionnaires allow you to capture information with a different quality from the numbers. So questionnaires allow you to:

(a) to improve the accuracy of the *Ergonomic Investigation*, and interpret the correspondence between the physical data measured and the perception of the users/occupants;

(b) add or expand research on cognitive rests linked to human-environment and welfare;
(c) guide the actions of corrections of the physical environment or of human behaviour;
(d) guide the strategies and corrective architectural solutions of the environment;
(e) expand medical research, environmental psychology,

in summary questionnaires provide information to be used for purposes outside of the construction of the questionnaire itself, and for situations where there are regulations, for a specific case it is not applicable. So for children in pre-school or schools for people with relationship difficulties or illnesses, situations of stress at work, situational or context, the use and interpretation of subjective data allows you to know information related to the physical and human aspects, that the measurement of the physical data does not allow you to do.

The scales of judgment that may be included in the questionnaire are:

- perceived or desired;
- related to the global context or localized (a particular portion of the room or building);
- relating to the present or the past, or referred to a moment of the response, to the day or seasons or previous situations, in this second case, the judgment refers to the memory of perception, so the uncertainty is greater, is no longer a judgment, but an opinion or a memory;
- refers to a single moment of a day or repeated in time, in the same day or more periods during the season, or year.

In the preparation and administration of the questionnaire attention should be paid to external climatic variables and/or the clothing of those who can affect the perception of comfort. In the case of children (or other vulnerable users, for example in those hospital or sick people) it is worth highlighting what is their degree of autonomy in choosing and changing their clothing during the day. In defining "*Principles of scale construction and condition of use*" ISO 10551 specifies that:

> Basic difficulties are encountered in any area which involves the use of language. In this regard, bias and variability in the data can results from inconsistencies and inappropriateness on accompanying instruction. Therefore, it become crucial to standardize preparatory instruction which explain the study, as well as the wording of the judgement scales.

In building the judgement scales the standard defines three types of questions:

(1) referred to the ASSESSMENT (*Scale Evaluative*) of feeling at the precise moment (*How are you feeling (at this precise moment)?*). For this type of specific question to define between the two poles of opinion, the extremes of feeling, Pole A (e.g. very cold) and pole B (e.g. very hot). This criterion can be applied to any judgement scale relative to the environment, it can go from "*dark*" to "*very bright*" or "*too bright*" or "*blinding*" or from "*quiet*" to "*too noisy*", etc.;
(2) referring to the PERCEPTION (*Perceptual Scale*) of feeling (Do you find this ...?) also expressed in percentage terms compared to an optimal (100% of comfort) worse or worst feeling (e.g. from perfect to not tolerable);

(3) referred to the PREFERENCE (*Scale of Preference*) of a feeling (*How would you prefer to be now*), which requires a judgment with respect to the previous answers, and the variability of environmental conditions with respect to itself, then the environment will be "*hotter*" or "*colder*", the question tends to evaluate a gradient, increase or decrease compared to the feeling valued or perceived.

In the questions for the *assessment* of feeling, the scale between Pole A and Pole B must be divided symmetrically in 7°, or possibly 9°, while maintaining the central degree (point of indifference) as the optimal situation, or neutral, as shown in Table 5.1.

Pole A, Polo B and the various Degrees may be associated with a word or matching image (e.g. Table 5.2).

In questions of the evaluation of *perception*, the scale increases from a Degree equal to 0 (zero) which corresponds to the situation "comfortable" or perfect, or outstanding, up to a Degree equal to 4 which makes the worst judgment "extremely uncomfortable". This type of scale can have more degrees of judgment (e.g. Table 5.3) and refers also to non-thermal conditions. Similarly the terminologies used for each degree can vary depending on the measurement; for example in the case of lighting can go from 0° correctly illuminated to 4°.

Questions related to *preference*, refer to the evaluation of preferable conditions compared to current conditions, and has the objective to evaluate the difference in degree with respect to that perceived. So this is a relative measurement mode, which allows you to evaluate the ability or willingness of people (e.g. Table 5.4).

These are criteria that can be used to assess the perception of comfort. The first criterion allows the relationship of the PMV evaluated with the questionnaire with

Table 5.1 Structure of the evaluation scale

Pole A	Degree of intensity	Point of indifference	Degree of intensity	Pole B
	(– 4) – 3 – 2 – 1	0	+ 1 + 2 + 3 (+ 4)	

Table 5.2 Example of scale of perceptual judgment on personal thermal state ("*How are you feeling now?*" ISO 10551)

Poles	Degree	Term wording
Hot	(+ 4)	Very hot
	+ 3	Hot
	+ 2	Warm
	+ 1	Slightly warm
Indifference	0	Neutral
Cold	– 1	Slightly cool
	– 2	Cool
	– 3	Cold
	(– 4)	Very cold

Table 5.3 Example scale of evalutative judgment on personal thermal state (Do you find this …? ISO 10551)

Pole	Degree	Wording of degrees
	0	Comfortable
	1	Slightly uncomfortable
	2	Uncomfortable
Discomfort	3	Very uncomfortable
	4	Extremely uncomfortable

Table 5.4 Example of thermal preference scale ("Please state how you would prefer to be now": ISO 10551)

Poles	Degree	Wording of degree for 7-degree scale
		I would prefer to be …
Warmer	+ 3	Much warmer
	+ 2	Warmer
	+ 1	A little warmer
	0	Neither warmer nor cooler
Cooler	− 1	Slightly cooler
	− 2	Cooler
	− 3	Much cooler

the PMV measured by the probes on site. The *Types of Judgements* are divided into two groups:

- The first group is related to the "*Personal thermal state*" of the person interviewed, in general these types of questions seeks to evaluate how the person perceives and evaluates the environment, and includes questions about:
  - *Perceptual*, judgment on the perception of comfort when you place the questionnaire (*How do you feel at this precise moment*?) and is rated with 7 or 9°;
  - *Affective evaluation*, opinion on the comfort expressed not as assessment but as perception between the extreme condition of the "*perfect*" and that of "*not tolerable*" or "*unbearable*" to "*harmful*" or considered such. This judgment in its formulation and can be understood relative to multiple variables related to comfort and care is needed that the entire questionnaire, or the question, are clearly understood as referring to the thermal comfort (or brightness, or indoor air quality, etc. …) and no judgments related to external factors such as the courtesy of the people, the difficulty of access to the site, the condition of hunger or pregnancy etc. all those psychological elements of context that may influence the judgment. In this regard, a situation common in buildings where there is an organization, more so if institutional (public offices, schools etc.) the risk of these questions is that they are intended as a judgment of the evaluator on the organization, or, conversely, a judgment of the organization on the evaluator. The subject must be in the condition of anonymity and tranquility

in order to express their opinion freely and in any case refer to the variable indoor climate, and nothing else;
  - *Thermal preference*, judgment concerning the conditions expected from the subject rather than that perceived, and this information is useful, both to guide design decisions and to evaluate the difference between what is perceived and desired, even with respect to the values of neutrality or standards required by law. Also in this case the judgment is reported at the time of administration of the questionnaire.

- The second group is related to *Thermal Ambience*, the judgment on the air (*exclusively*) and tolerability/acceptance of the perceived conditions, and includes questions about:
  - *Personal acceptability*, judgment on the acceptability of the level anticipated, expected, demanded from the subject with respect to the current one. The ISO 10551 provides 2°: "*Generally Acceptable*" and "*Generally Unacceptable*";
  - *Personal Tolerance*, opinions on the judgment preceding that indicate the tolerance of a condition is not acceptable in a 5° scale ranging from "*Perfectly tolerable*" to "*Intolerable*", in these kinds of questions the view can be bound on personal ratings, mood, to political considerations, or other factors that may depend on the social conditions of the place, the hierarchical relationships, the level of attendance or waiting as in the case of crowded waiting rooms etc., and each aspect that binds tolerance to non-environmental factors.

The judgments are expressed in relation to the time the was questionnaire submitted. In some cases it is possible to expand the questionnaire including questions related to feelings and judgments in other times of the day or seasons of the year or, in the past or future. In these cases it should be noted that judgment is not related to the judgment of feeling, but expressed "*the opinion that the subject with respect to possible judgement that could be expressed in the past or in the future*" (for example following actions): In this sense the randomness of the data is not comparable with data measured on site, but can be used to guide the design criteria if these include, as is the case of open spaces or spaces indoor collective, dynamics of social judgment.

Data analysis and the application of results to the standard ISO 10551 provides examples of statistical evaluation, highlighting how the representativeness of the data depends on the statistical method adopted, also "*This international Standard encourages practitioners and research with sufficient data, or those are able to collect data, to publish studies on the statistical characteristics of the sample distribution of such data.*"

The ISO 10551 Annex B (informative) "*Application of assessment procedures and judgment scales: Examples, including data analysis*" lists examples of statistical processing of the data.

Table 5.5 Example of calculations of central tendency and scatter

	Degree	Number (n.)		x_i^2	$x_i^2 * n$
	(a)	(b)	(c) = (a) * (b)	$(a)^2$	$(a)^2 * (b)$
Very cold	– 4	0	0	16	0
Cold	– 3	1	– 3	9	9
Cool	– 2	10	– 20	4	40
Slightly cool	– 1	22	– 22		22
Neutral	0	18	0	0	0
Slightly warm	+ 1	16	16	1	16
Warm	+ 2	5	10	4	20
Hot	+ 3	0	0	9	0
Very hot	+ 4	0	0	16	0
No reply		1			
Total		72	– 19		107

Central Tendency = 72/– 19 = – 0.26 (slightly, slightly cool)
Scatter (σ^2) or Standard Deviation = $\sqrt{}(107/72) - (-0.26)^2 = 1.19$

For the first evaluation team (*Personal thermal state*) statistical data to be reported are:

- for the central tendency, the median (second quartile);
- for the scatter (σ^2), the semi-interquartile distance (half the difference between the first and third quartiles).

For judgments regarding the acceptability statement, the statistical data to be reported are:

- for the central tendency, the mode (point of maximum frequency);
- for the scatter: the entropy of distribution.

Table 5.5 show and example of calculation of central tendency and scatter.

5.4 Statistical Representation of the Questionnaires

The processing of the results obtained from the questionnaires poses the problem of how to interpret the data from the statistical point of view, or what is the number of questionnaires, or samples (where the sample is defined as the number of answers to every single question), is it necessary to refer to, *to get a reliable result*?

In survey statistics two sampling techniques can be distinguished:

(a) the probabilistic techniques, which allow the calculation of statistical error,
(b) non-probabilistic techniques, which do not allow for the calculation of statistical error.

The techniques of type (a) apply when you have an exact list (or approximately exact) of all components of the population of the target universe. For example when you know all workers or students in the building. For this type of investigation it is possible to adopt survey systems with refined statistical errors (random, multistage, stratified, systematic), and in general, the purpose of the project, can be limited to the interpretation of the results obtained, as the sample is representative. In these cases, you can assess the representativeness of the sample, for example if you carry out 10 questionnaires to workers in a building where 100 people work, representativeness is 10% and you can match this 10% as if it were 100% (to put it crudely).

The techniques of type (b) apply when you do not have the exact list of the population of the target universe. For example in the case of interviews in public spaces, or in places frequented by different people with different types of influxes, you can not know what the number is, exactly, of people who frequent the place. In this case the reference population is "self selected", in that it refers only to people who are available to respond, and the self-selection is "not random" as it measures the rate of response to the propensity of the people interviewed. Such techniques are applied to marketing surveys and surveys to know the propensity for a given behavior.

In case (b) is therefore not possible to evaluate the error, or the representativeness of the sample, only the "propensity to answer", so 10 interviews represent all of those who are willing to answer questions.

In this type of investigation one method of data processing is to vote, for example if it is estimated that the population of a city is made up of 30% under 30, 40% between 30 and 60 years, and the remaining 30% over 61, a sample is representative even when there are only 10 questions including 3 relating to under 30, four between 40 and 60 years, and 3 over 60. This method is often adopted in telephone interviews.

The number of responses can reduce the randomness of the data but not its representativeness.

Non-probabilistic surveys are used when:

(a) we do not know the whole population (it is not known how large)
(b) it is a pilot surveys or exploratory type
(c) we do not have sufficient budget for other investigations.

5.4.1 *Questionnaire Design and the Relationship Between Questions and Variables Measurable by Instrumentation*

In the preparation of the questionnaires attention must be paid to four aspects:

- the *context* in which the questions are submitted, the time available, the fact that they are self-complied or interviews, as described in the preceding paragraphs;
- the *statistical representativeness* of the sample;

- *Types of Findings* of the questionnaire, assessment, perception, preference and the corresponding rating scales;
- The *Questionnaire design* or the layout of the questions, their sequence and the writing of the individual sentences that make up the questions and answers in relation to the cognitive levels of the subjects that have to respond, their availability and the time available;
- finally, in the event that is provided for a monitoring campaign, the *relationship between the physical variable* measured by instrumentation and/or calculated and the *Question* of the questionnaire.

In preparing the Questionnaire Design it can be useful to (check list):

- provide an introductory phrase in which you explain the purpose of the questionnaire, the proponent, the duration, the use and anonymity of data, etc. and any other matters which may clarify the contents of the experimental activities and facilitate the availability of the respondents;
- group questions for topics or issues, according to the increasing level of attention, bearing in mind that the concentration tends to increase and then decrease, depending on length of the questionnaire. So simple questions and/or personnel question "to break the ice" regarding their age, clothing etc., must be at the beginning, the more sophisticated and/or sensitive should be inserted to the middle of the questionnaire, while the final questions can be open or interlocutory;
- avoid nested questions related to the previous example, "*if you answered yes to the previous question which of these* ...", particularly if the questionnaire is self-compiled or online;
- feelings are measured with respect to the time during which the questionnaire is completed, to avoid questions about the past or the future, and if they are included always provide a question relating to the question as a "*control*";
- prefer short sentences with unique questions/answers, uninterpretable, avoiding long lists of multiple responses;
- avoid technical terms or questions related to technical solutions;
- prefer questions with enslaved answers, "feeling hot/cold";
- consider the availability and fatigue of the interviewee to answer the questions and the fatigue of the interviewer (if any);
- remember to collect data on the context: date, weather, sunshine, temperature, flow, time and number of questions, special situations, anecdotes, etc.

In the case in which there is a monitoring of physical variables by means of instrumentation, each question/answer must be able to evaluate a single physical variable, measured or processed, with which the results will be compared.

The questions on Thermal Comfort, such as those presented in Tables 5.2 and 5.3. can be directly compared with the value of PMV and PPD measured by means of the instrumentation and calculated according to ISO 7730.

The Air Temperature can be assessed by questions related to the temperature itself, or to the "feeling of temperature" with a ladder-type evaluation or preference "Thermal Preference scale", which report phrases like "I wish that in this environment it were + 2 °C, + 1 °C, …" or "I think the temperature is: according to the law, greater that the value of the law, lower that the value of the law, much lower …", etc.

The Relative Humidity can be expressed by the single term "Humidity" and the evaluation can go from the poles "Very Dry" to "Very Humid" and, in relative terms, such as "I want an environment drier, damper … etc." Data may be collected relating to considerations on the presence/absence of mold or the risk of it forming, the effect of the vegetation or other elements that can increase the concentration of water vapour in the environment. An example is shown in Table 5.6.

The air speed can be related to the evaluation of discomfort "*environment not-ventilated, properly ventilated, with a light breeze, with drafts, with strong air currents, etc.*" or to request feedback and opinions on the presence/absence of dust.

In the case of Indoor Air Quality it is necessary to identify a marker that can be perceived olfactory. Questions may be referred to specific odours from the outside: traffic, dust, gasoline or other pollutants, especially if are work activities present nearby or inside: detergents, odours from processing, etc. The other parameter that can be measured with the sense of smell is bad smells or "stale" air, two parameters related to the concentration of CO_2: the higher the concentration of CO_2 the greater the feeling of stale air. In some cases it is possible to correlate the CO_2 to other feelings.

Since the CO_2 concentration affects the amount of oxygen contained in the environment, and therefore with the amount of oxygen that reaches the brain, it is possible to correlate the concentration of CO_2 with feelings of fatigue, malaise, dizziness, redness of the eyes or difficulty concentrating. These phenomena can be coupled with an increase in temperature due to the plant/equipment or to the persistence of the employment of the people. In this sense there can be questions asked about the crowding of the room (very crowded, slightly crowded, etc.).

Table 5.6 Example questions related to relative humidity[a]

And if it should be warmer, what is your tolerance level?
[_] I do not have tolerance (perfect comfort condition should be ensured)
[_] I can tolerate a warmer thermal environment
[_] I can tolerate a hot environment
And if it should be more humid, what is your tolerance level?
[_] I do not have tolerance (perfect comfort condition should be ensured)
[_] I can tolerate a slightly more humid environment
[_] I can tolerate a humid environment
And if it should be more dry, what is your tolerance level?
[_] I do not have tolerance (perfect comfort condition should be ensured)
[_] I can tolerate a slightly more dry environment
[_] I can tolerate a dry environment

[a] From the questionnaire proposed within the on going Ph.D. research of Giovanni Litti; Faculty of Applied Science, EMIB Lab, University of Antwerp

The concentration of CO_2 is also a great indicator that lets you estimate the ventilation rate of the environment and thus the perception of the environment through questions like "do you think the environment is adequately ventilated, too ventilated, you must open the windows, the Air-Conditioning works/doesn't work properly, etc."

The evaluation of other parameters of comfort, such as Acoustic Comfort and Lighting/Visual comfort follows the same criteria relating to the perception of the phenomenon of sound and light as shown in Table 5.7.

In assessing the acoustic comfort, the element of discomfort is given by the noise coming from the outside or from other rooms in the building, so the Scale of Perception of the phenomenon may relate to the intensity of the noise (not Intense—Hardly

Table 5.7 Esempio domande relative a acoustic comfort and lighting comfort[a]

Acoustic comfort	Lighting comfort
In this moment, the noises you are hearing are	The lighting in this room is
• Not intense	• Too intense
• Slightly intense	• Intense
• Intense	• Rather intense
• Hardly intense	• Enough
The acoustic sensation, is causing	• Rather weak
• Well being	• Weak
• Slight annoyance	• Strongly weak
• Annoyance	In this moment, the lighting is causing
• Strong annoyance	• Well being
In this moment, you would like to hear	• Slight annoyance
• The same noise	• Annoyance
• Slightly less noise	• Strong annoyance
• Less noise	In this moment, you would like to have
According to the acoustic comfort, how do you consider this room?	• Much more intense light
	• More intense light
• Tolerable	• Slightly more intense light
• Not so tolerable	• The same light
• Hardly tolerable	• Slightly less intense light
	• Strongly less intense light
	According to the light intensity, how do you consider this room
	• Tolerable
	• Not so tolerable
	• Hardly tolerable

[a] From the questionnaire proposed within the on going Ph.D. research of Giovanni Litti; Faculty of Applied Science, EMIB Lab, University of Antwerp

intense) or to the Acoustic Sensation. Both definitions are kinds of generic character and co-related to the variable "noise", the questionnaires may refer to understanding of the text, the presence/absence of reverberation or rumble and other phenomena related to the acoustic quality of indoor environments, also in relation to the level of preparedness of the respondent.

For example in the case of theaters or music auditoriums, the perception of the musicians, and the amount of words they used to describe the spatial quality of the sound, will be greater than with respect to (for example, you can ask if the sound is "centered") to a user less prepared. In the same world in the case of classrooms, acoustic comfort may be related to the presence of noise, the noise due to people during class or during breaks, or to the intelligibility of the lesson (can you hear the words as the teacher explains?). Finally in the definition of the sounds you can use criteria synaesthetic, and associate sounds to colours, a yellow sound, may seem more "acute" and strident than a blue sound.

In assessing the brightness comfort elements of discomfort are two: (i) the correct lighting environments to perform the visual task, which can range from "properly illuminated" to "dark", and (ii) or the visual acuity or the presence of glare or other difficulties that may affect the clarity of vision. To these can be added (iii) the perception of colours, even in this case the level of accuracy and vocabulary depends on the familiarity of the respondent with respect to colours (Man/Woman, artist, Child/Adult, workers of the arts, the fashion industry, etc.). Colours can be associated as synaesthetic concepts referring to natural the environment (green frog, lemon yellow, etc.). The lighting level of the environment is directly correlated with the measurement of the lux, to an exact level or as an average, and you can use a Evaluative Scale from "strong light" to "insufficient light" or "dark". In Perceptual Scales poles ranging from environments properly illuminated to disturbed environments.

In both evaluations a scale of preference can be used with respect to the conditions of acoustic comfort and light, to have a quieter environment, more/less illuminated.

The results can be direct or indirect, derived from the interpretation of the que-questionnaires and activity monitoring.

Finally if you have children who answer the questionnaire it can be an opportunity for the interviewer and for the child to learn something new, so it's good to think of the Questionnaire Design, as an opportunity for learning and curiosity, leaving children a chance to give vent to their freedom and creativity.

References

Allison GT (1971) Essence of decision: explaining the Cuban missile crisis. Little, Brown, Boston

Andersen R, Fabi V, Totfum J, Corgnati SP, Olesen BW (2012) Window opening behaviour modelled from measurements in Danish dwellings. Build Environ 69:101–113

Andersson K (1998) Epidemiological approach to indoor air problems. Indoor Air 4:32–39

Andersson K, Stridth G, Fagerlund I, Larsson B (1993) The MM-questionnaire—a tool when solving indoor climate problems. Department of Occupational and Environmental Medicine, Sweden

ASHRAE 55. Thermal environmental conditions for human occupancy. American Society of Heating, Refrigerating and Air-conditioning Engineers (ASHRAE), Atlanta, Georgia
Ball SJ (1981) Beachside comprehensive: a case study of secondary schooling. Cambridge University Press
Bechhofer F, Elliot B, McCrone D (1984) Safety in numbers: one the use of multiple interviewers. Sociology 18:97–100
Bedford T (1936) The warmth factor in comfort at work. Medical Research Council Industrial Health Board, report no. 76. HMSO, London
Bluyssen PM, Cox C (2002) Indoor environment quality and upgrading of European office buildings. Energy Build 34:155–162
Bryman A (2004) Social research methods. Oxford University Press, Oxford
de Dear RJ (1998) A global database of thermal comfort experiments. ASHRAE Tech Data Bull 14(1):15–26
de Dear RJ, Brager GS (1998) Developing an adaptive model of thermal comfort and preference. ASHRAE Tech Data Bull 14(1):27–49
de Dear RJ, Brager G, Cooper D (1997) Developing an adaptive model of thermal comfort and preference. Final report on RP 884. Macquarie University, Sydney
Dili AD, Naseer MA, Varghese TZ (2010) Thermal comfort study of Kerala traditional residential buildings based on questionnaire survey among occupants of traditional and modern buildings. Energy Build 42:2139–2150
Fabi V, Andersen R, Corgnati SP, Olesen BW (2012) Occupants' window opening behaviour: a literature review of factors influencing occupant behavior and model. Build Environ 58:188–198
Hargreaves T, Nye M, Burgess J (2010) Making energy visible: a qualitative field study of how householders interact with feedback from smart energy monitors. Energy Policy 38:6111–6119
Hargreaves T, Nye M, Burgess J (2013) Keeping energy visible? Exploring how householder interact with feedback from smart energy monitors in the longer term. Energy Policy 52:126–134
Heinzerling D, Schiavon S, Webster T, Arens E (2013) Indoor environmental quality assessment models: a literature review and a proposed weighting and classification scheme. Build Environ 70:210–223
Humphreys MA (1975) Field studies of thermal comfort compared and applied. Current paper CP 76/75. Department of the Environment, Building Research Establishment, Watford, UK (also in Physiological requirements on the microclimate. Institute of Hygiene and Epidemiology, Prague, 1975, pp 115–181; J Inst Heat Vent Eng 44:5–27, 1976)
Humphreys MA, Nicol JO, Raja IA (2007) Field studies of indoor thermal comfort and the progress of the adaptive approach. Adv Build Energy Res 1(1):55–88
Hutter HP, Moshammer H, Wallner P, Damberger B, Tappler P, Kundi M (2006) Health complaints and annoyances after moving into a new office building: a multidisciplinary approach including analysis of questionnaires, air and house dust samples. Int J Hyg Environ Health 209:65–68
ISO 10551. Ergonomics of the thermal environment. Assessment of the influence of the thermal environment using subjective judgment scales
ISO 7730. Moderate thermal environments—determination of the PMV and PPD indices and specification of the condition for thermal comfort
Raw GJ, Roys MS, Whitehead C, Tong D (1996) Questionnaire design for sick building syndrome: an empirical comparison of options. Environ Int 22(1):61–72
Whyte WF (1955) Street corner society, 2nd edn. University of Chicago Press, Chicago

Chapter 6
The Thermal Comfort and Child Development Psychology

Abstract The study of Thermal Comfort as perceived by children requires us to know Thermal Cognitive Knowledge (TCK): children's ability to understand the feelings caused by iterations with the environment with respect to their own cognitive universe-world. Added to this is the need to define the questions that can be understood by children without the intermediation of the contents by adults. This chapter is a summary of the main issues affecting the study of Child Development and relations with the indoor environment and thermal comfort, and the information needed to prepare the contents of the survey questions to ask children.

6.1 Child Development and Indoor Environment

In previous chapters we have described the physical and physiological which define the conditions of Thermal Comfort, and the way we learn about the perception of comfort to the occupant, through the use of the occupants. Both the Standards, laboratory tests, and the use of questionnaires were processed and used mainly as user standards: men or women "healthy" and adult.

The literature reports some other cases in other contexts such as hospitals Verheyen et al. (2011), De Giuli et al. (2013), Khodakarami and Nasrollahi (2012), Lomas and Giridharan (2012), also with regard to the aspects of Environmental Psychology, as in Fornara et al. (2006), Mourshed and Zhao (2012), or in the concept of Healing Environment as in Huisman et al. (2012), as well as in the workplace Laurence et al. (2013). In hospitals there plays a certain perception of the quality of the environment also socially for patients (Campos et al. 2013) or in waiting rooms (Arneill and Devlin 2002).

In previous chapters we have described how to study Thermal Comfort in respect of the physical and physiological aspects, but there remains a third aspect to consider which is the *psychological and cultural* aspects. Thermal Comfort is expressed based on ratios of sensation (e.g. PMV or PPD), and only the extreme conditions (Heat Stroke, Sun Stroke, Hypothermia) are expressed exclusively by reference to physical and physiological factors (Human Body Heat Balance).

K. Fabbri, *Thermal Comfort Perception*, https://doi.org/10.1007/978-3-031-52610-7_6

The variables of judgment on Thermal Comfort due to psychological and cultural aspects include both the culture and the social context, the aspect of cognitive knowledge and cognitive process, or how the individual interprets the signals.

In the Standards and Studies of Fanger (1970) following, thanks to the development of a multitude of statistical data, the PMV index and other indexes of sensation are excluded, the aspect of "psychological-cognitive" is the most uncertain.

In the case of studies on subjects other than man, or woman, adult and healthy, we must also take into account aspects related to the Environmental Psychology (as in the previous chapter, account was taken of the studies of Sociology), especially as is used in the questionnaires.

In this chapter and in the second part of the book, our attention turns to a particular audience: children and young people.

Once the indicators of well-being are described, and how to make the questionnaires, we must describe how to use this tool, indexes and questionnaires to assess the feeling and understanding of the concept of comfort by children.

The study of Thermal Comfort and understanding of Environment by children and young people requires some brief reference to the notions of the discipline of *Developmental Psychology* and *Environmental Psychology* (Canter and Craik 1981). In the case of children and adolescents it can also influence the ability to learn, as, for example, described in Rivlin (1984), and Marchand et al. (2014).

The child and his understanding of the world varies according to the age, social and cultural context.

Developmental Psychology distinguishes the evolutionary development (according to the theory of J. Piaget) in several stages:

- The infancy, from 0 to 2 years;
- The early childhood, from 2 to 7 years;
- Primary school age/later childhood, from 7 to 11 years;
- Preadolescence, from 11 to 13 years;
- Adolescence, ages 13 and then up to 18–19 years.

This division has been reviewed by the authors and scholars and generally they regard the study on comfort and the educational system (at least the Italian one), to be a fair and realistic distribution.

Regarding Thermal Comfort and use of Questionnaires the following should be highlighted:

- *Infancy*, from zero to two years, a period in which there cannot be an interaction with the child as they are too young to understand the questions; the feeling of thermal comfort can be derived from physiological factors or by indirect experiments carried out by specialists in the field, as in Hodnett (1989);
- *Adolescence*, from thirteen years onwards, it is a period in which, as regards the perception of Thermal Comfort and questionnaires, the understanding of the questions is comparable with that of adults, if we exclude the problems related to the culture of adolescence, the social process that leads to the autonomy of the individual over society and the family;

these two aspects are excluded from the case studies reported in the second part of the book, which deals instead:

- *Early childhood*, two to six years, during which the child has yet to acquire the written language;
- *Primary school/later childhood*, from six to ten years, during which the child learns the concepts that allow them to describe the world and their own perceptions, but is still strongly anchored in the family context, cultural, is not yet an autonomous individual;
- *Preadolescence*, from ten to thirteen, during which the child begins the path of self-identification and autonomy from the social and cultural context, and the greater the influence of peers;

and for each period varies, in addition to cognitive aspects, the physical characteristics of the child (weight and height), the family and social environment, as well as the habits and autonomy in the choice of clothing, all of which impact on the parameters of Thermal Comfort as described in the previous chapters.

This chapter is a brief summary of the aspects of developmental psychology, to be taken into account in the preparation of questionnaires, in their administration and interpretation of results. The need to also study the cognitive aspects was necessary as a result of the research reported in the second part of the book. In dealing with children it was evident that there was a communicative gap with respect to the content of the questionnaires themselves.

During the research it was realized that for children the words are games (rhymes, puns) and have an evocative power greater than adults. They are interested in the new or strange words (drafts, infiltration, comfort, neutral, etc.) and for their significance both for what they evoke and how they sound.

In the case of pre-school children it was decided to use images to encode and represent the meaning of "hot", "very hot", "cold", etc., in this case the images chosen by the researchers did not correspond to the meaning which was attributed to the children.

*The question must be part of the language, or rather, the child's cognitive universe, which also includes the inner life of the child, for which the memory is linked to the story (*Piaget 1967*).*

The use of both these instruments express *words* (verbally and/or written) and *images*, along with the sharing and discussion of the meaning of each term by the children themselves, has proved the most effective solution.

In order to understand how and what to write in the questionnaires it has been very helpful to examine some aspects of developmental psychology, and it is good to point out that the author is not a scholar of this discipline. In order to ensure the interdisciplinary nature of this book, some concepts may be simplified for researchers in the field of human sciences, psychologists etc. But will be a new point of view for architects, doctors and those who work in comfort.

6.2 The Child Developmental Psychology (a Brief)

Developmental psychology includes the set of dynamics that bring a baby to become an adult, and which takes place during the entire span of life (Santrock 2013). Developmental psychology distinguishes between child development and adult-adolescent, identifying the evolutionary periods: infancy, early childhood, later childhood, preadolescence and adolescence that marks the transition to adulthood at around 18–19 years. The development includes biological, socio-emotional and cognitive processes in an inseparable continuum.

The psychoanalytical theories are different, in summary:

- *Erik Erikson* (1902–1994), adopts a psycho-social theory—the theory of Erikson (1950, 1968), which identifies eight stages of psychosocial development during the entire span of life, each of which consists of a dichotomy, that represents the transition from one phase to another by means of a "crisis", understood as a turning point. In early childhood the crisis between *trust/mistrust* between *autonomy/shame*, in the later childhood it is the crisis between *initiative/guilt*, and preadolescence between *industriousness/inferiority* when children come into contact with new initiatives, and then they pass to the crisis phase of adolescence between *identity/identity confusion*;
- *Jean Piaget* (1896–1980), developed a Cognitive Theory—Theory of Piaget—which is based on conscious thoughts of children, identifying different stages: sensory-motor for the first years of age, *pre-operative* and then *operators* for later childhood (up to 7 years). In the phase of pre-adolescence, it moves to a *concrete operational* stage in which the child is able to operate independently, and then it moves to the state of *formal operators* during adolescence, in which the subject constructs their identity compared to the social and cultural context;
- *Lev Vygotsky* (1896–1934) shares the view of Piaget highlighting the way in which culture and social interactions act and guide the cognitive development. His idea is that knowledge is *situated* (located, takes place in a precise physical, spatial and temporal context) and *collaborative* (collaborative, is through a dialogue with other people in a relationship of debit/credit), and is continuously processing information (Vygotskij, 1962).

These theories, especially those of Piaget and Vygotsky constitute a major contribution to the main concepts useful for well-being, we can add more fields of study, more related to research, such as:

- *The behavioral and social cognitive theories (or Behavioral Theories)*, from *Classical Conditioning or Respondent Conditioning* developed by Pavlov (1927) and Watson (1913) in which a stimulus corresponds to a reaction, the *Operant Conditioning* in which the behavior of the child is taught by the activities of reward or punishment in the wake of the actions taken by the child, until you get to Bandura (2001) observational learning or *Social Learning Theory* (or modelling) in which we learn by observing what others do;

- The Ethology (Ethological Theory) from research zoologist Konrad Lorenz (1903–1989), identified the *imprinting* and the rapid innate learning process of the animals, this concept was reproposed by Bowlby (1969) to human development;
- The Ecological Systems Theory of child development, by Bronfenbrenner (1979), in which the individual is located within the microsystem (family, school, etc.) in turn inserted into a meso-system, in an eco-system (family friends, neighbors) and finally in a macro system (society, culture reference).

6.2.1 The Research Instruments

Research in the field of child developmental psychology may adopt a method of *Descriptive Research*, similar to naturalistic observation, in which we observe the behavior of children in a given social context or environment, a method of *Correlation Research*, where they relate event and its consequences, through the interpretation of results.

When we plan to study the development evolved *over time*, searches can be carried out as *Longitudinal Studies*, if they study the same subjects over time, usually within a few years, or *Transversal studies*, if they study more subjects with different ages.

Experimental Research is set up like an Experiment in the sense understood by the scientific method, the scope and limits of the control experiment are defined with the creation of an *experimental group* and a *control group*, identifying the independent variables and independent. Typically, this type of testing takes place in a controlled environment, under laboratory conditions, for example the test of Povl O. Fanger is set up as an experiment in laboratory. Other tests may be performed outside of the laboratory in the Real World through naturalistic observation (field study).

Alternatively, we can adopt the *Quasi-Experiment or Quasi-Experimental Design*, in which not all the variables are under control, the context can be the normal school environment of the child, but it is possible to identify and measure the variables.

The experiments described in the second part of the book adopted the method of Quasi-Experiment, given that they were carried out in schools (Real World and field study), which was a monitored indoor microclimate to control the variables of the indoor microclimate and the PMV and PPD sensation/feelings indexes, according to the Standard, to be compared with the results of the texts.

The Child Developmental Psychology also takes account of the physical, motor and perceptive development, also from a neurological point of view, given that these factors affect the perception of the self by the subject, which varies with the evolution of the human body with the growth, with the possibilities of movement and aspects of neurological-perception. This chapter does not address the neurological aspects, although they presumably effect the exchange of energy on the body-environment, and further studies in this direction would be interesting.

The discipline that studies comfort, uses as tools for evaluation the indexes or indicators linked to the perception or sensation (feeling indexes PMV and PPD). It is therefore the indicators which require an interaction with the subject, and therefore are considered as already encoded by neurological processes the variables related to the metabolism and clothing, aspects of physical development, motor and perception. Regarding neurological encoding please refer to the Standard and the—limited—studies on the sick and in the hospital environment, reported in Khodakarami and Nasrollahi (2012).

From the point of view of the Discipline that deals with thermal comfort there are excluded—or at least the present author believes to be missing—studies that relate to cognitive-psychological variables, compared to the equations to determine the PMV comfort indexes.

This book seeks to contribute to the subject area in this direction, bringing some examples and methods of investigation.

6.2.2 The Development Process

The development process, according to Piaget, follows the patterns of actions to interact with the environment and maintain balance in the relationship between their internal environment (body and mind) with the external environment (physical and social). In this process of maintaining equilibrium there are *functional invariants*, i.e. the biological mechanisms that ensure the functioning of the body, and the processes of *assimilation and accommodation.*

Assimilation is the process by which the child incorporates information on the context, such as the ability to dress and to recognize with "heavier" clothing (with higher thermal resistance) when you feel cold (in contrast to the physiological which wants an increase in metabolism by mechanical work), or the ability to recognize the plant devices that provide the heat or the cold. An *arrangement*, or *adaptation*, is the mechanism by which children change their behavior patterns to fit new experiences.

The development is divided into stages, each of which corresponds to a development and a process of assimilation and accommodation, put in place by the child.

The first stage (0–2 years) is the *Sensorimotor Stage*, in which the child is related to the environment based on physical activities and the senses (sight, hearing, touch, taste, smell). The child reacts to stimuli through patterns, such as shelter from a harassing sound, or finding clothes in case of feeling cold, such patterns are internalized. In this period the child begins to understand the limits of their body, *Proprioception*, and the permanence of objects beyond their own sight.

In the following period (2–7 years) there is the *Preoperational Stage*, during which the child performs operations to get or do something, and begin to represent the world through words, pictures, drawings and games ("*pretend* …"), and is characterized by a marked egocentrism intellectual thought, in which the child is at the centre and the explanations are finalized.

It is a phase characterized by *symbolic* explanations (*Symbolic Function Substage*), where the nature of human activity is predominant, the explanations are intended to *animism*—it is believed that inanimate objects have life or vital qualities and are able to do things on their own—*finality*—for which all phenomena are goal-oriented, and the *artificialism*—in which the child is convinced that everything has been built by man or by a divine entity that regulates everything. Until the age of four years the child does not conceive that reality can be different from what is presented to him (Rollo 2005).

In the *Pre-operational Stage* the child develops an intuitive thought: they know things without using rational thinking. The child's thinking is irreversible and focuses its attention on a single characteristic of what is being viewed and draws conclusions without proceeding according to a particular reasoning.

The research carried out at the Diana kindergarten (Chap. 8), with children aged between 4 and 5 years old, has shown the ability of children to have an explanation for every judgment of feelings of comfort, and that explanation was often correlated with natural phenomena (e.g. Very Hot = Volcano) or causes aimed (Cold = Snow).

The *Concrete Operational Stage* (7–11 years) is where intuitive thinking is replaced by reasoning. The child (boy) is able to perform logical operations, related associations between them and compare different points of view, to classify a series of elements on the basis of a sequence or series, or according to the process ("*if A happens then B happens*", "*if A happens C can't happen*").

The last stage is the *Formal Operations Stage* (11–18 years) during which the subject (teenager) is related to the environmental context attributing to himself his own identity he has in this case adolescent egocentrism, in which the adolescent is assumed to have an "*imaginary audience*" and to build their personal fable. In this stage he develops abstract thinking, ideal and logical, and hypothetical-deductive reasoning.

In relation to the study of comfort (and in the school) the definition of the Stages introduced by Piaget is a useful guide to set the contents of the quiz and the route of administration of the questionnaire. The same theory was revised and deepened in the following decades, taking into account culture and education, considering the child as an *Active Thinker*.

In Vygostkij's theory the evolution of the mind takes place through the acquisition of psychic structures of increasing complexity, due to the use of language and dialogue (with others and with himself) through which the child is able to build his own symbolic reality (inner language of thought as a guide).

The signs (words, numbers) are symbolic mediators and depend on the culture in which the child belongs. These mediators have a symbolic function "*sign*" (indicate something) and an "*evocative*" function (tell something). Cognitive development is the result of a continuous dialogue between the child and what is happening in his proximal area (ZPD—*Zone of Proximal Development*).

The role of language in cognitive development is also emphasized by Bruner (1996), describing in the narrative the representation by means of a cultural approach to the instrument of codification of knowledge of the child by the representation, which can be an *Executive Representation* for a 2 years old (the child tells of the action he intends to make, e.g. take an object), between 2 and 7 years, the child uses the I*conic Representation*, the reality is represented through images, between 7 and

11 years the child adopts *Symbolic Representation*, the reality is codified through language and symbolic systems (e.g. numbers).

6.2.3 How Children Think. Information Processing

Children interpret the world in the world differently from adults, it is a process that depends on *Capacity and Speed* of information processing that the child receives. Children use three mechanisms (Siegler 1998) of *Information Processing*:

- *Coding*, with which information passes from the senses to the memory of the child, the increased ability to encode information coincides with a greater cognitive development;
- *Automaticity*, the ability to process information independently, as happens at the end of the process of learning to read;
- *Strategy Construction*, to continue learning through self-editing, or the acquisition of knowledge to know the acting world with an active role.

The process requires an attention span to the world and the knowledge to be encoded. Attention is an activity that requires effort and concentration, therefore It lasts, the child must get used to concentrating, from early childhood. Attention can be *Selective*, when the child focuses on a specific aspect of the environment, or *Uniform* when it focuses on more tasks simultaneously. With increasing complexity of cognitive functions, in later childhood, the child adopts forms of attention *Sustained (or Focused)* that help maintain the attention over an extended period of time, to reach an *Executive* focus, that is designed to accomplish an action.

The focus depends on the memory, the ability to retain information over time. The memory is divided into:

- *Short-Term Memory* (15–30 s) that maintains the information only for the time necessary to encode the environmental stimuli that are not deemed useful for the future;
- *Long-Term Memory* (no-limits), or memory commonly known, that of personal memories, notions, instructions, etc.;
- *Working Memory*, the memory required to perform the operations in which we take decisions or resolve problems.

Attention and memory depend on the developmental stage and are the factors that are worth keeping in mind in the purpose of the administration of the questionnaire, both for the duration of the questionnaire (number of questions) and in the phase of construction or understanding of the questions on the basis of the children. Children need to recognize and remember the meaning of the terms of judgment on Thermal Comfort and relate their feeling momentarily (short-term memory and sustained attention) with the judging criteria (working memory and executive attention), without distraction.

Memory has a special role as it contains the memories of children who must draw on it to be able to associate the feelings to something they remember. The construction process of memories is what makes the child relate to the feeling of very hot to a volcano. The child must remember what "hot" and "very", mean and connect this to an element of nature that has already been stored, allowing them to consolidate the memory, in this case the "volcano". The process is automatic.

The process of construction of the memories is carried out for patterns (for mental structures) or, according to another approach, according to the *Fuzzy Trace Theory* that distinguishes between memories that consist of the precise details and blurry traces in which there is only the central idea of the event.

The memories of infants (0–2 years) are episodic and tied to big emotional events; it is a memory that stores the contents of the main events. From early childhood (2–7 years) when children start primary education and then and in later childhood (7–11 years) with the construction of rational thought school education, children begin to structure their memory, in particular the long-term memory, even adopting some strategies such as repetition of concepts, their organization (such as the sequence of months), the development of a meaningful structure (image, story etc.) thanks to the *mental images*, that the child constructs to remember mental information.

In later childhood begins the process of elaboration of *Critical Thinking*, the child thinks productively and evaluates the facts. It is a process called *Mindfulness*, to be aware of their own thinking (Langer 2005), which follows the scientific thinking of the child—i.e. asking questions about the reality for answers—and *Problem Solving* to achieve a goal.

6.2.4 The Language

The language of children is the way in which they perceive and understand their world. The construction of language, words, grammar, syntax, and the process leading to the construction of the language of the child, follows the stages of cognitive development.

When using tools to ask the children their feelings it must be taken into account the level of understanding and structuring of sentences and meaning (semantic role) of the words used.

The language, as a form of communication, is characterized by having its own P*honology* (the words have a sound, thanks to the composition of several phonemes), the *Morphology* of the words, the Syntax of the language reference, and the *Semantics* which consists of assigning meaning to words and phrases. Finally, the language has its own purpose, linked to its use. The appropriate use of language (e.g. depending on the context) is called *Pragmatics*.

In infancy, children do not have a real language, are able to process sentences composed of two words, and with early childhood, aged three years, the child is able to put together more than three words, make a complete sentence and apply the rules of syntax. Between 4 and 5 years old they are able to formulate complex utterances and express semantic relations.

School age once acquired the knowledge of the language, the child is equipped with a broader vocabulary and grammar and acquires the ability to read and write.

6.2.5 *Child and Environment*

The cognitive development of the child plays a role also in other aspects related to the environment, and the relationship between child and environment. The child is in the world and learns to relate to the world through the emotions, affections and acquiring its own identity.

The Emotions

Emotions are a mode of interaction with the world, to express the sense of unease or approval, of joy or anger, enthusiasm, etc. The definition of emotion is not unique; the scientific description (Kagan 2010) correlates with both the neuro-biological basis and the experience and culture.

In the case of children emotion plays an important role in the relationship with the world, with his mother, parents, social context. Like adults, children are provided with their own emotional life, including in relation to overcoming purposes (for example, a child has a thrill of joy when he manages to overcome an obstacle or a game, and a feeling of anger when he doesn't get what he wants).

The "*emotional competence*" varies with the development and the acquisition of social skills of the individual, and affects the expression of emotions, their understanding and adjustment.

Emotions can be expressed through verbal language (words, tone of voice, etc.) and non-verbal (behavior of the body) and are expressed by means of the *rules of expression (Display Rules)* recognized on the basis of socio-cultural conventions.

Understanding the emotions of another person, on the part of the child, starts from early childhood (2–7 years), using the *Adaptive Coping*, and in childhood the child is able to attach to a correlated moral emotion or the cause of the emotion itself.

The emotions are regulated, and in the course of their emotional development, the child acquires the ability to control their emotions, using external processes (from another person) or internal (self). In the dynamics of emotion regulation children can react to external or internal stimuli (e.g. in the case of physical pain, stomach ache, an expression of emotions with pain). From later childhood children start to adopt cognitive strategies to control and regulate their emotions, for example controlling the expression of anger in relation to the social context and relations, and to manage stress situations.

For Thermal Comfort, emotion (excitement, indifference, joy, etc.) plays an important role in completing the questionnaires. In the event that the child interprets the questionnaire as a time of testing and assessment (very high risk in schools) there will live answers with emotional anxiety for the evaluation, the child will be concerned about his personal judgment relative to Thermal Comfort of the classroom, the effect on their school life, or, if emotionally linked to other students or the teacher, has a consequence for the teacher or the school itself. For this reason, when administering the questionnaire it has to do done very carefully so that it is not perceived by the children as at the valuation, on the contrary, the researcher will have to administer the questionnaire in a game atmosphere and dialogue, trying to attract their attention (with the help educators) and stimulating emotions of participation, joy, curiosity.

Another aspect to take into consideration with regard to children's emotions, from that which occurred from the case studies described in the second part, is that concerning the relationships with other children, with adults, and also with objects and places.

Children respond with a strong emotional component, membership or repulsion, criticism, joy, anger, etc. also in relation to the places and the space or spaces in the building where they are staying for the most part of the day.

Emotions have a role in the construction of the judgment of feeling of Thermal Comfort. If the experience was carried out on young children (3–5 years) it was verified that the association between thermal sensation, and its term, symbol or image (e.g. volcano or sun for heat, snow or ice for cold) had with the child all the emotions that the child attributed to that association. A little girl (4 years) who liked the winter, was excited when she saw the snow, and attributed emotionally positive judgment to the cold.

Participation in the questionnaire should take into account the ways in which, in that context, children regulate their emotions for two reasons:

- The emotion acts on judgment and assessment of sensation, a child who is angry or unfriendly, will tend to give negative feedback;
- Also the emotion alters the metabolic activity of the child, usually increasing it, emotion of joy, curiosity and anger results in an increase in heartbeat and metabolism.

Also, you can not expect (or attempt) a neutral stance with respect to emotions, otherwise you may get the opposite effect of boredom and indifference to the questionnaire.

During the research you will have to register if the level of "*emotional alteration*", caused by the performance of the test, is likely to significantly alter the metabolism. Typically this happens only if the emotions during the test are much more intense than during the rest of the day.

The researcher must also manage their emotions to avoid that, as a consequence, they are misinterpreted by the children. The activity of the questionnaire is situation of *light stress*, since it departs from the ordinary, and requires concentration and superior skills for children, since they pose intimate questions (the *personal* opinion of the child on the Thermal Comfort).

In stressful situations, in later childhood, children use the strategy of *Coping*, adaptive or passive, which consists of a number of strategies to assess and react to situations of stress. If the researcher has an emotionally positive attitude, for example smiles when administering the test, the children will have a positive attitude.

Conversely, if the researcher is bored, stressed or adopts repulsive facial expressions or doesn't participate, children can interpret such expressions in the repulsive sense and negatively altering or not working to complete the questionnaire.

Identity Development

The construction of identity, designed by Erikson (1968), is constructed through exchanges that the individual has with society: the family context (uniqueness and reciprocity), gender and sexuality, culture and ethnicity (ethnic identity).

The construction of identity is a process that lasts a lifetime, and starts from infancy when the child acquires the "*self-understanding*" of themselves and their bodies.

From later childhood understanding of the self takes place, as well as from a visual standpoint, including by verbal description, recognizing the differences between themselves and others (height, hair color, eyes, etc.). In addition, children define themselves based on the task carried out and are extremely positive, for example, a child may claim to "*never be afraid*", "*never feel hot*" and "*not be afraid of the cold*", etc.

During childhood (7–11 years) the child acquires greater realism and self-awareness, even in terms of character and psychological profile, giving more concrete descriptions of themselves in relation to the social context. The process ends in adolescence, when triggering other processes related to the social context and judgment of latria. At this stage, the identity is fluctuating, the person questions himself, and the distance between how they feel and how they are perceived by others.

The other aspect related to the construction of identity is given by the mode of understanding of the other, starting from early childhood children are able to interact with other children, to understand the emotions of others, children and adults, they are also able to construct abstract concepts, like lies or believe in Santa Clause, the Tooth Fairy, etc. In childhood such a capacity increases and the child adopts *Perspective Taking*, i.e. the ability to take the perspective of another person so as to create pro-social or anti-social behavior.

In the research field, with regard to the evaluation of Thermal Comfort, the construction of identity can affect the mode of administration of the questionnaire. The classrooms are micro-companies, with subject leaders, rebels, social dynamics, etc. and it is in this context that individuals construct their identity.

During the search there were different behaviors in asking questions concerning the content or purpose of the questionnaire, to the researcher:

- In the stage of early childhood (3–7 years) the child does not pose questions directly to the adult, and intervenes to tell their own personal experience, it is a monologue addressed to adults, there is an answer or a dialogue with what has been said by other children; the child brings their own personal account;
- In later childhood (7–11 years), the child is able to have a logical thought and is able to answer questions, as well as to propose solutions, associations, and ask questions according to their own notions (notions such as geographical, climate-related, personal experience, etc.); the child tells their own point of view or his own

opinion, and interacts with other children, even talking to each other to discuss or argue;
- In pre-adolescence (11–14 years) the subject has their opinions and experience, and is able to give direct answers, in these cases the social context is crucial, individuals are characterized by something they do, they have already drafted their own personality and individuality, when they tell their own point of view or their own personal narrative, but in this case, unlike the children, the discussion turned to the other members of the class or social group, rather than the adult.

To conclude the paragraph, it emphasizes that the research concerning the *Child Developmental Psychology*, in studying the relationship between the child and environment, are not concerned with how children encode their feelings (not emotions) they are concerned with the microclimate of the indoor and outdoor environment.

Understanding the concepts of hot/cold, wet/dry, etc. are not investigated, although we have an influence in the construction of the identity of the child. Children use expressions related to the microclimate is to define your own feeling, and to define themselves.

A child is defined as "sensitive to the cold" if he or she is sensitive to the feeling of cold or can be defined "warm" in the sense that he is often hot, and in the sense of being very emotional, who demonstrates their emotions and expressions of affection towards others.

Also, the microclimate of indoor environments has an effect on the psychological development, cognitive and moral development of children as shown by several studies on Fuel Poverty (Boardman 1991, 2010; Ormandy and Ezratty 2012).

6.2.6 *Child and Social Context*

Finally, a role to the social aspects of development, understood as the relationship between the children and the others: family, peers and school.

The Family

The family is a system, a micro-society, where relations between the individual take place in several directions, in a context of mutual socialization and mutual support. During infancy and early childhood, the parent–child relationship is centered on daily nourishment, it is the parent who takes care of nutrition, the phases of the day (lunch, dinner, time to go to sleep) and clothing; and is the same parent who chooses the type of clothes to buy and wear daily.

In childhood the child is more independent on their parents, which lay down *the rules* (lunch and dinner times, time to return home, to go to sleep) and discipline, obligations and discussions with the children, and build a "*scaffold*" (scaffolding) that increases the responsibility of the children themselves. The child is able to dress himself and choose the clothes to wear.

In the adolescent phase the process is characterized by a greater autonomy of the subject than the parents, that depends on the type of attachment that is made in relation to the family, an attachment that can be safe/autonomous/distant/avoidant, in which the adolescent develops an attitude of rejection, or worry/ambivalence, in which the teenager has an insecure attitude towards the family and/or the outside.

Peers

Peers are the people who are the same age as the child, the friends and acquaintances, classmates, and the like. Peer relationships are horizontal and not hierarchical. The peer group influences the individual and the construction of one's personality in a positive or negative sense, when there are attitudes of rejection of the context, and its popularity.

In infancy peers are playmates, distinguished between "*friends*" and "*non-friends*" and are divided by genre: boys prefer to play with boys and girls with girls.

In the context of peers' children can have different statuses:

- Popular children, liked most of the children;
- Average children;
- Neglected or shy children;
- Controversial children neither friendly nor unfriendly;
- Rejected children, judged unpleasant by most children.

In the administration of the questionnaires these types of individuals are evident and may cause difficulties or obstacles in the conduct of research. The refused or controversial children, if not involved, may boycott the questionnaire cause annoyance and introduce elements that alter the results, neglected or timid children, can make untrue or non-autonomous judgments, or refrain from communicating their views about their contentment of comfort. The researcher in order to avoid situations of disorder or boycott will have to try to involve all stakeholders in an equal manner, also reducing the possibility of protagonism: in the framing of the questionnaires all subjects are equal, and in the phase of each response every opinion is important.

The School

The school, in the various school levels, is the environment in which the child spends most of his time, meaning both "social environment" and physical environment, the school building. The two aspects, social and physical, are not separable; the space has an educational role on the individual and on social relations. The classroom, the furniture and the organization of the building may depend on the type of the didactic and pedagogic model adopted (OECD 2011).

The teaching method (how and what is taught) depends on the pedagogical method adopted. Pedagogy is the discipline that studies the problems related to education and training of man, in order to indicate the methods of educational practice.

The approaches to teaching are different:

- In the *Constructivism Approach*, which refers to the psychology of Piaget and Vygotskij the teacher helps the child to discover the world, the relationship

Teacher ↔ Student is bi-univocal, to learn learning. In these classes you can expect to build the questionnaire for Thermal Comfort together with the children;
- In the *Direct Instruction* (authoritarian), teaching is central and the teacher does not have an interaction with the child, it is a unique relationship Teacher → Student;
- The *Montessori* method, derived by the Italian educator Maria Montessori, plans to leave a lot of freedom in the learning of the child, even in the organization of furniture and classroom. The teacher is a facilitator of learning.

In this chapter we want to emphasize the need to identify, before the test, any subjects or critical elements that can affect the test.

Disabilities may affect physical and psychological aspects, or learning. Learning disabilities includes various disorders such as dyslexia, difficulty writing or calculating, emotional or behavioral disorders, and *Attention Deficit Hyperactivity Disorder (ADHA)*, and other more serious problems such as *Autism Spectrum Disorder* and *Asperger Syndrome.*

In the classrooms are small children or young people with problems physical (handicap) psychological or behavioral. The researcher must take note of such situations and administer the questionnaire to such entities. It is also important to avoid any situation of discrimination within the classroom, for proper and smooth conduct of the questionnaires. Any cognitive difficulties of the subjects or learning delays are not an impediment to the execution of the questionnaire, on the contrary you will have to explain to such persons the contents and get their response.

This is important, the test cannot alter the integration process in place, the feeling of Thermal Comfort is common to all subjects and deciding to administer the questionnaire only to persons deemed "healthy" is not a good outcome of the test.

6.3 Children's Sensitivity of Thermal Comfort, a Child's Point of View

The point of view of children with respect to the Thermal Comfort, with regard to the physiological and energy exchanges, does not vary with respect to the *Body Heat Balance Equation*, it is only to identify how to calibrate the physical and physiological parameters compared to the physical characteristics of children. Here the unknowns may be different, the first is the different metabolism of children than adults, and the role clothing and adaptive behavior can play implemented unintentionally by children.

In the study of Thermal Comfort the object is the thermal sensation that depends both on the energy exchanges from both the judgment given by the subject in such an exchange, and judgment on Thermal Comfort both perceived and desired, depends on the individual, in this case by the construction of the Environment by the child. As mentioned, the Standard ISO 7730 and the PMV well-being indexes are the result of statistical questionnaires administered to more subjects, a process that has allowed standardizing of the individual variables.

In the case of children it is necessary to keep in mind that the cognitive aspect is different from that of adults. Children build their relationship with the outside world, including the Thermal Sensation, as a personal narrative, also influenced by the emotional aspects and the level of autonomy of language and person.

Also, regarding a questionnaire in which it is asked what the feelings are, they may have difficulty in understanding the questions. The relationship between the children and others (parents, adults, children, society) or the method of learning the language. The study obviously touches both but relates the perception of the physical world, the feelings that involve the child and the representation that the child does to himself, and of the others (questionnaire) of these sensations.

It is not possible to assess the feelings of comfort of these subjects without taking into account the physiological and psychological-perceptual including cognitive aspects.

How to measure the cognitive variable and its incidence in the expression of the opinion of Thermal Comfort, compared to the standard behavior of adults?

The first thing to clarify is that the cognitive pattern (the construction and understanding of the world) by the child is different than adults, as mentioned the child, depending on age tends to attribute an animism, it is a finality of things and events. The connection with nature is a natural aptitude for children (Cheng and Monroe 2010). In the cognitive pattern of the adult, as shown in Fig. 6.1, the survey questions, which are formulated as sentences, oral or written, are understood: at the end hot, cold, etc. the adult gives a common meaning to that of other adult's culture reference. Likewise, they can understand that the questions are part of a questionnaire and asking about feelings and not opinion (this must always be clear and explained during the course of the research, it also needs to be explained that the questionnaire is anonymous).

In the child's cognitive pattern, evaluation questions, posed as sentences, oral or written, are not included: the child does not understand the meaning of the questionnaire, both in the terminology and in the very purpose of the questionnaire.

In addition, if the single child is able to say if he feels hot or cold, it is difficult to be sure of the meaning which is attached to hot and cold is the same as other children. Added to this is the environment in which the questionnaire occurs, which depends on the age, the school environment, and the fact that, in general, the person who put the questionnaire to the child is an adult and therefore can be seen as an activity of judgment of the adult to the child. This may result in an alteration of the judgment; the child may tend to express an opinion to have a favorable impression to the adult.

To be sure that the terminologies and the scale of judgment—*even the concept of the grading scale, of accounting values, is an aspect not always clear or known to the children*—you need to ask and compare every single degree of judgment, you need to ask: Hot like? Cold like? and give to this a Symbolic Code: a word, a design, a color, an association with natural phenomena, geographical etc.

The case studies of the second part describe some strategies adopted to define a common questionnaire to different classes, able to formulate questions comparable

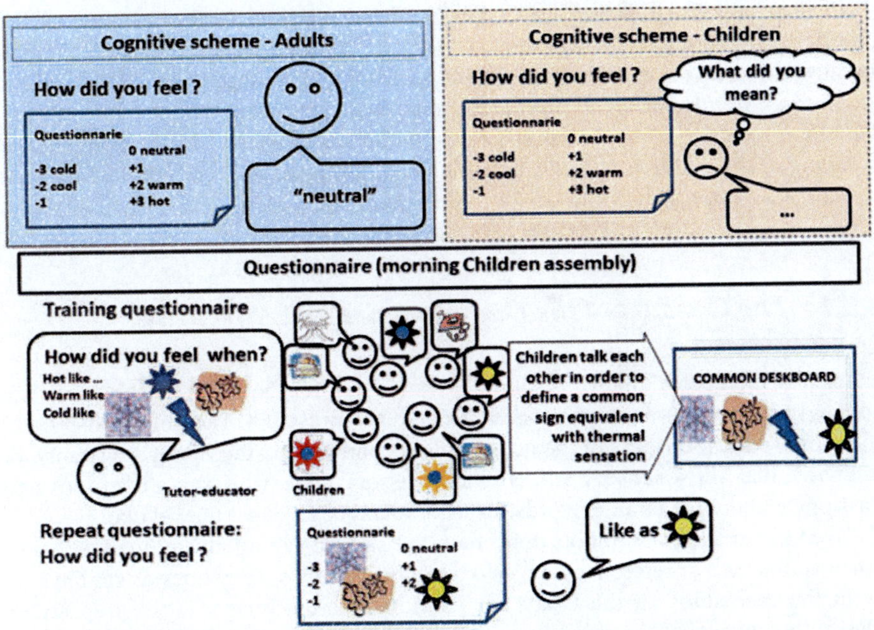

Fig. 6.1 Example of how to construct the questionnaire to the child's cognitive pattern

to the questionnaires of adults. So, we are able to analyze the data and compare them with that measured by the instrumentation or the standard approach.

The questions are:

- The assessment of the current condition of the environment compared to a scale of values (hot, cold, etc.);
- The judgment of Thermal Comfort compared to their own opinion (perfect, tolerable, etc.);
- The attribution of preferences (hotter, colder) than the ability to change the conditions of the environment in the future.

Regarding the first aspect we have made clear the need to clarify the different values and degrees of the scale of Thermal Comfort, the same degrees and terms can also be used for other applications, keeping in mind that:

- Judgment, assumes that the child is able to express his related opinion (tolerable, just tolerable, intolerable, etc.) only to Thermal Comfort and/or IAQ, namely the questions of the questionnaire, without expressing judgment related to social aspects (relations with the class, peers, adults, etc.) or emotional (love/hate toward the building, the school in general) or other moods or emotional variables;
- Preference, presupposes, in addition to what has already been said, that the child is able to define a future, or the ability to change the current conditions.

In both cases it is possible that the child will respond to the question of the questionnaire with a story, also those which are collected during the questionnaire, because they can be an important tool in the development of the results.

Thermal Comfort is the discipline that measures, statistically, the feelings of individuals. From the scientific point of view this "randomness" of the content and meaning of the question, must be identified and corrected in a manner that does not affect the assessment.

6.3.1 The Conduct of the Questionnaire and Affectivity

In making the questionnaires it should be noted that for the children, the words are sounds to play with—*nursery rhymes, puns*—that represent worlds and curiosity, and therefore so are the words of the questionnaire, particularly those that are strange or unknown that allow sensory and emotional associations. Children are curious and passionate about the strange words for their sound or for what they evoke.

Even the images or symbols don't have the same meaning given by adults, even when adults take as reference images of cartoons, comics, fairy tales or similar.

In the case study of the Diana nursery/kindergarten images were prepared to clarify the content of the questionnaires and it was a mistake: the interpretation of the images by the children had nothing to do with the provisions of adults; their associations were trying to build a story between the characters/comics shown in our diagrams. Since the question has to be part of the cognitive universe, of the language of the children, only children are able to express, speak, write and describe their universe.

In addition, children have a strong inner life, the language—the question—intersects with their internal representation

> With language, the child in fact discovers unsuspected riches of a world of reality superior to him: his parents and the adults around him appear as being big and strong, the origin of unforeseen and often mysterious activities, but now these beings reveal their thoughts and their will, and this new universe will begin to dominate with an incomparable aura of seduction and prestige. (Piaget 1964, English Edition, 1967, p. 27)

In submitting questionnaires, we must avoid the "*coercion of the adult on the child*". In the administration of the questionnaire we must pay attention to the role of the teacher, as it is the subject that "*imposes things*" (orders, duties, discipline, etc.), or better, depending on what the educational and pedagogical model says/asks of children during the day. For this reason it is useful to gather information on the didactic and pedagogic model adopted.

The questionnaire must collect the child's opinion, free from possible constraints, we should try to eliminate any impression that the questionnaire involves a judgment or an assessment of the child, the teacher, the school, or building.

The memory of the children in the age of later childhood (2–7 years), according to J. Piaget, is linked to story and they struggle to see the point of view of others, tending to bring a collective monologue. In the questionnaire it is best to use symbols rather than words. The symbolic game or imaginative words, symbols drawn (or chosen) directly by the children, allows us to be sure that, in the various individual monologues, all children have the same scale of judgment; while it is proposed to avoid the symbol directly from the adults.

Children make statements, they do not argue, and they always have an answer (consistent) to everything. They affirms and don't show. They tend to find a finalist or animist explanation to the questions, including those on comfort: the heat is that of the sun, the super heat of the volcano (and how do they know about the volcano?), the cold to the snow.

Attention should be paid to the analysis of the questions you intend to submit to the children, and the questions that children ask to the adults, or the analysis of the "why" placed directly by the child. Asking the "why" is the communication tool of the child, when he asks questions and is not learning the basics of other children or adults.

The questionnaire is an anomalous situation because the child is asked to find ways to express his personal opinion. Those of the questionnaire are not questions asked by the child (the why), but become so if the child constructs for himself and the other children.

The interest of the child, to the question or to the questionnaire or to the assessor that puts the questionnaire, regulates the system of values. The child, particularly in second infancy and childhood, tends to try to please—*or seek attention*—the teacher, or the assessor as it is a new subject.

L'interesse del bambino, verso il quesito o verso il questionario o verso il valutatore che pone il questionario, regola il sistema di valori. Il bambino, in particolare nella seconda infanzia e nella fanciullezza, tende a cercare di compiacere—*ovvero cercare l'attenzione*—l'insegnate, o il valutatore in quanto è un soggetto nuovo.

> The transformation of the action resulting from sketches of socialization not only affect the intelligence and thought, but also profoundly affect the emotional life. As we have already mentioned, in fact, from the preverbal period there is a close parallel between the development of affect and that of intellectual functions, as they are two inseparable aspects of every action: in any conduct in fact the motivation and dynamism energy dependent from affection, while the techniques and the adaptation of the means employed are the cognitive aspect (sensorimotor or rational). There is therefore a purely intellectual (in the solution of a mathematical problem there are many such feelings: interests, values, impressions of harmony, etc.) And even purely affective acts (love always implies understanding), but always and in any case, both in the pipelines relative to the objects and in those relating to individuals, involved both elements, as one presupposes the other. (Piaget 1964, English Edition, 1967, p. 41)

We also need to decide (and record) if the assessor and/or the teacher should be present in the room during the questionnaire, if the evaluator it is only one interface with the teacher, if they have to be present only in the stage of development of the questionnaire, if they have to answer the questions asked by the children during the completion of the questionnaire, etc. The solution to the criteria adopted during the experiment must be registered in the same climatic conditions of outdoor and indoor microclimate. During the processing and interpretation of the data it is necessary to assess whether, and how, socio-environmental factors could affect the metabolism, clothing, and if they have an impact on statistical answers.

6.3.2 *The Educational Context: The Space that Educates*

The study is precisely the difficulty in linking these typical aspects of human sciences and medicine, with the physical ones typical of physical technique and description of the indoor microclimate variables. Add to that the educational environment and the school system: nursery/kindergarten, elementary, and middle school. The educational model is crucial in the preparation and administration of the questionnaire, and must be coordinated with the teachers, professors and executives, in the complex bureaucracy of the school, which involves difficulties that should not be underestimated.

The educational environment covers two aspects:

- The *role of the educational space*, given by the physical and geometrical characteristics of the space of the building;
- The *pedagogical model* and the teaching method adopted.

Educational Role of Space

The educating space, as it is an "*element of quality teaching*" for the purposes of learning teaching, the space is also a place of affection, the bearer of values. In this sense also the Environmental Building, or the feeling tied to the perceived comfort in environments helps to qualify the building, to give it a value, and vice versa, the features of the building have an influence on the perception of comfort by children.

The relationship between space and the educational role, as the "*third teacher*" (Cannella 2013) was highlighted by Loris Malaguzzi, the teacher of Reggio Emilia, founder of the Reggio Emilia method applied in the Diana kindergarten as the "third teacher." On the side of architecture it is reminiscent of the experience of Herman Hertzberger, and other architects (Giancarlo De Carlo, Aldo van Eyck, Team X, etc.) who in the fifties and seventies of the twentieth century have designed testing new architectures in relation to their educational purpose, architecture fit for the child.

The study of Nair and Fielding (2005) provides for a division into six broad categories based on the teaching functions:

- Parts of the Whole,
- Spatial Quality,
- Brain-Based,
- High Performance,
- Community Connected,
- Higher Order.

For example, the term Learning Studio (L-Shaped) refers to a class that has the shape of an L, with the entrance on the long side of the class, an Active Zone (entry, storage, project work) a Flex Space (seminars, individual or collaborative) and one space, the bottom of the L, is intended as a Breakout Area (soft seating).

This model was introduced for the first time in 1940 in Crow Island School in Winnetka, Illinois (USA) and is also used in the educational model of Loriz Malaguzzi at the Diana kindergarten and other kindergartens of Reggio Emilia. Studies of Fielding Nair and also deal with matters relating to the Environmental Building, such as Daylight and Solar Energy, to be introduced in the school, shielding the affections of glare or heat (although not reported in the text references to the comfort, or the effect of comfort/Discomfort learning), the Natural Ventilation, following design criteria "bioclimatic" or, in general, sustainability criteria.

The OECD introduced the concept of "*Innovative Learning Environment*" (CELE 2011), in which they analyze the effectiveness of learning environments on the dynamics that take place inside, to the social and collaborative, inclusion/exclusion, to the motivation of the student, and to the enhancement of individual differences.

Thermal Comfort, since it depends so much by the architectural shape of the plant, when its technological characteristics, plant, and how they are perceived, is among the factors to be taken into account in the study of the comfort and the perception of comfort by children living in a school setting.

The pedagogical approach defines the "narrative space", the place where the stories take place, the lives of the children. The movement in the room by children is a way to measure them in the appropriate environments. Children measured as units themselves and not the Metric System, it is so large and small in relation to its size, warm and cold is not related to temperature but to their perception, in the space, of their perception of calculating cold.

The relationship between space and what space teaches (mode of use, available inputs and openings, available control devices and interface, temperature, color, odor, etc.) still has a psychological and cognitive effect on people affecting the perception of comfort. By way of example, a thermally neutral environment, with an index PMV equal to zero, of red color may give a different perception of the sense of comfort with respect to a blue color, and examples are similar in relation to the size of the space (airy/narrow) or the state of deterioration and maintenance (new/dilapidated).

The current approach is to make the educational environment, teaching environments (Learning Environments), promote adaptability and flexibility of space.

Innovative learning environments.

Embrace pedagogical approaches and create innovative contemporary learning environments that:

(i) Promote learning for students, professional and the wider community through active investigation, social interactions and collaboration;

(ii) Support a full range of learning and teaching strategies from direct explicit instruction to facilitation of inquiry to virtual connection and communication;

(iii) Support disciplinary and interdisciplinary learning;

(iv) Move beyond the simplicity of flexile open spaces to integrate resource rich, special purpose spaces with flexible, adaptable multipurpose spaces to provide a dynamic workshop environment for learning;
(v) Support to individual, 1 to 1, small group and larger group learning;
(vi) Are age-stage appropriate;
(vii) Facilitate learning anywhere, anytime, by any means, through seamless access to ICT distribution of learning resources for ease of access in learning spaces and accessibility beyond the traditionally defined school day;
(viii) Activate and invigorate learning space – indoor and outdoor;
(ix) Inspire participation in, and responsibility for the learner's community;
(x) Enable all aspects of the buildings, building design and outdoor spaces to be learning tools in themselves. (CELE 2011, p. 26)

In this approach the Thermal Comfort, and the feeling and understanding of Thermal Comfort by children, is a challenge for different disciplines: architecture, technical facilities, pedagogy, developmental psychology, health and medicine, etc. The use of Thermal Comfort is described in detail in van Hoof et al. (2010) and applied to a case of homes for older people with dementia, in which is highlighted the relationship between cognitive abilities of users (Dementia and Thermal Comfort) and elements of Building Design. The article emphasizes the role of the technological professions, the ethical, economic, and even the use of Building systems (facade systems, heating, HVAC, ventilation, control systems and individual control).

The article is based on ISO 10551

> "According to this standard, subject should rate the environment on a perceptual scale first, the an evaluative scale, followed by a preference scale, concluding with rating for personal acceptability and personal tolerance. Since the validity of rating and answers given by people with dementia is poor (some suffer from aphasia, other are happy to give any answer to the researcher in order to comply), family carers should be asked to rate the thermal perception of their partner/spouse with dementia as an additional measure, based on observation and knowledge of their partner/spouse with dementia." (…) "Integrated building calls for an integrated approach of the building process by considering the needs of the various stakeholders involved This is not an easy task, as the need of others may be overlooked in the complex process of construction or refurbishing buildings, in particular the needs of stakeholders that do not reverberate strongly."

As for the Architectural Point of view, to know the feeling of comfort perceived by children, through the use of questionnaires, together with the onsite measurement of microclimatic parameters, allows you to drive the design process (or re-design) of architecture, the technical components of the building (walls, windows, etc.), in the design of the Open Space (Nikolopoulou and Steemers 2003) for the management of HVAC systems and the use of the environments in which children live according to a sustainable optic (e.g. Holopainen et al. 2014) in the broadest sense of the term sustainable.

6.4 The Pedagogical Models

Evaluation should take into account the pedagogical-didactic method, assessing, where possible, to be accompanied and supported by education experts and educators according to the teaching method of the context in which it is carried out the questionnaire: kindergartens, elementary schools, middle schools, or non-institutional institutional setting, in a lecture or in a recreation, etc.

With respect to the Standard ISO 10551 requires introduced "*communicative corrections*".

The execution experiment involves two steps:

(1) identification of educational terms with which children are presented to the questionnaire, with child psychologists and children as required by the pedagogical model;

(2.a) questionnaire responses in successive phases;

(2.b) (simultaneously) Measurement of microclimatic parameters using sensors located in the environment.

The evaluation of thermal comfort is accomplished by an on-site measuring campaign and in response test by the children, with subsequent comparison between the results obtained by the two different types of investigation.

Pedagogy is the discipline that studies the problems related to education and training of man, in order to indicate the methods of educational practice. In dealing with the preparation and administration of the questionnaire it is useful to have some brief notions of different pedagogies, keeping in mind that the same pedagogy is a multidisciplinary science that employs more skills. What is the context in which the questionnaire is carried out, school or non-school, in the moment in which we compare and explain the content questions to another subject you are teaching, and then you are having a "*pedagogical approach*", in general such teaching is between an adult and a child.

The twentieth century saw the emergence of pedagogy, in parallel with mass culture and the rise of literacy and science education.

The pedagogical models are distinguished (also) according to social and economic models and political development of the country, in relation to the role that the school and education have in forming citizens.

The first pedagogical model is defined "Active Education" or "activism", whose main theorists is John Dewey (1859–1952) and other educators of the early twentieth century. In this model, the child is spontaneously active and learning must be in contact with the external environment. The educational model foresees that the child learns the concepts and sociability, through their own experience and the comparison with the teacher and with the other children. Activism focuses on education the child with an active role, highlights the "doing" as a way of learning, the manual and the game, the motivation understood as interest on the part of the child, interest emotional, practical and cognitive, and the relationship with the physical and social environment in which the child lives. In this model, according to John Dewey the child is educated through democracy and through direct experience.

The model is anti-authoritarian has had several changes over time, including Ovide Decroly (1871–1932) and Maria Montessori (1807–1952). The Montessori method has had wide international spread and is more widely adopted in Italian schools. This method requires that the child carries out their activities freely, and the role of the teacher is to help you learn, the child must learn how to learn, the child must receive stimuli and stress especially from the environment and not directly by the adult. What Is More:

> The same school furniture should be designed to fit the child, so that he can handle it directly, and also includes those materials through the use of which allows the self-education of the child, although under close monitoring, but not overwhelming, the teacher. (Cambi 2005, p. 33)

The "traditional" pedagogical models in most cases no longer used, are those of *authoritarianism*, in which discipline plays an important part, and who see the teacher as the person who educates and the scholar who learns. In this model the role of the teacher is central, as well as the relationship between teacher and pupil, who is passive, who must learn the concepts provided by the teacher.

In pedagogical Marxist models, the school is at the centre of society, and education work is at the center of society. The school is a miniature society in which reigns organization, discipline, education and work.

In pedagogical models of Christian origin the good of the person, as an individual, and family is at the centre. In this model the relationship is unique of the kind teacher-student.

In the second half of the twentieth century the approach of didactic lectures, teacher-students, are gradually abandoned or transformed, thanks to the emergence of *Science Education*, which focuses on learning methodologies rather than on the aspects of philosophical pedagogy character.

In this context, the approach is closer to the anti-authoritarian models of activism, a pragmatic model that is focused on training doing and learning through the scientific method or by the artistic experience.

The approach of Piaget and Vygotsky (cognitive pedagogy), which sees in education an instrument to act and produce practical effects of education, results in teaching models in which the child has an active role, learning occurs through the action, the representation, the visual organization and the symbolic language. The teacher guides the understanding of things starting from questions, comments or actions of children.

Jerome S. Bruner's approach (psycho-pedagogical structuralism) the symbolic function of language education moves towards forms of scientific knowledge not such as art, in this approach, any idea of childhood may be covered in teaching. In both approaches the cognitivist pedagogy moves the axis on the learning process (as the child learns) rather than on the notion learned. The schools, in recent decades have been confronted with a mass education, in a complex environment and with multiple sources and sources of information from the child. The school is not the only place where the child can learn concepts and is not the only source of education, it plays a role in the "*open*" family contexts, peers and the media.

In most of the case studies in this book we compare schools that adopt an updated version of the Montessori Method (Middle School) or its educational and pedagogical methods derived from pedagogies that are always part of activism. In this sense it was easier to interact with faculty and with children to build the questionnaire.

6.5 How to Make a Questionnaire for Children

The construction of the questionnaire must be consistent with the pedagogical method adopted. The questionnaire should be at a time similar to that of the lessons or days normally carried out by children. In the preparation of the questionnaire, you can follow the criteria adopted by the social sciences (Bryman 2004) in the case of self-completion questionnaire:

> In many ways, the self-completion questionnaire and the structured interview are very similar methods of social research. (…) However, because, there is no interviewer in the administration of the self-completion questionnaire, the research instrument has to be especially easy to follow and its questions have to be particular easy to answer. (…) as a result, self-completion questionnaires, as compared to structured interviews, tend to:
>
> - have fewer open questions, since closed ones then to be easier to answer;
> - have easy to follow design to minimize the risk that the respondent will fail to follow filter questions or will inadvertently omit a question;
> - be shorter to reduce the risk of "respondent fatigue", since it is manifestly easier for respondent who become tired of answering questions in a long questionnaire to consign it to a waste paper bin than to terminate an interview. (Bryman 2004, p. 133)

In preparing the questionnaire it is good to pay attention to the number of questions (a good criterion is to use a maximum of 10 questions) and must be divided in such a way to check the results, some questions have only one answer, others multiple questions.

In the construction of a survey it is necessary to draw attention to the order of asking questions. For example, the first questions may relate to the characteristics of the building, so as to familiarize the test interviewee, then, when the interviewee is still careful, questions of order "personal" (questions) and their opinion on the Thermal Comfort finally the last question, when the interviewee is tired, with simple questions to end the test. In using the images when they are not encoded in a univocal way you may fall in a situation of "tilt" by the person who is responding.

The questions must be clear about their purpose, if the goal is to re-study the general opinion of the people about the comfort, without specifying where, then it is possible to refer to different moments. On the other hand, the objective is to verify and validate the comfort of the environment, and then compare it to monitoring activity the question is too specific to be able to give valid results, equivale polls on voting intentions, or to the exit pool, which collide with the real data of the polls.

The questionnaire should arouse curiosity in children and enable them to respond independently and with their own time, without accelerating the answers. We must be sure that the questions are clear, and the answers are objective, and build the questionnaire in order to avoid ambiguous terms, mixed, to distinguish the notions of measurement and judgment, from where you can get an opinion. It is also necessary to avoid any influence of the teachers in the answers.

During the administration of the questionnaire the microclimate must be monitored, the physical characteristics of the children (age, height, weight) and type of clothing.

For the subjective evaluation it was necessary to define the pedagogical approach and the method by which to define and submit questionnaires to children in order to get answers comparable with those of the instrument. For example, the preschool children are not able to independently assess the quality of the air, having as an indicator the concentration of CO_2, so are not good indicators, until the level of CO_2 concentration does not reach the threshold values or high discomfort. They still manage to express a perception of air quality with respect to a rating scale.

The ventilation of the rooms is delegated to adults, both as regards the execution of the manual or mechanical ventilation, for the autonomous detection of a threshold limit for which it is necessary to open the window to ventilate.

The gap between the understanding of air quality of children with respect to adults and instrumentation deserves to be studied further.

To qualitative and quantitative evaluation, you can add the didactic-pedagogical, with the support of child psychologists and educators of the kindergartens, according to the pedagogical method, in itself a topic that deserves in-depth psychological and pedagogical analysis.

Another aspect to consider, especially if you use a questionnaire with only written questions, is whether to read the contents of the question, misunderstandings arise regarding a word, or the meaning of a sentence, it is useful to find a synonym or an example the explanation of "teaching" is likely to be ineffective. For example, the term "*bad*", can be translated as "*with bad smell*" or, "*the smell of the air when you don't open the windows for a long time.*"

The questions should relate to the perception of thermal comfort at the moment when you carry out the questionnaire and avoid asking sensations related to the perception of comfort at other times of the year. If it is deemed useful to see how changes over time are necessary to provide more questionnaires in time and/or a longitudinal study.

The children of nursery and primary schools have an age which, because of their cognitive development, does not allow him to understand all the questions provided in the questionnaire of adults.

In writing of the questionnaire some expressions may not be clear, such terms seemingly trivial as "*draft*" ("*air-leakage*") common by technicians, cannot be understood by children, who may not know the term, use slang expressions or their language of origin (if foreign) call it in another way, or do not consider it as a problem.

Or the term "*isolated*" ("*Insulated*", "*Well insulated*", "*Badly insulated*") that in the ergo technical means the presence of insulation but which cannot be understood, it is also not clear what can be understood as *well insulated* or *little isolated*, if you have no technical knowledge.

Added to this is that, in Italian isolated, means both "*insulated with*" and "*alone without anyone about*" (from island). Less clear is to use a more specific term such as "thermal insulation" or "shields", not known either by children or, in many cases, by adults.

The questions which ask for confirmation of an opinion, for example, "*The classroom is well ventilated? [_] Yes [_] No*", puts the respondent in the difficulty of having to assign themselves a quality level of judgment and then see if this is respected.

Other questions can ask the respondent preferences or advice on the adoption of possible solutions. Even in these cases, in defining multiple answers, avoid formulations too defined ("*do you think it would be right to color the wall yellow ocher?*"), with technical solutions that include specific terminology or complex solutions.

The same concept of comfort is not separable from environmental contexts, present in the indoor environment. For the evaluator comfort is what has been explained in the previous parameters, and it is possible to evaluate (and measure) the various types of comfort: Thermal Comfort, Visual Comfort, Acoustic Comfort, Indoor Air Quality. For those not technical these different types of amenities are present together, synesthetic, and so is the judgment.

The IAQ is correlated to the temperature due to concentration of people in the same environment for a long time, therefore, the increase of CO_2. The light, illuminance and radiation (*thermal phenomenon*) are strongly correlated, the visual comfort given by a greater illumination can be influenced by the warm feeling due to the higher irradiation.

The term environment also includes other aspects not physical or physiological, the same environment, or socializing with other visitors of the environment. The notion of environment and comfort is different if you interview a person who is a visitor who has a momentary perception, compared to those who work every day in that environment, and thus also social relations, relationships, moods that change during the seasons and days, and with the other workers. The memory of the social relations that are established within the environment affects the perception of comfort, and the response to the questions. This happens for children and for adults.

Those who answer tend to associate the history to the question, so the question on the presence of noise, can be linked to a noisy neighbor, or deemed to be noisy because it has different habits than their own, so the presence, in the vicinity of an industry, can increase their perception of negative phenomena related to air quality. Those who answers builds stories, it is not a lab rat that responds to neurological impulses, however, they tend to associate anything of their own personal experiences related to the claim or to the boredom of answering the question. Always bear in mind the evaluator, when present is always seen as an examiner.

For example, during a trial a child, did not know which option to answer the question "*what do you think the climate and comfort in the classroom is?*". The options ranged from "*perfect*" to "*tolerable*", to "*not acceptable*", but the child's observation was that he "*he found it bad because there was a child who was unkind who didn't speak to him, and so then did not speak with the others.*"

This relationship affects the social climate in the classroom, which was understood not so much as a "*physical climate*", but as "*social climate*". In clarifying this aspect, we said "yes, but if you look at only the appearance of the temperature, the light, the room, the atmosphere of the room, as you think", the answer was "I like the room, I feel good inside, not too hot", and then he then replied, confidently, "perfect".

The difficulty is to construct a questionnaire that normalize these variables, where possible reporting them as a disturbing element in the response, and then interpreting the results from the statistical point of view; so it is important to collect the narrative aspects related to construction of the questionnaire and the responses.

References

Arneill AB, Devlin AS (2002) Perceived quality of care: the influence of the waiting room environment. J Environ Psychol 22:345–360

Bandura A (2001) Social cognitive theory. Annu Rev Psychol 52

Boardman B (1991) Fuel poverty: from cold homes to affordable warmth. Belhaven Press

Boardman B (2010) Fixing fuel poverty: challenges and solutions. Earthscan

Bowlby J (1969) Attachment and loss. Hogart Press, London

Bronfenbrenner U (1979) The ecology of human development. Harvard University Press, Cambridge, MA

Bruner JS (1996) The culture of education. Harvard University Press, Cambridge, MA

Bryman A (2004) Social research methods. Oxford University Press, Oxford

Cambi F (2005) Le pedagogie del Novecento (Pedagogies of the twentieth century). Editori Laterza, Bari

Campos AC, Lima NL, Pereira CR, Fornara F, Buonaiuto M (2013) Inpatients' and outpatients' satisfaction: the mediating role of perceived quality of physical and social environment. Health Place 21:122–132

Cannella G (2013) Architetture pedagogiche, 28 Nov 2013 (Pedagogical architecture, 28 Nov 2013)

Canter DV, Craik KH (1981) Environmental psychology. J Environ Psychol 1(1):1–11

Centre for Effective Learning Environments (CELE) (2011) Designing for education. Compendium of exemplary education facilities. OECD Centre for Effective Learning Environments, OECD Publishing

Cheng JC-H, Monroe MC (2010) Connection to nature: children's affective attitude toward nature. Environ Behav 44(1):31–49

De Giuli V, Zecchin R, Salmaso L, Corain L, De Carli M (2013) Measured and perceived indoor environmental quality: Padua Hospital case study. Build Environ 59:211–226

Erikson EH (1950) Childhood and society. W. W. Norton, New York

Erikson EH (1968) Identity: youth and crisis. W. W. Norton, New York

Fanger PO (1970) Thermal comfort-analysis and applications in environmental engineering. Danish Technical Press, Copenhagen

Fornara F, Bonaiuto M, Bonnes M (2006) Perceived hospital environment quality indicators: a study of orthopaedic units. J Environ Psychol 26(4):321–334

Hodnett ED (1989) Personal control and the birth environment: comparisons between home and hospital settings. J Environ Psychol 9(3):207–216

Holopainen R, Tuomaala P, Hernandez P, Hakkinen T, Piira PJ (2014) Comfort assessment in the context of sustainable buildings: comparison of simplified and detailed human thermal sensation methods. Build Environ 71:60–70

Huisman ERCM, Morales E, van Hoof J, Kort HSM (2012) Healing environment: a review of the impact of physical environmental factors on users. Build Environ 58:70–80

Kagan J (2010) Emotions and temperament. In: Bornstein MH (ed) Handbook of cultural developmental science. Psychology Press, New York

Khodakarami J, Nasrollahi N (2012) Thermal comfort in hospital—a literature review. Renew Sustain Energy Rev 16:4071–4077
Langer EJ (2005) On becoming an artist. Ballantine, New York
Laurence GA, Yitzhak FY, Linda H, Slowik LH (2013) "My space": a moderated mediation model of the effect of architectural and experienced privacy and workspace personalization on emotional exhaustion at work. J Environ Psychol 36:144–152
Lomas KJ, Giridharan R (2012) Thermal comfort standards, measured internal temperatures and thermal resilience to climate change of free-running buildings: a case-study of hospital wards. Build Environ 55:57–72
Marchand GC, Nardi NM, Reynolds D, Pamoukov S (2014) The impact of the classroom built environment on student perceptions and learning. J Environ Psychol 40:187–197
Mourshed M, Zhao Y (2012) Healthcare providers' perception of design factors related to physical environments in hospitals. J Environ Psychol 32(4):362–370
Nair P, Fielding R (2005) The language of school design: design patterns for 21st century schools, 2nd edn. DesignShare, Inc., Minneapolis, MN
Nikolopoulou M, Steemers K (2003) Thermal comfort and psychological adaptation as a guide for designing urban spaces. Energy Build 35:95–101
OECD (2011) Designing for education. Compendium of exemplary education facilities. OECD Publishing, Paris
Ormandy D, Ezratty V (2012) Health and thermal comfort: from WHO guidance to housing strategies. Energy Policy 49:116–121
Pavlov IP (1927) In: Anrep GV (trans) Conditioned reflex. Oxford University Press, London
Piaget J (1967) Lo sviluppo mentale del bambino. E altri studi di psicologia. Einaudi, Torino, p 38 (Piaget J, Six èuteds de Psychologie. Editions Gonthier, 1964) (English: Six psychological studies. Random House, New York, 1967)
Rivlin LG (1984) Educational issues, school settings, and environmental psychology. J Environ Psychol 4(4):347–364
Rollo D (2005) Breve dizionario di psicologia dello sviluppo e dell'educazione (Brief dictionary on development psychology and education). Carocci, Roma
Santrock JW (2013) Child development, 13th edn. McGraw-Hill Companies Inc., New York (Italian edition by Rollo D, 2013)
Siegler RS (1998) Children's thinking, 3rd edn. Prentice Hall, Upper Saddle River, NJ
van Hoof J, Kort HSM, Hensen JLM, Duijnstee MSH, Rutten PGS (2010) Thermal comfort and the integrated design of homes for older people with dementia. Build Environ 45:358–370
Verheyen J, Theys N, Allonsius L, Descamps F (2011) Thermal comfort of patients: objective and subjective measurements in patient rooms of a Belgian healthcare facility. Build Environ 46:1195–2120
Vygotskij LS (1962) Though and language. MIT Press, Cambridge, MA (original edition 1934)
Watson JB (1913) Psychology as the behaviorist views it. Psychol Rev 20:158–177

Chapter 7
New Classroom: Indoor Air Quality and Distance Learning

Abstract The first edition of book was in 2015, so many things have happened that have changed research in indoor thermal comfort, primarily, the effect of the pandemic on the classroom. Classrooms changed place: not only inside a school building, but also other space became school, such as a room in a home, thanks to DAT and outdoor space. This upgraded new chapter presents a brief survey of the scientific literature on thermal comfort and the classroom, their measurements, how COVID changed indoor air ventilation in the classroom.

7.1 Literature Review Upgrade

The first edition of book was in 2015, so many things have happened that have changed research in indoor thermal comfort, primarily, the effect of the pandemic on the classroom. In updating the book and reviewing the studies from 2015 to the present day, referring to the research that this book cites, I became aware that it is divided into two distinct moments: before and after the COVID-19 pandemic.

The literature review of Zomorodian et al. (2016) identifies two categories of thermal comfort models:

(a) *rational*, following Fanger's approach (Fanger 1970), and
(b) *adaptive*, following Nicol et al. (2012).

The chapter presents a review of research carried out in schools of different levels of education—primary, lower and upper secondary school and university—in different countries, some with evaluation during the entire year, others only in one season, with environments equipped with Natural Ventilation or Controlled Mechanical Ventilation (CMV), with or without questionnaire. The references report a wide variety of cases and survey methodologies, demonstrating how very alive research is in this specific sector, even if limited to indoor studies that use ISO 7730 (ISO 2005) and ASHRAE 55 (ASHRAE 2010) for the evaluation of the Predicted Mean Vote (PMV, see Chaps. 3 and 4).

The literature review of Lamberti et al. (2021) reports the synthesis of about a thousand articles distributed in all continents, climatic zones and levels of school

K. Fabbri, *Thermal Comfort Perception*, https://doi.org/10.1007/978-3-031-52610-7_7

education: kindergarten, primary, lower and upper secondary school and university. The article is a collection of research on multiple topics and shows the plurality of approaches in the study of thermal comfort in school buildings. Some research focuses on aspects related to Building Energy Performance or Energy Consumption, on the activation or automation of Heating, Ventilation and Air Conditioning (HVAC) systems, Building Automation and Control Systems (BACS) referring to all uses of the building or non-specific components such as shielding elements, automation of ventilation and window opening, turning on lights, and so forth. In this line of research, thermal comfort is the set-point condition, according to regulatory parameters, while the opinion or perceptions of children, teenagers and students is not taken into consideration.

The methodology proposed in this book has been applied in buildings with uses other than as a school, such as modern architecture (Galiano-Garrigós et al. 2022), mosques (Ahriz et al. 2021), Malay house (Hassin and Misni 2022), Passive Houses (Wąs et al. 2022), kindergartens (Gajić et al. 2019), comfort and energy efficiency in the case of Arctic buildings (Abazari et al. 2022), dwellings in the Brazilian Amazon (Celuppi et al. 2019).

Other articles cite research concerning *outdoor space* to evaluate pedestrian comfort (Zhang et al. 2023), the relationship between thermal comfort and day degree, such as in the case study in Jordan (Oroud 2022), in the evaluation of the Standard Effective Temperature adapted for human walking (Fang et al. 2021), and in the outdoor shaded space in Tehran, Iran (Hadianpour et al. 2018).

The distinction is made between both *Architectural characteristics and thermal comfort* and *Health and student productivity and thermal comfort*.

In regard to architectural characteristics, indoor Thermal Comfort depends on (a) place, (b) building and (c) local culture and/or local habits of people who work, study or inhabit the indoor environments. The geometry and technology of the buildings depend on the local architectural, distribution, construction and plant engineering characteristics. Thermal Comfort is influenced, in its physical component, by the envelope (e.g. level of thermal insulation, type of windows and related shielding systems), by the equipment of the Heating, Ventilation and Air-Conditioning systems (HVACs) and Building Automation and Control Systems (BACS), including the possibility people have of being able to control the microclimatic conditions (e.g. the possibility of adjusting the thermostat of an environment).

In regard to health and student productivity, research confirms that understanding the indoor environment and comfort varies greatly with the age of the students. Therefore, a distinction must be made between:

(i) research carried out involving adolescents (lower and upper secondary school) and/or young adults (university) who can also be given questionnaires with well-established concepts; and
(ii) research related to children, from kindergarten to primary school, for which, depending on the age and the developmental and cognitive phase, the environment takes on a different meaning, therefore the questions and answers depend on the '*internal narrative*' of the child.

In the field of research on children, Karyono's research (Karyono et al. 2020) reports a very useful structure with a timeline diagram for development of the methods in thermal comfort from 1920 until the twenty-first century. The timeline distinguishes between the first studies of the 1920s and 1930s up to the article by A. P. Gagge of 1967 (Gagge et al. 1967). In the 1970s, the research is divided into two approaches, one adopts the Thermal Physiology approach, starting from P. O. Fanger, the other adopts the Human Behavior approach, starting from the article by M. A. Humphreys and F. Nicol on the self-regulation of the human body (Nicol and Humphreys 1973). In the last decade, the research has expanded to different categories of subjects: the young and the elderly, male and female, people with Alzheimer's, and so forth. All subjects that do not correspond to the model of a healthy man of 35 years. Finally, it reports a series of research that focuses on the relationship between the adaptive approach and improvement by means of Artificial Intelligence.

The article by Djamila (2017) focuses on *thermal comfort models*, following the approach provided by the ASHRAE 55 and ISO 7730 standards. The literature focuses on the variables of the indoor environment used in the research and the age of the subjects.

The questionnaire is always a useful tool for the evaluation of thermal comfort or thermal sensation vote, in indoor environments and in school buildings. The study of Elsayed et al. proposes the Actual Mean Vote (AMV) (Elsayed et al. 2023), in Watchman et al. the *biophilic school* in cold climates is proposed (Watchman et al. 2021).

The COVID-19 pandemic has changed the ways that school buildings are managed as well as the research on the subject, including Thermal Comfort and ventilation.

Prior to 2020, studies concentrated on air pollution (Salthammer et al. 2016) or smart homes and the control of indoor air quality (Schieweck et al. 2018). Other research concentrated on Indoor Air Quality, indoor pollutants and Thermal Comfort. The literature shows research on the relationship between air quality and student performance, including the impact of the concentration of CO_2 on the number of pauses/breaks during lessons for ventilation of the room (Al-Rashidi et al. 2012) correlated also to the ventilation rates, to the shielding systems, or to the internal temperature (Wargocki and Wyon 2007). The research of Barbic et al. (2019) compares the microclimate of the class and the measurement of cardiac autonomic control (electrocardiogram ECG) and cognitive performance of the students.

After 2020, the research is concentrated on ventilation and Indoor Air Quality, proposing new standards of ventilation, superior to the values laid down by the current standards, also in the case of natural ventilation, in addition to the research on the perception of IAQ in naturally ventilated primary schools (Korsavi et al. 2021) and nursery schools (Yuan and Ryu 2022). Flanking research on IAQ is research on the Indoor Environmental Quality (IEQ) in school buildings (Korsavi et al. 2020).

The need to increase ventilation in environments leads to dealing with arguments that have rarely been object of research in the literature, for example, the effect that the classroom air temperature has on people, how it affects student behavior (Korsavi and Montazami 2020) and the mood of the teachers (Boix-Vilella et al. 2021).

The article published in 2023 by Zapata-Lancaster *Carbon Dioxide Concentration Levels and Thermal Comfort in Primary School Classrooms: What Pupils and Teachers Do* (Zapata-Lancaster et al. 2023) deals with the relationship between the CO_2 concentration, ventilation and didactic activities in light of the COVID-19 pandemic, monitoring six classes of a primary school in South Wales (UK). The focus of the research is on the strategies adopted by teachers for ventilation and the corresponding response of students.

7.2 COVID-19 and Classroom Ventilation

The coronavirus pandemic, in different ways from country to country, has modified the classroom by moving it from the school building to other places, in particular into the children's and students' homes. The class is split into two parts:

(a) on the one hand, the *class* understood as a community of students and the teacher (one or more teachers depending on the teaching model), and
(b) on the other hand, the *classroom* intended as a physical place to go to and remain for a specific number of hours of the day, consisting of walls, floors, desks, furniture, etc., with its indoor microclimate.

The idea of the class as a community remained even during the period of isolation due to the pandemic. Distance learning (e-learning) has allowed the student–teacher relationship to be maintained as well as, in part and with the difficulties of connections, the relationship between students. The climate of the class given by these relationships, although frayed and with difficulty, has remained (almost) intact. The relationship through video (smartphone, computer, tablet) and connection involves only the senses of sight and hearing, with greater difficulty in concentration and tendency toward distraction in regard to the didactic task.

At the same time, the students experienced a different relationship with the place where they were having lessons: *they were no longer in class with others but at home.*

Referring to the subject of this book, the Thermal Comfort was different for each student, as well as the color of the room, the furnishings, the windows and the outdoor spaces.

The school and the classroom as a physical place became, for a few weeks, a different, distant and isolated (off-site) place. The classroom was abandoned, for months, as commonly happens in the for several months of the year during school break.

Upon returning to the physical space of the classroom (on-site), the management of the opening of the windows and the ventilation of the classrooms became an obligatory gesture even during the lesson. For this reason, it was necessary to adapt the environments to increase the Air Change per Hour (ACH) and—where possible or provided for by new laws—to introduce Controlled Mechanical Ventilation (CMV) or Heating, Ventilation and Air-Conditioning (HVAC) systems or Heat Recovery Ventilation (HRV), or Mechanical Ventilation Heat Recovery (MVHR).

Children and young people spend a lot of time in classrooms that have a density of crowding (number of people per square meter) higher than that of any other building, including offices.

The COVID-19 pandemic required the revision of indoor air quality and thermal comfort standards of school buildings.

The concentration of children leads to an increase in the concentration of indoor pollutants, in particular CO_2 (Zheng et al. 2022) which can also influence the thermal sensation (Shin et al. 2021) and the concentration of fine particles (PM 10, PM 2.5, etc.) condition that increases both the risk of diseases related to breathing (colds, flu, rhinitis, asthma, etc.) and the performance and level of attention, an aspect studied in the scientific literature (Mendell and Heath 2005; Andrea et al. 2014; Nazaroff 2022).

The EN 16798-1:2019 standard, Energy performance of buildings—Ventilation for buildings (CEN EN 16798-1 2019) together with the ASHRAE 62.1 standard, Ventilation for Acceptable Indoor Air Quality (ASHRAE 62.1) define the criteria for calculating and sizing natural ventilation systems or Mechanical Air Ventilation or Heating, Ventilation and Air-Conditioning systems or Heat Recovery Air Ventilation.

In particular, EN 16798-1 provides for four categories of Indoor Environmental Quality (IEQ) from IV (low) to I (high), each defined according to the rate of ventilation and concentration of pollutants. The parameter that allows one to design a good IAQ is the Ventilation Rate, which can be calculated based on perceived air quality, with the formula:

$$q_{tot} = n \cdot q_p \cdot A_R \cdot q_b \quad \left(\mathrm{l\ s^{-1}}\right) \tag{7.1}$$

where q_{tot} is the total ventilation rate for the breathing zone ($\mathrm{l\ s^{-1}}$); n is design value for the number of the persons in the room; q_p is ventilation rate for occupancy per person ($\mathrm{l\ s^{-1}\ person^{-1}}$); A_R = floor area (m^2); q_B is ventilation rate for emissions from building ($\mathrm{l\ s^{-1}\ m^{-2}}$).

Alternatively, it can be calculated on the basis of the concentration of the individual substances (e.g. CO_2, VOC, PM, etc.) with the formula:

$$Q_h = \frac{G_h}{C_{h,i} - C_{h,0}} \cdot \frac{1}{\varepsilon_v} \quad \left(\mathrm{m^3\ s^{-1}}\right) \tag{7.2}$$

where Q_h is the ventilation rate required for dilution ($\mathrm{m^3\ s^{-1}}$); G_h is the generation rate of the substance ($\mathrm{\mu g\ s^{-1}}$); $C_{h,i}$ is the guideline value of the substance ($\mathrm{\mu g\ s^{-1}}$); $C_{h,0}$ is the concentration of the substance of the supply air ($\mathrm{\mu g\ s^{-1}}$); εv is the ventilation effectiveness (–).

In the case of schools and child care facilities, it is possible to increase the air flow rate according to the EN 16798-7 standard (CEN EN 16798-7 2017) which shows the formulas for calculating, in detail, airflow due to windows opening, whose values are defined by the standards for the calculation of building energy performance and expressed as Air-Change per Hour (ACH).

The research of Babich et al. (2023) reports that following the EN 15251 standard (EN 15251 2008) the minimum ventilation rate to ensure the IAQ level I is 5 L s^{-1} $person^{-1}$, value that must be increased by 0.6 L s^{-1} $person^{-1}$, based on the density of persons in the room, a number obtained by the effective number of students or by the number of students per square meter required by legislation.

Children are climate-sensitive subjects and more prone to getting sick, since their immune system is in a phase of growth. In a similar way, the elderly tend to get sick more often because their immune system is weaker (Sheffield and Landrigan 2011). Both are more sensitive to the consequences of indoor and outdoor pollution, including phenomena related to climate change (pollution and heat waves). The time spent in climate-controlled indoor environments has effects on the physiological level (breathing, temperature, acclimatization of the body, adaptation of the body during transition from environments with different microclimatic conditions) and on the psychological level, due to the time spent in circumscribed or confined spaces (i.e. with a physical confine, as is precisely the classroom surrounded by walls and floors).

In particular, children tend to have greater climate-related diseases in the lower respiratory tract, allergic or non-allergic reactions due to air pollution that may be related to climate change, such as asthma (Biagioni et al. 2023) also in school environments (Mortelliti et al. 2022).

The World Health Organization (WHO) has drafted a series of indicators of children's environmental health (World Health Organization 2009). The report defines the MEME (Multiple Exposure—Multiple Effect) framework in which the context determines the causes of diseases affecting children (health outcome) based on exposure to pollutants that can be distal or proximal, and on which preventive action must be taken.

In the last decade, research has also been interested in the role played in the development of children in the city, in the built environment and in natural contexts (nature-based), a condition that has been confirmed during the restrictions due to COVID-19 (Martin et al. 2023).

COVID-19 has had a decisive impact on Indoor Air Quality and on the natural and mechanical ventilation systems in all sectors and uses of buildings (Agarwal et al. 2021). The strategies for reducing the risk of contagion of the COVID-19 virus has seen the use of several instruments:

(a) the first was the lockdown and the reduction of the presence of people in places, which was followed by
(b) the use of the protective mask, in presence, which led to the omission of expressions, difficulty breathing and during play among children, which was accompanied by
(c) social distancing (ECDC 2020; Cao et al. 2020; Agarwal et al. 2021), strategies that have reduced the risk of contagion from COVID-19 and other typical childhood diseases such as colds, measles, asthma, and so forth.

These strategies have changed the way school spaces are used, in particular, the frequency of the air ventilation exchange. The WHO recommends, in general, a ventilation rate of 288 m^3/h/person in case of natural (Dong et al. 2019) or mechanical ventilation.

Air can be filtered using an Activated Carbon and HEPA (High-Efficiency Particulate Air) filter to reduce the rate of pollutants including NO_2 and VOCs, as well as Photo-Catalytic Oxidation, Ultraviolet radiation and germicidal UV radiation, or Ionisation air purification.

Among the studies on the sizing of control mechanical ventilation to reduce the risk of COVID-19 infection, the research of Sha et al. (2021) follows the *Wells–Riley (WR) Model* for the reduction of the risks of infection. The WR model is a model proposed by William F. Wells and Richard L. Riley to estimate the transmission of tuberculosis and measles infections (Wells 1934; Riley et al. 1978, 1995; Riley 2001; Noakes et al. 2006; Sze To and Chao 2010).

These studies show that the ventilation rate can be calculated with the following formulas:

$$Q_{DV} = -\frac{P_d \cdot E_m \cdot B \cdot Npqt}{\ln(1 - P_I)} \quad (\text{m}^3/\text{h}) \text{ or } (\text{l/s}) \tag{7.3}$$

$$ACH = -\frac{P_d \cdot E_m \cdot B \cdot Nqpt}{V \cdot \ln(1 - P_I)} \quad (\text{h}^{-1}) \tag{7.4}$$

where P_d is social distance index, E_m is mask index, B is initial infection rate, N is number of occupants, V is space volume (m^3), P_I is probability of infection, p is pulmonary ventilation rate of susceptible individuals (m^3/h), q is quantum generation rate (h^{-1}) and t is exposure time (h).

What has the pandemic made evident?

To date, in 2024, these strategies seem distant, even in memory, but it must be borne in mind that these strategies have had and will have an effect on the behavior, training and psychological growth of people who were school-age children and young people in the years 2020–2022; effects that, at present, we are not able to assess.

During and after the pandemic, outdoor activities have been the subject of several research studies confirming the role that the outdoor environment, natural or of the city, can play in children's development. Despite the restrictions, for example, children were given more time to play outdoors. Outdoor activities allow children to acquire skills related to (i) their relationship with the world and nature, and (ii) to play itself.

Playing, for children, is the way through which they experience the world and social relationships, increasing their creative abilities (Nussbaum 2013). The pandemic has highlighted the negative effects of confined living at home, both in terms of repercussions on health; for example, the absence of outdoor activities increases the risk of obesity in children (Patterson et al. 2021) and psychological and relational problems (socialization difficulties, fear or phobic sense in social interactions, etc.). In this period, the importance of the role of parents as a role model has grown: reactions to COVID restrictions by adults, including anxieties or concerns, have been imitated and internalized by children (Dodd et al. 2021) at home and upon returning to the school environment (Dodd et al. 2021).

The pandemic has changed the relationship with the environment, in particular with external activities. It has changed, sometimes in a pervasive way, especially in the sectors of the media and politics, what is called the 'narrative'.

What have we learned thanks to the COVID-19 pandemic?

Nothing, although something has changed: we have learned the role of relationships and the characteristics of the place or environment in which they occur, the advantages and disadvantages, the shortcomings of the city, of the built environment and of nature. Urban planners, mayors and real estate developers are reviewing the dynamics that determine the choices of living in a given place. In particular, the importance of relationships with distances no farther than 15 or 30 min, the neighborhood, gardens and public parks. These dynamics also influence the school. The places of learning can no longer be limited to a room, and not only in order to ensure air quality or the concentration of pollutants.

The relationship with the place where we live, play and learn influences the growth of the child, starting from three years of age. Being outdoors improves relational skills with others and the understanding and definition of the self.

Children (see Part III) interpret the environment, including the thermal sensation, through narration, the self-narration of the world. The climatic conditions or the feeling of heat or cold do not have a specific, absolute value, but acquire it in that they are linked to a relationship. This relationship takes place through the narration that occurs in a place, so the snow and the associated cold can correspond to positive sensations if, for example, the first snow seen is associated with positive experiences, play, fun, affection, making new friends or pets with which to play.

Space, place and environment

Those who have studied architecture know how important the debate on the subject of space and time is, and are aware of the influence in the cultural context of modern architecture and urbanism of Siegfried Giedion's book first published in 1941, '*Space: Time and Architecture: The Growth of a New Tradition*' (Giedion 2009). Giedion espouses the theories of physics of the early twentieth century, from the theory of relativity of Albert Einstein (1905), he distinguishes between the notion of space and time in architecture, attributing to time the fourth dimension of architecture. Giedion reinterprets the architecture of the great masters such as Wright, Mies van der Rohe and Le Corbusier in this light, attributing to them the notion of '*not discrete space*'.

The twenty-first century requires a redefinition of terms, at least in my view, in order to address the issues partly dealt with in the third part of this book and anticipated here.

Space, understood as an area or boundary, can refer to the geometric, dimensional datum where the body can move, walk, run. Space can be small, smaller than the human body, or large, huge beyond the dimensions of the earth and the cosmos. Between the infinitely large and the infinitely small is the human scale, ranging from 10^{-3} (the millimeter) to 10^3 (the kilometer) meters. These are the space dimensions that the human body can experience directly. The body area (dimension of the human body and interpersonal relationships (Hall 1966)), is within that scale and the actions in space have an above and a below, a front and a back. Space is geometric dimensions, with no other quality than dimensions, useful in defining the ergonomic standards of school desks, chairs, teachers' desks, blackboards, dimensions of the windows, volumes and ventilation rates.

The *place*, also termed location, includes the affective and sensory data, it is endowed with unique characteristics to itself, which affect the emotional state, allowing the sedimentation of notions, memories and the definition of feelings. The place has a name: house, park, school, room, classroom, garden, and is always the subject of specific experiences (running, jumping, falling, sitting) and emotional relationships (fear, joy, boredom). The place can have a boundary, confines, as it is a room or building, or it can be out in the open, such as a garden or a park, and is characterised by containing one or more objects, things or trees, people and animals. The place also is endowed with colors, light, smells, tastes and sounds, and both the school and the classroom are a place in which, and thanks to which, one learns.

Finally, the *environment* is the social and community datum. It includes all the precedents to which is added the relationship with others, with nature and with the city, memory and history.

The COVID-19 pandemic has changed classroom ventilation habits when children and young people are in the classroom. In addition, it was a sort of acceleration of two teaching methods already present in the pedagogical field also with some pilot experiences: e-Learning or Distance Learning, briefly presented in the next paragraph, and Outdoor Education, to which the third part of the book is dedicated.

7.3 A New Class Outside the Classroom—Distance Learning

It is not easy to evaluate the effects of e-learning lessons and tools compared to the traditional lesson, and it is not the subject of this section, since the author does not have the competence to respond to that task. I'd like instead to bring to the reader's attention to the fact that the experience of the pandemic required one to '*learn by doing*', as if a person had to learn how to fly an airplane while building the plane itself. In the didactic and pedagogical field, the teaching mode has changed, today many more people and institutions are aware of new ways of being-in-class and these new ways include adopting a different relationship with *the classroom*: its space *may not be an indoor one*. The place where you learn, where the lesson is held, may not be the school as a building, the school is also outside the school and can include the courtyard, the urban park, the entire neighborhood, the city or in nature, even in the wilderness.

In all cases, children and young people will have the opportunity to understand the context of which Thermal Comfort is an integral and, in some cases, even decisive part.

Parts I and II of the book deal with theoretical notions and case studies related to the '*construction-of-the-world*' and '*comprehension-of-the-world*' by children aged three to eleven, starting from thermal sensation and thermal comfort.

Thermal Comfort can also affect learning performance even in the case of online lessons (online e-learning performance). The study of Wong et al. (2023) reports an experiment during which the skin temperature of students who had to attend an online lesson were measured along with the environmental parameters of the room. At the end of the lesson (75 min) a questionnaire was given to them to evaluate the Thermal Sensation Vote and perceived learning performance. The results show that these factors had an influence on reading and numeracy skills. The scientific literature reports similar results referring to indoor environmental quality (Savelieva et al. 2020; Wang et al. 2021).

Following the approval of the restrictive measures (stay-at-home), different countries have adopted different mitigation strategies described in the study published by Anderson et al. (2020) published in the journal Lancet and reported in the literature review Learning from a Pandemic around the World by Reimers (2022).

COVID-19, as mentioned, has been a sort of acceleration in the use of distance learning or e-learning, as it has been in the world of work: today, in 2024, many more meetings are carried out with online platforms and software webinars (Zoom, Teams, etc., as reported in Chick et al. (2020)). What has been '*discovered*' is that lessons can also be held online, even university exams or the discussion of degree theses; and if, on the one hand, there are obvious limits compared to the on-site lesson (discussion between peers, socialization, questions and answers, longer levels of attention in presence, etc.), on the other hand, it constitutes a new way to do things as well as an opportunity.

In respect to the subject of this book, the pandemic has brought a novelty related to the perception of the classroom and its comfort. During the teaching activity in e-learning mode a double phenomenon occurs: on the one hand, one's own *room* becomes the *classroom*, where, for example, it is no longer mandatory to 'tolerate' the same air temperature that everyone else does. The same applies to the position of the child in his or her *room*, when the child is alone, he or she is the '*master*' of the space, while in the classroom the space must always be shared. This is true, however, only in the case that the student has his or her own room dedicated only to study, because this is not a condition common to everyone, especially in the poorest sections of the population, in families with several children or houses with few rooms, where it is necessary to attend the lesson in the kitchen or in the living room.

On the other hand, the *classroom* enters into the child's *room*, in the sense that, thanks to the video, the other classmates and/or the teacher are able to enter into the house, even if virtually. This second aspect expands the range of information on individual students: classmates see how the room other students have is furnished and everyone sees the teacher's room, except if the video is obscured.

One last aspect, not to be underestimated, the concentration and attention during the online e-learning activity has a shorter duration than the on-site learning activity, perhaps ten times shorter in duration, for three reasons: (a) being in one's own room, (b) the greater amount of distractions (and no one who is there to point out when a student is distracted), (c) the inability of teachers to change their teaching model, that is, the model that is used in the frontal lesson.

The concept of comfort expands beyond thermal comfort or visual comfort, the child is in an environment that he or she codes as *his or her own* and *personal*, adaptable and adapted to his or her needs.

The scientific literature about Digital Education or E-Learning has had a significant growth as a result of the pandemic, precisely because it has become the only way, in many countries, to guarantee the continuity of lessons. There has been an increase and rapid development of the online lesson tool and the various meeting platforms Zoom, Microsoft Teams, and Google Meet as their virtual classrooms where teachers synchronously meet their students (Kansal et al. 2021).

To this were added, in progress and subsequently, at the end of the pandemic, reflections on totally online or mixed teaching methods (blending). Even teachers were not prepared for this change of method and continued to use online the teaching methods that work in presence. In other words, the frontal lesson was carried out in front of the camera and the students were in front of a video. The lesson turned into a lecture of one hour or more with repercussions on the concentration of the students. In online teaching and, in general, in front of a screen, the level of attention is reduced compared to the frontal lesson in person: after ten minutes or a quarter of an hour of listening to the online lesson, the student tends to become distracted (Wilson and Korn 2007). Attention decreases, and the student starts looking for other footholds to continue stimulating the brain, distracting himself or herself. These dynamics are common in subjects with attention deficit (Attention Deficit/Hyperactivity Disorder, ADHD) and it can be considered that they have increased even in subjects without attentional deficits, precisely due to the teaching tool.

Hybrid Teaching (e.g. Blended Learning) is a different lesson model than the classic frontal lesson, similar to the pedagogical models that reject the frontal lesson, preferring teaching methods that favor direct student–teacher relationships (from one-to-one) or between peers (from a few to a few) as in Montessori schools, or with the Steiner method, to name two pedagogical models that are still current. For example, flipped classrooms are a chance to reverse the trend towards frontal lessons (Awidi and Paynter 2019; Cheng et al. 2019; Strelan et al. 2020; Cao et al. 2023; Fischer et al. 2023), as well as Outdoor Education (see Chap. 13).

Classroom environment

The UNESCO Report "Reimagining our futures together: a new social contract for education" highlights the impact that the COVID pandemic has had on schools. "*The pandemic has proved that the school cannot be entirely displaced into virtual spaces.*" UNESCO Report (UNESCO 2021) Schools must build collective capacity, while digital technologies should aim to support—and not replace—schools.

The Classroom Environment (different from the Class Environment that determines the indoor microclimate) can improve students' behavior and willingness to learn. The classroom environment can refer to the classroom as relationships of people when they are in the classroom and to the classroom as architecture.

The teaching of the twenty-first century has been oriented towards overcoming the frontal lesson, in particular for the age range of 6–11 years, so that the child can occupy different spaces in the classroom, not only his or her desk. Sometimes it is the desk itself to move or it is the child to change classroom, moving in the laboratories where specific educational activities are carried out (music, art, dance, crafts, etc.). The classrooms must guarantee the conditions of thermal comfort, both those provided for by regulations and standards, and those desired by the students, so that the environment is stimulating for the senses as well as being emotionally interesting (see Part III).

The Classroom Environment, understood as a relationship of people, is influenced by: (a) the didactic model (frontal lesson, students at desks, flipped room, discussion, etc.) or pedagogical model (authoritarian, Montessorian, Stenierian schools of thought, Dewey approach, etc.); (b) the relationship between teacher and students of a direct type or between peers; (c) the size and crowding of the classroom; (d) the windows and level of illuminance and distribution of light and colors; (e) the arrangement of the desks, quality and quantity of the furnishings, posters, etc., in addition to the cleanliness of the places. In addition, the Classroom Environment can be a:

(a) *developmental environment*, often used in kindergarten and preschool, in this case, the teacher interacts with everyone, together or one by one, stimulating basic skills and social interaction;
(b) *competitive environment*, typical of the high school classroom, the lesson is structured, most often frontal or with group discussion, there are tests, intermediate or final exams, and students are invited to challenge each other, improve with each other, so as to improve their high skills and also deal with conflict from the relationship point of view (similar to competitive games: running, chess, etc.);
(c) *collaborative classroom* where students work in a group or in multiple groups with a common goal to achieve all together (similar to role-playing or team sports).

In conclusion, one may wonder what all this has to do with children's perception of Thermal Comfort, as it can be said that the understanding of thermal comfort has not changed, and that what is written about it in the first edition remains valid. What has changed is the *classroom well-being*, because the school is no longer (only) at school: the lesson can be held at home, in other buildings or outside, the Thermal Comfort modifies the *emotional* response of the child that varies each time. Not only that, in addition to what has already been said in other chapters of the book, the emotional response (see Part III) is part of the journey—not necessarily rational—of learning: one learns through emotions, and the feeling of hot/cold, dry/wet, fragrant/smelly contribute to the construction of emotions.

References

Abazari T, Potvin A, Demers CMH, Gosselin L (2022) A biophilic wellbeing framework for positive indoor-outdoor connections in energy-efficient Arctic buildings. Build Environ 226:109773. https://doi.org/10.1016/j.buildenv.2022.109773

Agarwal N, Meena CS, Raj BP et al (2021) Indoor air quality improvement in COVID-19 pandemic: review. Sustain Cities Soc 70:102942. https://doi.org/10.1016/j.scs.2021.102942

Ahriz A, Mesloub A, Elkhayat K et al (2021) Development of a mosque design for a hot, dry climate based on a holistic bioclimatic vision. Sustainability 13:6254. https://doi.org/10.3390/su13116254

Al-Rashidi K, Loveday D, Al-Mutawa N (2012) Impact of ventilation modes on carbon dioxide concentration levels in Kuwait classrooms. Energy Build 47:540–549. https://doi.org/10.1016/j.enbuild.2011.12.030

Anderson RM, Heesterbeek H, Klinkenberg D, Hollingsworth TD (2020) How will country-based mitigation measures influence the course of the COVID-19 epidemic? The Lancet 395:931–934. https://doi.org/10.1016/S0140-6736(20)30567-5

Andrea B, Fabbri K, Boeri A (2014) IAQ evaluation in kindergarten: the Italian case of Asilo Diana. Adv Build Energy Res 8:1–18. https://doi.org/10.1080/17512549.2014.890532

ASHRAE (2010) ASHRAE 55 thermal environmental conditions for human occupancy

ASHRAE 62.1. Ventilation for acceptable indoor air quality

Awidi IT, Paynter M (2019) The impact of a flipped classroom approach on student learning experience. Comput Educ 128:269–283. https://doi.org/10.1016/j.compedu.2018.09.013

Babich F, Torriani G, Corona J, Lara-Ibeas I (2023) Comparison of indoor air quality and thermal comfort standards and variations in exceedance for school buildings. J Build Eng 71:106405. https://doi.org/10.1016/j.jobe.2023.106405

Barbic F, Minonzio M, Cairo B et al (2019) Effects of different classroom temperatures on cardiac autonomic control and cognitive performances in undergraduate students. Physiol Meas 40:054005. https://doi.org/10.1088/1361-6579/ab1816

Biagioni B, Cecchi L, D'Amato G, Annesi-Maesano I (2023) Environmental influences on childhood asthma: climate change. Pediatr Allergy Immunol 34. https://doi.org/10.1111/pai.13961

Boix-Vilella S, Saiz-Clar E, León-Zarceño E, Serrano MA (2021) Influence of air temperature on school teachers' mood and the perception of students' behavior. Sustainability 13:9707. https://doi.org/10.3390/su13179707

Cao W, Fang Z, Hou G et al (2020) The psychological impact of the COVID-19 epidemic on college students in China. Psychiatry Res 287:112934. https://doi.org/10.1016/j.psychres.2020.112934

Cao Y-X, Xia S-L, Zhu Z-Y et al (2023) Exploring lemology teaching with "internet plus" flipped classroom pedagogy. BMC Med Educ 23:341. https://doi.org/10.1186/s12909-023-04309-x

Celuppi MC, Meirelles CRM, Cymrot R et al (2019) Preliminary approach to the analysis of climate perception and human thermal comfort for riverside dwellings the Brazilian Amazon. J Build Eng 23:77–89. https://doi.org/10.1016/j.jobe.2019.01.019

CEN EN 16798-1 (2019) EN 16798-1, energy performance of buildings—ventilation for buildings—part 1: indoor environmental input parameters for design and assessment of energy performance of buildings addressing indoor air quality, thermal environment, lighting and acoustics

CEN EN 16798-7 (2017) EN 16798-7, energy performance of buildings—ventilation—part 7: calculation methods for the determination of air flow rates in buildings including infiltration

Cheng L, Ritzhaupt AD, Antonenko P (2019) Effects of the flipped classroom instructional strategy on students' learning outcomes: a meta-analysis. Educ Technol Res Dev 67:793–824. https://doi.org/10.1007/s11423-018-9633-7

Chick RC, Clifton GT, Peace KM et al (2020) Using technology to maintain the education of residents during the COVID-19 pandemic. J Surg Educ 77:729–732. https://doi.org/10.1016/j.jsurg.2020.03.018

Djamila H (2017) Indoor thermal comfort predictions: selected issues and trends. Renew Sustain Energy Rev 74:569–580. https://doi.org/10.1016/j.rser.2017.02.076

Dodd HF, FitzGibbon L, Watson BE, Nesbit RJ (2021) Children's play and independent mobility in 2020: results from the British children's play survey. Int J Environ Res Public Health 18:4334. https://doi.org/10.3390/ijerph18084334

Dong W, Liu S, Chu M et al (2019) Different cardiorespiratory effects of indoor air pollution intervention with ionization air purifier: findings from a randomized, double-blind crossover study among school children in Beijing. Environ Pollut 254:113054. https://doi.org/10.1016/j.envpol.2019.113054

ECDC ECFDPAC (2020) Considerations relating to social distancing measures in response to COVID-19—second update. Stockholm

Elsayed M, Romagnoni P, Pelsmakers S et al (2023) The actual performance of retrofitted residential apartments: post-occupancy evaluation study in Italy. Build Res Inf 51:411–429. https://doi.org/10.1080/09613218.2022.2121908

EN 15251 (2008) Indoor environmental input parameters for design and assessment of energy performance of buildings addressing indoor air quality, thermal environment, lighting and acoustic. UNI

Fang Z, Zhang Y, Yang Y et al (2021) Experimental investigation of standard effective temperature (SET*) adapted for human walking in an indoor and transitional thermal environment. Sci Total Environ 793:148421. https://doi.org/10.1016/j.scitotenv.2021.148421

Fanger PO (1970) Thermal comfort. Analysis and applications in environmental engineering

Fischer K, Sullivan AM, Cohen AP et al (2023) Using cognitive load theory to evaluate and improve preparatory materials and study time for the flipped classroom. BMC Med Educ 23:345. https://doi.org/10.1186/s12909-023-04325-x

Gagge AP, Stolwijk JAJ, Hardy JD (1967) Comfort and thermal sensations and associated physiological responses at various ambient temperatures. Environ Res 1:1–20. https://doi.org/10.1016/0013-9351(67)90002-3

Gajić D, Stupar D, Antunović B, Janković A (2019) Determination of the energy performance indicator of kindergartens through design, measured and recommended parameters. Energy Build 204:109511. https://doi.org/10.1016/j.enbuild.2019.109511

Galiano-Garrigós A, Marcos CL, Kouider T, Juan Gutiérrez PJ (2022) Reassessing thermal comfort in modern architecture: E.1027 as a case study. Build Res Inf 50:230–254. https://doi.org/10.1080/09613218.2021.1971060

Giedion S (2009) Space, time and architecture. Harvard University Press

Hadianpour M, Mahdavinejad M, Bemanian M, Nasrollahi F (2018) Seasonal differences of subjective thermal sensation and neutral temperature in an outdoor shaded space in Tehran, Iran. Sustain Cities Soc 39:751–764. https://doi.org/10.1016/j.scs.2018.03.003

Hall ET (1966) The hidden dimension. Garden City, New York

Hassin NSFN, Misni A (2022) Developing methodology to assess the thermal comfort of traditional Malay house. IOP Conf Ser Earth Environ Sci 1067:012023. https://doi.org/10.1088/1755-1315/1067/1/012023

ISO (2005) ISO 7730 ergonomics of the thermal environment—analytical determination and interpretation of thermal comfort using calculation of the PMV and PPD indices and local thermal comfort criteria. ISO 2005

Kansal AK, Gautam J, Chintalapudi N et al (2021) Google trend analysis and paradigm shift of online education platforms during the COVID-19 pandemic. Infect Dis Rep 13:418–428. https://doi.org/10.3390/idr13020040

Karyono K, Abdullah BM, Cotgrave AJ, Bras A (2020) The adaptive thermal comfort review from the 1920s, the present, and the future. Dev Built Environ 4:100032. https://doi.org/10.1016/j.dibe.2020.100032

Korsavi SS, Montazami A (2020) Children's thermal comfort and adaptive behaviours; UK primary schools during non-heating and heating seasons. Energy Build 214:109857. https://doi.org/10.1016/j.enbuild.2020.109857

Korsavi SS, Montazami A, Mumovic D (2020) The impact of indoor environment quality (IEQ) on school children's overall comfort in the UK; a regression approach. Build Environ 185:107309. https://doi.org/10.1016/j.buildenv.2020.107309

Korsavi SS, Montazami A, Mumovic D (2021) Perceived indoor air quality in naturally ventilated primary schools in the UK: impact of environmental variables and thermal sensation. Indoor Air 31:480–501. https://doi.org/10.1111/ina.12740

Lamberti G, Salvadori G, Leccese F et al (2021) Advancement on thermal comfort in educational buildings: current issues and way forward. Sustainability 13:10315. https://doi.org/10.3390/su131810315

Martin A, Clarke J, Johnstone A et al (2023) A qualitative study of parental strategies to enable pre-school children's outdoor and nature experiences during COVID-19 restrictions. Health Place 79:102967. https://doi.org/10.1016/j.healthplace.2023.102967

Mendell MJ, Heath GA (2005) Do indoor pollutants and thermal conditions in schools influence student performance? A critical review of the literature. Indoor Air 15:27–52. https://doi.org/10.1111/j.1600-0668.2004.00320.x

Mortelliti CL, Banzon TM, Vieira CZ, Phipatanakul W (2022) Environmental exposures impact pediatric asthma within the school environment. Immunol Allergy Clin North Am 42:743–760. https://doi.org/10.1016/j.iac.2022.05.005

Nazaroff WW (2022) 30+ years of knowledge creation: indoor air 1991–2021. Indoor Air 32. https://doi.org/10.1111/ina.13074

Nicol JF, Humphreys MA (1973) Thermal comfort as part of a self-regulating system. Build Res Pract 1:174–179. https://doi.org/10.1080/09613217308550237

Nicol F, Humphreys M, Roaf S (2012) Adaptive thermal comfort: principles and practice. Routledge

Noakes CJ, Beggs CB, Sleigh PA, Kerr KG (2006) Modelling the transmission of airborne infections in enclosed spaces. Epidemiol Infect 134:1082–1091. https://doi.org/10.1017/S0950268806005875

Nussbaum MC (2013) Creating capabilities. Harvard University Press

Oroud IM (2022) Integration of GIS and remote sensing to derive spatially continuous thermal comfort and degree days across the populated areas in Jordan. Int J Biometeorol 66:2273–2285. https://doi.org/10.1007/s00484-022-02355-6

Patterson RR, Sornalingam S, Cooper M (2021) Consequences of COVID-19 on the childhood obesity epidemic. BMJ n953. https://doi.org/10.1136/bmj.n953

Reimers FM (2022) Learning from a pandemic. The impact of COVID-19 on education around the world. In: Primary and secondary education during COVID-19. Springer International Publishing, Cham, pp 1–37

Riley RL (2001) What nobody needs to know about airborne infection. Am J Respir Crit Care Med 163:7–8. https://doi.org/10.1164/ajrccm.163.1.hh11-00

Riley EC, Murphy G, Riley RL (1978) Airborne spread of measles in a suburban elementary school. Am J Epidemiol 107:421–432. https://doi.org/10.1093/oxfordjournals.aje.a112560

Riley RL, Mills CC, Nyka W et al (1995) Aerial dissemination of pulmonary tuberculosis: a two-year study of contagion in a tuberculosis ward. Am J Epidemiol 142:3–14. https://doi.org/10.1093/oxfordjournals.aje.a117542

Salthammer T, Uhde E, Schripp T et al (2016) Children's well-being at schools: impact of climatic conditions and air pollution. Environ Int 94:196–210. https://doi.org/10.1016/j.envint.2016.05.009

Savelieva K, Elovainio M, Lampi J et al (2020) Psychosocial factors and indoor environmental quality in respiratory symptom reports of pupils: a cross-sectional study in Finnish schools. BMJ Open 10:e036873. https://doi.org/10.1136/bmjopen-2020-036873

Schieweck A, Uhde E, Salthammer T et al (2018) Smart homes and the control of indoor air quality. Renew Sustain Energy Rev 94:705–718. https://doi.org/10.1016/j.rser.2018.05.057

Sha H, Zhang X, Qi D (2021) Optimal control of high-rise building mechanical ventilation system for achieving low risk of COVID-19 transmission and ventilative cooling. Sustain Cities Soc 74:103256. https://doi.org/10.1016/j.scs.2021.103256

Sheffield PE, Landrigan PJ (2011) Global climate change and children's health: threats and strategies for prevention. Environ Health Perspect 119:291–298. https://doi.org/10.1289/ehp.1002233

Shin H, Kang M, Mun S-H et al (2021) A study on changes in occupants' thermal sensation owing to CO_2 concentration using PMV and TSV. Build Environ 187:107413. https://doi.org/10.1016/j.buildenv.2020.107413

Strelan P, Osborn A, Palmer E (2020) The flipped classroom: a meta-analysis of effects on student performance across disciplines and education levels. Educ Res Rev 30:100314. https://doi.org/10.1016/j.edurev.2020.100314

Sze To GN, Chao CYH (2010) Review and comparison between the Wells–Riley and dose-response approaches to risk assessment of infectious respiratory diseases. Indoor Air 20:2–16. https://doi.org/10.1111/j.1600-0668.2009.00621.x

UNESCO (2021) Reimagining our futures together: a new social contract for education. UNESCO

Wang C, Zhang F, Wang J et al (2021) How indoor environmental quality affects occupants' cognitive functions: a systematic review. Build Environ 193:107647. https://doi.org/10.1016/j.buildenv.2021.107647

Wargocki P, Wyon D (2007) The effects of moderately raised classroom temperatures and classroom ventilation rate on the performance of schoolwork by children (RP-1257). HVAC&R Res 13:193–220. https://doi.org/10.1080/10789669.2007.10390951

Wąs K, Radoń J, Sadłowska-Sałęga A (2022) Thermal comfort—case study in a lightweight passive house. Energies (Basel) 15:4687. https://doi.org/10.3390/en15134687

Watchman M, Demers CMH, Potvin A (2021) Biophilic school architecture in cold climates. Indoor Built Environ 30:585–605. https://doi.org/10.1177/1420326X20908308

Wells WF (1934) On air-borne infection. Am J Epidemiol 20:611–618. https://doi.org/10.1093/oxfordjournals.aje.a118097

Wilson K, Korn JH (2007) Attention during lectures: beyond ten minutes. Teach Psychol 34:85–89. https://doi.org/10.1080/00986280701291291

Wong L-T, Chan MT, Zhang D, Mui K-W (2023) Impact of thermal comfort on online learning performance. Build Environ 236:110291. https://doi.org/10.1016/j.buildenv.2023.110291

World Health Organization (2009) Children's environmental health indicators (CEHI)

Yuan X, Ryu Y (2022) Evaluation of children's thermal environment in nursery school: through the questionnaire and measurement of wearable sensors approach. Int J Environ Res Public Health 19:2866. https://doi.org/10.3390/ijerph19052866

Zapata-Lancaster MG, Ionas M, Toyinbo O, Smith TA (2023) Carbon dioxide concentration levels and thermal comfort in primary school classrooms: what pupils and teachers do. Sustainability 15:4803. https://doi.org/10.3390/su15064803

Zhang Y, He H, Dai K et al (2023) Thermal responses of face-masked pedestrians during summer: an outdoor investigation under tree-shaded areas. Build Environ 233:110058. https://doi.org/10.1016/j.buildenv.2023.110058

Zheng J, Guo X, Hu S et al (2022) Study on long-term variation characteristics of indoor CO_2 concentrations based on a data-driven method. Energy Build 256:111760. https://doi.org/10.1016/j.enbuild.2021.111760

Zomorodian ZS, Tahsildoost M, Hafezi M (2016) Thermal comfort in educational buildings: a review article. Renew Sustain Energy Rev 59:895–906. https://doi.org/10.1016/j.rser.2016.01.033

Part II
Case Studies

Chapter 8
Field Research

Abstract In Part Two we report the results of research in four case studies carried out at kindergartens and Italian schools of various levels. This chapter discusses the objectives, methodology and equipment common to all the research and the Italian educational system.

8.1 Objectives of the Case Studies of the Second Part

In the last decade several studies have formulated different approaches and indicators about adaptive comfort (adaptive approach), such as the "*indoor comfort operating temperature*" according to De Dear and Brager (1997) for ASHRAE the Smart Control and Thermal Comfort (SCAT) updated in EN 15251, for the European approach McCartney and Nicol (2002), as described by Ferrari and Zanotto (2012) and Nicol and Humphreys (2010) and the possible developments of this approach are discussed in Halawa and van Hoof (2012).

The study of Adaptive Thermal comfort has also been applied in the primary school classroom (ter Mors et al. 2011).

The studies are related to buildings with several categories of uses such as offices Goto et al. (2007), Kuchen and Fisch (2009), Ricciardi and Buratti (2012) or in hospitals Verheyen et al. (2011), De Giuli et al. (2013) and the review of Khodakarami and Nasrollahi (2012).

Other studies address the measurement of well-being in a school environment, also according to the adaptive approach by comparing the experimental data with mathematical models, such as university classes Buratti and Ricciardi (2009), in which students are subjected to questionnaires provided by the standard EN ISO 10551. These are therefore adults perfectly capable of understanding the content of the questions, and the collection (collect data) of the results of the PMV and PPD compared with that measured by the instrumentation according to EN ISO 7730. The analysis consists of the comparison between the indexes of the questionnaire and those of the on-site measurement.

K. Fabbri, *Thermal Comfort Perception*, https://doi.org/10.1007/978-3-031-52610-7_8

The study by De Giuli et al. (2012) in Italy refers to a primary school and children already in school, in which he deals with the aspect of the *Pupils' perception of the indoor environment* (IEA) to children who underwent questionnaires translated from English into Italian and were compared to the results of the questionnaire measured in situ from 7 different schools for a total of 614 pupils.

The studies of Pereria et al. (2014) in Portugal and the research of Yun et al. (2014) adopts a similar approach to that of this book, by comparing the results of a questionnaire with the instrumental, the contents of the questionnaire are expressed through "childish" images like emoticons, anime and koran-cartoons.

The study of comfort within the preschool is tackled by Eusebio et al. (2012) for a nursery (kindergarten), in which is addressed only the numerical approach, in particular linked to the effect of ventilation, but there is not an assessment of the response of children.

In facing a testing activity it is useful to understand and describe the building in which the activities take place, in terms of spatial dimensions, the dimension of-and the number of classrooms, the provision of more levels, the green space.

The spaces are inhabited by the children during the entire span of school, their understandings of space, as well as the finishing, the windows, the instrumental equipment of the classroom, the level of maintenance of the school, has an influence on their perception of the environment, including the condition of comfort. If the building and classrooms are in a state of disrepair, with plaster that is detached or broken parts and elements, the bad opinion that the child has of the school, is reflected also on the opinion that they will have on the building.

The subsequent chapters describe the following case studies:

- Kindergarten Diana in Reggio-Emilia (early childhood 0–7 years);
- Elementary School (CEIS - Educational Center Italio Swiss) in Rimini (later childhood 8–11 years);
- Elementary School Leonardo da Vinci in Carpi Modena (later childhood 8–11 years);
- Middle School in Cesena (pre-adolescence 11–14 years).

For all we used the same methodology of the research described in the next section.

8.2 The Research Methodology

The method used to evaluate the perception of Thermal Comfort in preschoolers consists of comparing the results of the evaluation by two approaches:

- on the one hand, the "*Objective and Quantitative Approach*" (QOS), which is based on the measurement of indoor microclimate and in the evaluation of the feeling of comfort according to ISO 7730 (Chap. 4) and related Standards;
- on the other hand, the "*Qualitative and Subjective Approach*" (QSA), based on subjective questionnaires and according to the Standard ISO 10551 (Chap. 5).

8.2.1 *Quantitative and Objective Approach (QOA)*

For the evaluation of Thermal Comfort a Microclimate Station was positioned in the environment for the duration of the administration of the questionnaire.

In each case study, we used the Microclimate Station "HD 32.3 Instrument for the analysis of the indices: WBGT—PMV—PPD index analysis in moderate environment" (DeltaOhm srl) and results were Elaborated with DeltaLog 10 v.0.1.5.28 software in accordance with the "moderate environments" standards prescribed by ISO 7730, ISO 7726 and ISO 8996 regulations. We used Instrument HD—WBGT—PMV HD 32.3 for monitoring campaign.

The instrument is designed for analysis of moderate environment and it can contemporarily detect: the Globe thermometer temperature (T_g in °C), Wet bulb temperature with natural ventilation (T_w in °C), Ambient temperature (T_a in °C), Relative Humidity (RH in %) and Air speed (w in m/s). Instrument HD are used to evaluate thermal comfort and PMV PPD indexes in moderate environmental, for these reason is suitable for our research. Data download and data export are possible with software Deltalog 10, with export data in excel. Data logging interval is every 5 min.

For the evaluation of *Indoor Air Quality*, the concentration of carbon dioxide (CO_2) was chosen as an indicator as a parameter related to breathing and, therefore, with the oxygenation of the brain, a factor that allows us to relate the feeling of tiredness or fatigue (or headache) with the level of concentration of a indoor pollutant.

The environments in which the experiments were carried out, are such as not to have sources of particular indoor pollution; if not chalk dust, cleaning products, all pollutants whose level of concentration and dilution in the indoor air, can be assimilated with those of CO_2.

Furthermore we adopted the CO_2 level as an IAQ marker. We decided to use the CO_2 marker for three reasons:

(1) it can be measured without interrupting normal school activity;
(2) CO_2 concentration has an influence on children's weariness (if CO_2 concentration increases, they can have lapses in concentration and feel tired);
(3) Empirical evaluation of the sense of smell (air smell).

Some research (Chithra and Nagendra 2012) correlate the IAQ with the ventilation conditions of outdoor pollution.

The CO_2 concentration measurement with a probe and a data logger, including PPD (Percentage People Dissatisfied) evaluation;

- EN 15251 standard adopts CO_2 as a marker to evaluate IAQ and Indoor Environmental Quality (IEQ) classification. (In Table 8.1 we show the recommended level of CO_2 concentration).
- EN 15251 uses the CO_2 marker when people are the main pollution source, as it was in the case of the kindergarten in our study.

Table 8.1 Recommended CO_2 concentration above outdoor concentration by Standard EN 15251

Category	Corresponding CO_2 above outdoor in ppm for energy calculations
I	350
II	500
III	800
IV	> 800

Table 8.2 IAQ Category and CO_2 Level by Standard EN 13,799

Category	Description	CO_2 (ppm) level in rooms	
		Typical range	Default value
IDA 1	High indoor air quality	≤ 400	350
IDA 2	Medium indoor air quality	400–600	500
IDA 3	Moderate indoor air quality	600–1000	800
IDA 4	Low indoor air quality	> 1000	1200

– EN 13779 standard defines the classification level of indoor air (see IDA in Table 8.2) with a CO_2 marker, as CO_2 is a good indicator for human bio-effluent emission.

In our research, we used an HD 37 AB 17 D (DeltaOhm) data logger to measure Relative Humidity RH (%), Environmental Temperature T (°C), Carbon Monoxide CO (ppm) and Carbon Dioxide CO_2 (ppm). The CO_2 measurement was obtained with a special infrared sensor (NDIR technology: Non-Dispersive Infrared Technology) which, thanks to the use of a double filter and special measurement techniques, guaranteed accurate and stable measurements over time.

8.2.2 *Qualitative and Subjective Approach (QSA)*

The questionnaires are processed in such a way to keep under control and be assured—*by researchers*—the understanding of the meaning of the terms and concepts of the questionnaire by children depending on their age, cognitive development and the level of education.

The construction phase and processing of the content of the questionnaire is the element most subject to variables of the research. For this reason, in the description of the case studies it pays particular attention to the preparation of the questionnaires.

The subjective judgment is strongly influenced by personal opinions, judgment, test conditions etc. even when using the methods of social research developed according to the survey statistics. This variability or uncertainty of the response increases as the object of study is related to opinion and concepts to employees dependent also on the mood of the person and of the ability to the understand terms that referred to.

What is asked of the respondents is not, for example, a political opinion (*would you vote X or Y?*), social (*what do you think of immigration?) or a general preference (do you like most yellow or red?*); questions which ask directly the opinion of the person, but we ask for a "*judgment regarding their own feeling compared to the microclimatic conditions of the environment*". Even in this sentence there are "ambiguous" terms:

- "*Judgment*" and not opinion, we ask the respondent to assign a value, this raises doubts about the availability of the subjects on the consequences from this allocation of value; in the case of one of the children one of the aspects is to pay attention and to emphasize that the questionnaire is not an assessment on the respondent or on the building or school organization, it is not a classwork task;
- "*Feeling*" is a term that can include the sensation of the skin, the smell, the light and the sound, it is a synesthetic phenomenon, which also includes what has been experienced in the day, their own vision of the world etc.;
- "*Microclimate*" is a very specific expression, the weather is often understood as the outdoor weather conditions and not as a condition of physical parameters in indoor spaces;
- "*Environment*" is an even broader term that includes ecology, pollution, social environments, working, etc.

And if these are difficulties we have with adults it is even more difficult in the case of children where the "***evocative power***" of words and images is greater. For this in the research particular attention was given to the construction of the questionnaires ***with the children***.

8.2.3 Comparison of the Results of the Two Approaches

The question remains if the above considerations have to do with the physical-technical approach and the study of comfort? The moment we study the physically measurable variables (microclimate indoor outdoor) and correlate to a judgment that depends on:

(a) physiological variables related to the variability of the characteristics of the human body, age and health status;
(b) behaviour and cultural;
(c) personal opinions;
(d) the understanding of the questions for subjects using a non-verbal language, in this case children.

Each of these variables has been the subject of study and scientific literature and ISO 7730, the Adaptive Thermal Comfort, etc. described in the previous paragraphs, which tries to evaluate, the behaviours (physical activity and work) while personal views are included in the statistical treatment of results.

The approach taken in this book and in the case study focuses on the study variables of type (d) in the case of children. The "objective" and quantifiable results must be comparable with the subjective results, and this requires a standardization work by subjective data during the development and administration of the questionnaire. More so, where the number of responses, and the randomness of each set of responses, does not allow us to have a high number of responses to which to apply the statistical error.

The results obtained from the monitoring (QOA) are compared with those of the questionnaire (QSA) in order to identify what are the variables that depend on the children. In particular, the results show a different metabolic activity to be given to children and a greater propensity to the feeling of "hot" or "very hot".

The following chapters describe the application of the methodology and the results for each case study, divided by grade level:

- kindergarten,
- primary school,
- middle school,

each characterized by a level of cognitive development and different school organization, conditions that required the adaptation of the methodological approach to each specific case.

8.3 The Italian Educational System

The OECD Centre for Effective Learning Environments (OECD 2011)reports the Classification of levels of education based on the revised International Standard Classification of Education (ISCED-97). ISCED is an instrument for compiling statistics on education international.

A summary of the levels (Annex 1 OECD 2011):

"*Pre-primary education* (ISCED 0) is defined as the initial stage of organized instruction, designed primarily to introduce very young children to a school-type environment, that is, to provide a bridge between home and a school-based atmosphere;

Primary education (ISCED 1) usually begins at age five, six or seven and last for four to six years (the mode of the DECO countries being sic years). Programme at the primary level generally require no previous formal education, although it is becoming increasingly common for children to have attended a pre-primary programme before entering primary education;

Secondary education (ISCED 2) generally continues the basic programme of the primary level, although teaching is typically more subject-focused, often employing more specialized teachers who conduct classes in their field of specialization;

Upper secondary education (ISCED 3) corresponds of the final stage of secondary education in most OECD countries. Instruction is often more organised along subject-matter lines that at ISCED level 2 and teachers usually need to have a higher level of, or more subject-specific, qualifications than at ISCED 2. The entrance age to this level Is typically 15 ore 16 years.

Tertiary education (ISCED 5/6) programme have and educational content more advanced than those offered of theory levels.

The Italian school system is divided into two cycles of study:

- The first cycle is divided into two paths:
  - Primary School, from six years, mandatory for all young Italians and foreigners who have completed six years of age,
  - Secondary School First Grade (Middle School), lasting three years, mandatory for all Italian and resident children who have completed primary school.

The entire first cycle of education ends with a State examination, after which it is possible to reach higher levels of education (Decree of the President of the Republic no. 89 of March 20, 2009 "Revision of the organization and teaching of school childhood and the first cycle of education ").

- The Second Cycle of education concerns the Secondary School Second Grade, and is divided into a course of five years, which may have different locations grouped into three modules, high schools, technical schools and vocational schools, all consist of two two-year classes and one-fifth year, and a final state examination.

The first and the second cycle are the levels of education, the first and secondary are *compulsory*. Higher education is excluded, and the following level of university education (Master of Science, Specialist, Ph.D.), the Higher level art, music and dance, technical institutes, Master etc.

Similarly there remain excluded levels of *pre-school education*, which includes *non mandatory* pre-schools, which are not included in the education systems, and where children mainly practice play and socialization.

The whole mandatory and not mandatory training, can be carried out in public or private structures. In the case of mandatory training and high schools, most of the schools are public, and therefore the employees are public as well as the school regulations made on the basis of ministerial decrees. Private schools may be different in nature and receive state contributions, and unlike public schools, while guaranteeing the provisions of the ministerial programs, adopt their own pedagogical models or learning, whether secular or religious schools.

In the case of pre-schools, the percentage of public facilities is lower, and there are different types and pedagogical models, Steiner schools (Steiner method), Montessori schools (Montessori Method), etc. Pres-chools are an economic equalizer for many families that having both parents employed in the work, would find it difficult to look after the children. Nevertheless, there may be instances where parents decide to entrust children of preschool age to maternal care or grandparents.

Finally, to conclude the framework of the school system, the political role that the public school has played in the unification of the Italian language, especially since World War II should be emphasized, a role still played with the economic difficulties (and political) dictated by the public administration structure.

It is useful to keep this in mind for conducting research as part of educational facilities, because the availability of teachers and the school administration with their hierarchy, may constitute an obstacle to the preparation and administration of the questionnaires, making the activity occasional or one-off, or making it impossible to complete the questionnaires administered due to unavailability of a teacher. Finally, schools consist of teaching staff, those that work for the administration (and construction) of the questionnaires, administrative and non-teaching staff, and faculty support for disadvantaged students with physical or psychological illnesses.

References

Buratti C, Ricciardi P (2009) Adaptive analysis of thermal comfort in university classrooms: correlation between experimental data and mathematical models. Build Environ 44:674–687
Centre for Effective Learning Environments (CELE) (2011) Designing for education. compendium of exemplary education facilities 2011. OECD Centre for Effective Learning Environments, OECD Publishing
Chithra VS, Nagendra S (2012) Indoor air quality investigation in a naturally ventilated school building located close to an urban roadway in Chennai, India. Build Environ 54:159–167
De Dear RJ, Brager GS (1997) ASHRAE RP-884, Final report: developing an adaptive model of thermal comfort and preference. American Society of Heating, Refrigerating and Air-Conditioning Engineers, Atlanta
De Giuli V, Da Pos O, De Carli M (2012) Indoor environmental quality and pupil perception in Italian primary schools. Build Environ 56:335–345
De Giuli V, Zecchin R, Salmaso L, Corain L, De Carli M (2013) Measured and perceived indoor environmental quality: Padua Hospital case study. Build Environ 59:211–226
EN ISO 10551 Ergonomics of the thermal environment. Assessment of the influence of the thermal environment using subjective judgement scales
Eusébio ZE, Conceição JMM, Gomes ANH, Manuela M, Lúcio JR (2012) Application of a developed adaptive model in the evaluation of thermal comfort in ventilated kindergarten occupied spaces. Build Environ 50:190–201
Ferrari S, Zanotto V (2012) Adaptive comfort: analysis and application of the main indices. Build Environ 49:25–32
Goto T, Mitamura T, Yoshino H, Tamura A, Inomata E (2007) Long-term field survey on thermal adaptation in office buildings in Japan. Build Environ 42:3944–3954
Halawa E, van Hoof J (2012) The adaptive approach to thermal comfort: a critical overview. Energy Build 51:101–110
Khodakarami J, Nasrollahi N (2012) Thermal comfort in hospital—a literature review. Renew Sustain Energy Rev 16:4071–4077
Kuchen E, Fisch MN (2009) Spot monitoring: thermal comfort evaluation in 25 office buildings in winter. Build Environ 44:839–847
McCartney KJ, Nicol JF (2002) Developing an adaptive algorithm for Europe. Energy Build 34(6):623–635
Nicol F, Humphreys M (2010) Derivation of the adaptive equations for thermal comfort in free-running buildings in European standard EN 15251. Build Environ 45:11–17
Pereira LD, Raimondo D, Corgnati SP, da Silva MG (2014) Assessment of indoor air quality and thermal comfort in Portuguese secondary classrooms: methodology and results. Build Environ 81:69–80
Ricciardi P, Buratti C (2012) Thermal comfort in open plan offices in northern Italy: an adaptive approach. Build Environ 56:314–320

ter Mors S, Hensen JLM, Loomans MGLC, Boestra AC (2011) Adaptive thermal comfort in primary school classroom: creating and validating PMV-based comfort charts. Build Environ 46:2454–2461

Verheyen J, Theys N, Allonsius L, Descamps L (2011) Thermal comfort of patients: objective and subjective measurements in patient rooms of a Belgian healthcare facility. Build Environ 46:1195–2120

Yun H, Nam I, Kim J, Yang J, Lee K, Sohn J (2014) A field study of thermal comfort for kindergarten children in Korea: an assessment of existing models and preference of children. Build Environ 75:182–189

Chapter 9
Kindergarten Case Study

Abstract In this chapter we report a case study of the Diana Kindergarten in Reggio-Emilia (Italy). The research concerns the assessment of the Thermal Comfort and Indoor Environmental Quality through questionnaires administered to children aged 4 and 5 years of age. The questionnaires were processed with the children with the support of teachers and education experts, according to the pedagogical model of the kindergarten (the Reggio-Emilia Approach). The results show a different cognitive universe of the children with regard to the understanding of the indoor and outdoor environment, and a metabolic rate slightly higher than that of adults. In addition, research has provided useful insights into the cognitive aspects of children and changes to the research approach.

9.1 The Kindergarten Diana Case Study in Reggio Emilia

The case study described in this paragraph refers to the activity carried out at the experimental kindergarten "*Asilo Diana*" in Reggio-Emilia, a city in northern Italy (44.69 N–10.64 E). The building is home to the first location where the method of education and teaching called the "*Model Reggio-Emilia Children*", was applied. It is internationally known, and therefore has a value both historically and symbolically. The same architectural shape of the classrooms was chosen by the founder of the method "Reggio Children" (Reggio-Emilia pedagogical model): Loris Malaguzzi (1920–1994).

The building was built in the sixties of the twentieth century by the Municipal Architecture Department (Municipal Department of Architecture) and is now under the direction of Reggio Emilia Municipal Instruction Office (Office of Education, Schools and kindergarten in the Municipality of Reggio Emilia).

The kindergarten has been the subject of pedagogical studies at an international level and, in the past, has been defined as the "*most wonderful kindergarten in the world*". Figures 9.1, 9.2 and 9.3 show building plan and views.

The building consists of only one floor of about 785 m^2, and a flat roof (pitched roof), and there are two open courtyards, where there are trees and shrubs from

K. Fabbri, *Thermal Comfort Perception*, https://doi.org/10.1007/978-3-031-52610-7_9

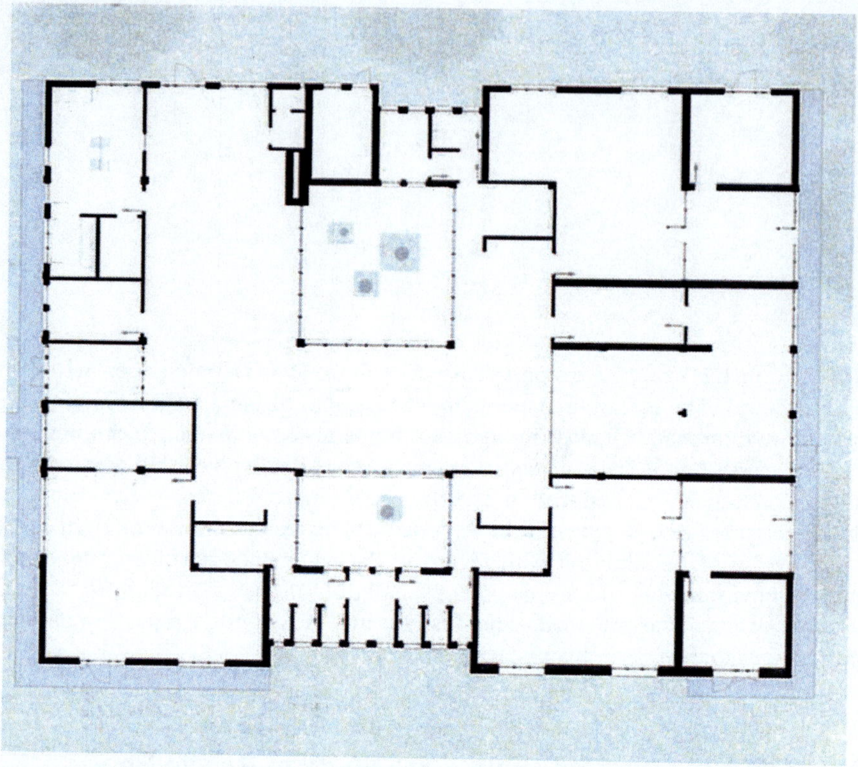

Fig. 9.1 Ground floor plan Diana kindergarten

which directly enters the natural light. The building is located in the city centre and the main urban park in the city, and is surrounded by a garden for the exclusive use of kindergarten, enclosed by a fence and shrubs to protect children from external intrusions. The outdoor and inner areas form a unique pedagogic, and in both places educational activities are held.

The internal layout follows the pedagogical criteria defined by Loris Malaguzzi, the classrooms constitute one space: a Flexible L, where there are present social and common spaces and spaces more "*private*" for children. There are also two small adjoining spaces called "*mini-workshops*" where they play the artistic activities, and a large central common room, called the "piazza" (Square). Inside the building there are spaces for the toilet, the kitchen and dining room that, in fact, are part of the square.

The building is made of a reinforced concrete structure and a masonry half-full brick exterior (without thermal insulation), window frames with wood frame and single glass.

Fig. 9.2 View of an internal classroom

Despite the crisis, Reggio-Emilia is a city with a flourishing economy, mainly in the manufacturing sector, and a high income. The city is among the first places nation-wide for income and quality of life, there is a high rate of immigration, compared to other Italian cities, and there are no phenomena of petty crime or degradation. Finally it is worth highlighting the active role played by the Municipality of Reggio-Emilia concerning school and social building, education, urban décor and spatial planning. The city has a network of heating systems which also serve the building under study. The city has always been characterized by a progressive politics, communist, socialist or social democratic.

The Diana Kindergarten is public, municipal although a registration fee must be paid, as most of kindergartens in Reggio Emilia and all adopt the pedagogical Reggio-Emilia method; there is also, in the city, the institute *'Reggio Children'* a private facility that conducts research, training and dissemination.

Fig. 9.3 Transformation of the classroom in the dormitory for an afternoon rest

9.1.1 The Children Age 3–5 (Later Infancy)

The Diana kindergarten is a public structure, the employees, educators, pedagogues and assistants are therefore public employees and trained to apply the Reggio Children teaching method. At the end of the day, and periodically staff compare the teaching activities carried out, any problems encountered and solutions to solve them. All this takes place with an atmosphere highly participatory by educators and pedagogues; this has contributed greatly to the research, as educators have supported the researchers in the administration of the questionnaire, immediately understanding the purpose of the trial. Also because the kindergarten itself was often involved itself in pedagogical research activities.

The nursery has three classes divided by age: 3, 4 and 5 year-old, for a total of about 78–80 students per year; and is open from September to June from 8 am to 4 pm. During the summer months the building is closed and is not involved in any activities.

The children spend most of their day with their peers in their own class, in the studio, except during lunch, where they mingle together, or when they perform common tasks in the square or in the garden.

The day will include three main phases:

- The arrival of the children in the morning and in the entrance hall, followed immediately by an "*Assembly*" (Figs. 9.5 and 9.15) confirming the active-participatory method of the school, where, in each class, the children gather in a circle to discuss, among themselves and with the teacher, what will be the activities that they are going to play during the day;
- The morning until the stages of preparation for lunch, constitute moment s of meeting with the children of other classes;
- After lunch is the time of the afternoon nap, during lunch classrooms are prepared by the educators with mattresses, children sleep for about 1 h after lunch in the same room in which they carried out the morning activities, on waking there is another hour of play until 4.pm when parents come to collect them.

Not all children remain until 4 pm, some parents prefer to take them after lunch at 1 pm. Parents are involved in the activities and are regularly invited to attend meetings with child psychologists and educators.

9.1.2 The Pedagogical Model of the Diana Kindergarten

The nursery Diana adopts an innovative pedagogical model, derived from the will of the founder Loris Malaguzzi. The school refers to the values, methods, practices and teaching of active education and educational cooperation, of scientific psychology, following the approaches respectively.

The pedagogical model of Loris Malaguzzi and of Reggio-Emilia, are part of the series of the pedagogical Piaget and Dewey in a version updated and revised: the child learns by doing things, mainly with an artistic approach. This therefore differs from the active pedagogical model that adopts the scientific method: learn to discover. Also thanks to the work of Loris Malaguzzi (1920–1994) (Copple 2003; MacDonald 2007), founder of Reggio Emilia pedagogical model, this kindergarten has become an object of study for international pedagogues and, in the past, has sometimes been defined the "most wonderful kindergarten in the world".

9.2 The Research Activity

The research has focused on the development and administration of the questionnaire, in line with the pedagogical model and kindergarten in order to ensure the complete understanding by children of the contents of the questionnaire. The educators chose to administer the questionnaire only to children aged 4 and 5, because the children of 3 years, provide multiple answers to the questionnaires tending to respond by choosing the last option which is listed (based on the observation declaration of teachers and education experts at the Diana kindergarten).

This factor is probably due to the memory of the children of three years, if a question is posed that does not refer to objects or other items that can occur directly, but that refers to something inherent to the memories or the opinion of the child, they tend not to remember all of the options. To the question:

1. you prefer the colour "yellow", "green" o "blue", or
2. you prefer the colour "green", "blue" o "yellow",

the children of 3 tend to respond "blue" tot he first question and "yellow" to the second question.

The questions related to Thermal Comfort were:

(1) "*What is your perception of the environment in this precise moment*?" with possible answers:

- (+3) hot,
- (+2) warm,
- (+1) slightly warm,
- (0) neutral,
- (−1) slightly cool,
- (−2) cool,
- (−3) cold

(2) "*At this time would you prefer to be…?*" in this case, the preferential judgment scale included the answers:

- (−2) much cooler,
- (−1) cooler,
- (0) without change,
- (+1) warmer,
- (+2) much warmer

The first question allowed us to determine PMV and PPD indexes, and the second to evaluate information about adaptive thermal comfort.

The first series of questions for the children were "translated", also in symbols as reported in the following paragraphs, even with a terminology more effective and evocative for them; so for the question (1) "*What is your perception of the environment in this precise moment?*" the names of the children's responses are:

Italian version	English Version
(+3) caldo bestiale megatravolgente,	(+3) super boiling hot,
(+2) caldo bestiale,	(+2) boiling hot,
(+1) caldo,	(+1) hot,
(0) caldino,	(0) warm,
(−1) freddo,	(−1) cold,

(continued)

(continued)

Italian version	English Version
(−2) freddo freddo,	(−2) very cold,
(−3) freddo congelante	(−3) freezing cold

The second question "*At this time would you prefer to be…?*" was asked orally, but answers did not demonstrate a significant repetitiveness, and turned out to be too different and subjective.

At first children didn't understand the second question, they believe it was a not a question on thermal sensation, but about anything, for example some answers at question "At this moment would you prefer". Was: "at home", "to play" or "I am feeling tired, or hungry". So these answers confirm difficulties to do questionnaire, because we need to clarify we are talking about thermal feeling, for each question, and we're not talking about other things. In any case children have some difficulties to recognize between present thermal sensation and a desire to do another activity or other aspects that we cannot evaluate. In spite of these difficulties, we obtained the correct answers, asking them which thermal comfort sign, on those reported on desk board they preferred at the moment.

In the experiment, we decided to evaluate whether the children had notions and perceptions related to Indoor Air Quality (IAQ). In this case, the questions raised are two:

(3) "*What does the air smell like now?*", which provides the answers:

- (0) without odour
- (+1) with slightly bad odour
- (+2) with bad odour
- (+3) with offensive odour
- (+4) with a very bad offensive odour
- Children have preferred "translate" the answers in their own language:

- (0) great air
- (+1) very good air
- (+2) good air
- (+3) bad air
- (+4) repugnant air.

(4) "*How do you feel right now?*" which does not provide answers to a list, but free responses by children, as if it were an interview about their opinions.

Answers to "What does the air smell like now?" were directly written down, without using graphic representation.

The children answered individually, without any influence from their schoolmates or adults. The graphic elaboration was a complex psycho-pedagogical work. Educators reported some of the children's funny judgments about air quality. For the first query, 5 year-old children's qualitative judgment was mainly related to food perceptions (T: "There is a smell of chocolate"), other people (G: "This is Tommaso's smell") and allergy (A: "When there is dust in the air, I sneeze").

The 4 year-old children's qualitative judgments were very similar and related to school activities (A: "This is the smell of toys/When I come into the classroom there is a good smell/The smell changes because things change/In the kitchen there is a good smell, while in the bathroom there is a bad smell")(A: "Some smells exist and others do not exist/There is a smell of table/The smell changes if you touch different objects").

The questionnaires were developed and administered according to the Pedagogical method of the Diana kindergarten. In particular, it was decided to describe to children the processing tasks of the questionnaire at the arrival of children, in the morning during the morning assembly.

The steps of the experiment were:

(1) the definition of the pedagogical presentation and processing of the questionnaires with the children, the teachers and the researchers, a step necessary because children aged 4 and 5 were not able to understand the questions as they are described (and administered) according to UNI 10551;
(2) the administration of the questionnaire in three different times of the day;
(3) monitoring, throughout the day and in conjunction with the administration of the questionnaires, the indoor microclimate parameters and the concentration of CO_2 probes and data loggers placed in site.

The research took place over two days, the first day in the classroom for 5 year-olds and the second in the classroom for the 4 year-olds.

The first approach was to submit the questionnaire for children with images prepared independently by researchers before the meeting. This approach was revealed to be neither useful nor effective because the children could not understand the contents of the questionnaire questions or attributed different meanings from those expected. For example in order to explain the feeling of cold the researchers had prepared a series of images (Fig. 9.4) where he represented (as a comic) a "*trembling-man*" (a man who was trembling), to indicate the feeling of "very cold" and designed (as a comic) a "*sweaty-man*" (a man who was sweating) for feeling "*very hot*". The children did not understand the meaning of the images and attributed a different meaning to the images: to the "*trembling-man*" the children attributed the feeling of fear, according to them the man was frightened by something and trembled from fear (think of the drawings *Scooby Doo*) and children have imagined and started discussing about why he could be so scared. The image of the "*sweaty-man*" has been interpreted as a lazy man, who had no desire to work.

In summary of this first meeting it was noted that the images which were prepared in order to translate words and thermal sensations and smell, had nothing in common with the "*Cognitive Universe*" of the child. We eliminated all the work done, and proceeded with a new construction of the questionnaires.

So we decided to follow kindergarten pedagogue model. The Loris Malaguzzi pedagogue models assert children learn automatically without a linear relationship

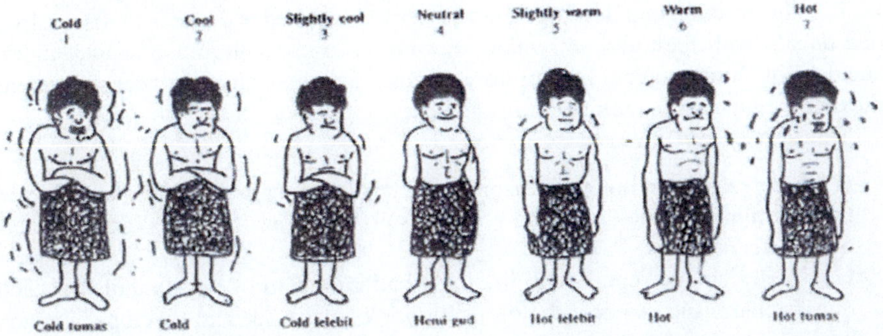

Fig. 9.4 The designs for the evaluation of thermal sensation developed by researchers but not involving children

Fig. 9.5 The morning assembly and the answer to the questions, on the bottom board with symbols and envelopes

cause-effect, but children are active subject of learning world construction process, so education is an auto-learning process. In this way "pedagogical approach" we debated with children about thermal concept through questions like "is hot like…" or "is cold like …".

The construction and development activity has resulted in a series of discussions and debate with the children, which allowed us to know the opinion and understanding of children about the comfort, climate and sophisticated concepts such as the distinction between temperature and heat.

The activity took place in three stages:

(1) the presentation and discussion activity with the children of the meaning of the questions and the terms and symbols suitable to describe the different levels of each feeling;
(2) the design and representation on the bulletin board of the symbols for each questionnaire and response, below each symbol was placed an envelope in which each child put a small stick in the envelope below the symbol that corresponded to their feeling;
(3) the effective administration of the questionnaire at three different times of the day.

The aspect of research related to the pedagogical approach was to identify the most appropriate pedagogical methods to obtain from the children the answers corresponding to the questions asked. To do this, there were a series of meetings together with kindergarten educators to achieve a result graphically and conceptually understandable for children and then on reliable ratings.

Together with educators the duration of the questionnaires was discussed and on how to submit questions to the children. It was decided to act directly, or asking the question orally and proposing the possible alternative answers graphically through symbols of different shape and colour, designed by the children themselves. The oral question involved discussion with the children, activities that involved organizing the children, for the two classes, in groups of three or five children.

9.2.1 The Discussion Stage with the Children to Find Matching Symbols

The discussion activity was held for each classroom of the children's ages, and was moderated by the educators, who described the activities of the day and led the discussion.

In order to understand how we got to the processing of queries it is useful to recap the more interesting and curious observations of the children. The children were placed in a disorderly fashion. Below are the answers of the children (the initials of the children are used) as well as transcribed by teachers before and during the research of the symbols.

In the room of children of 4 years old and the teacher asked the question:
"*What is the temperature?*"

Original Italian Version	English translation
A: La temperatura è quando c'è caldo fuori potete mettere il fresco, quando c'è il fresco fuori potete mettere il caldo;	A: The temperature is when it's hot out, and you can put on the cool, when it is cool outside you can put on the heat;
C: Perché poi se c'è freddo come si fa a riscaldarsi?	C: How then if it's cold, do you warm up?
F: La temperatura sono le cose che, sono tipo, adesso forse c'è una temperatura che c'è il vento La temperatura che c'è troppo caldo, freddo, delle volte piove viene il sole	F: The temperature are the things that, they are that kinds, now maybe there is a temperature that there is the wind the temperature that is too hot, cold, sometimes it rains and out comes the sun

Here the educator intervened to clarify the relationship between temperature and climate, asking "*When it rains or when the sun returns, the temperature is always the same or it changes*"?

A: When it rains it is cold when there's the sun it gets hot.

Educator: "*with what do you feel the cold*?"

Original Italian Version	English translation
F: io non lo sento molto perché mi metto una giacca pesante e ho anche l'ombrello	F: I do not feel it much because I wear a heavy jacket and I also have an umbrella
A: oppure può venire freddo quando c'è la neve	A: or it can be cold when there's snow
F: per è il contrario invece di avere freddo ho caldo, mi metto una giacca troppo pesante	F: for it's the other way round, instead being cold I'm hot, I wear a jacket too heavy

[Here we see the self-centeredness of the children who express their feeling, in relation to their own clothing]

Teacher: "*with what do you feel the heat and cold*?"

Original Italian Version	*English translation*
F: io riesco a capire	F: I understand
A: ci abbiamo il cervello	A: We have a brain
F: forse è la nostra pelle che sente freddo o caldo	F: perhaps it is our skin that feels cold or heat
A: quando la pelle ci ha freddo vengono tutti i peli su e quando c'è caldo la pelle diventa tuta liscia	A: when the skin is cold all the hairs are up and when it's warm, the skin becomes smooth
F: poi anche vengono già delle gocce d'acqua. Sai qual è l'acqua del sudore? Quella che beviamo noi. Quando sudiamo viene fuori. La disegnamo?	F: then there are also water droplets. You know the water of sweat? What we drink. When we sweat comes out. Shall we draw it?

The answers are related to the opinion of that child at that time compared to their notions of climate, temperature and their own feeling. The answers do not correspond directly to questions, but they try to find an analogy or an experience that can match the subject matter of the question.

The discussion was useful to understand the imagination of children and what were the symbols and images that they use, to express the feelings of comfort and air quality. Also by participating in the discussion and in the design in an active manner you are sure that the symbols correspond exactly to the type of thermal sensation or opinion on IAQ.

The teachers collected some phrases/children's views about the Thermal Comfort, during the debate of the early morning.

To questions like "What is temperature?" 5 year-old children answered with qualitative judgments about climate (e.g. A. said: "When it rains, it is cold and when the sun appears, it is hot"), or fantastic statements (L.: "If you are near a volcano, you feel sizzling hot"). Someone answered getting closer to the scientific discipline of comfort (F.: "Maybe it is our skin that feels cold or warm"). Answers of 4 year-old children were strictly related to outside climate (M.: "When it is sunny, it is hot. In Winter, it is cold because the sun is not in sky"; or A.: "In the morning it is hot. At the gym you feel much warmer because you run.") and some of the answers were very funny (A.: "I feel excited when I talk about temperature, because I think about Winter and I really love it.").

Answers to "What does the air smell like now?" were directly written down, without using graphic representation. The children answered individually, without any influence from their schoolmates or adults. The graphic elaboration was a complex psycho-pedagogical work. Educators reported some children's funny judgments about air quality. For the first query, 5 year-old children's qualitative judgment was mainly related to food perceptions (T: "There is a smell of chocolate"), other people (G: "This is Tommaso's smell") and allergy (A: "When there is dust in the air, I sneeze"). 4 year-old children's qualitative judgments were very similar and related to school activities (A: "This is the smell of toys/When I come into the classroom there is a good smell/The smell changes because things change/In the kitchen there is a good smell, while in the bathroom there is a bad smell") (A: "Some smells exist and others do not exist/There is a smell of table/The smell changes if you touch different objects").

Annex shows discussions with the children during the morning assembly.

9.2.2 *The Design and Representation on the Board*

In the joint discussion with the children, each of them drew the symbols corresponding to the feeling. Then we proceeded to the realization of a board of 1 m × 1 m, in which was reported a single graduated scale of symbols shared by the children and at the bottom of each symbol was placed an envelope.

In the three times of the day in which the survey was done, the children were asked to insert a wooden stick in the envelope which corresponded to the symbol and the sensation perceived by the child. For example in the case of the first application, the child who perceived the environment "very hot", inserted the stick inside the envelope positioned below the symbol that corresponded for the class, to the feeling of "very hot".

In this way the children were autonomous in expressing their opinion and, as was required in groups of 3–5 students, it was possible to minimize the risk of mutual influence in the response.

At the end of each session of questions the researchers collected the number of sticks for each answer, and cataloged the results of the experiment. The experiment was performed for the two classes with 3 questionnaires per day. This criterion was decided in the first meeting with only the educators, where they performed a complex activity with the children, in which they undertook to propose the theme of environmental comfort within their section to understand how children conceive the concepts of temperature, smell, etc.

Figures 9.6, 9.7, 9.8, 9.9, 9.10, 9.11, 9.12, 9.13 and 9.14 show above activities and children symbols.

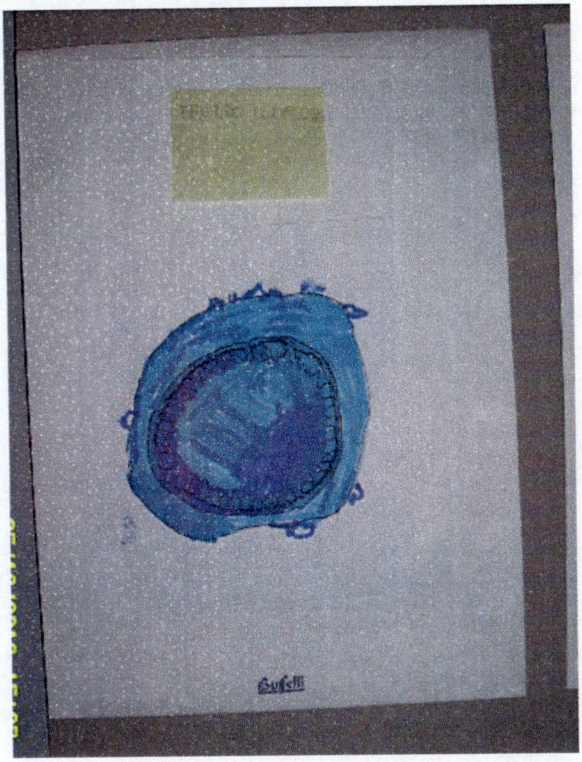

Fig. 9.6 Sun-cold-ice corresponds to the score (−3) "*Very cold*"

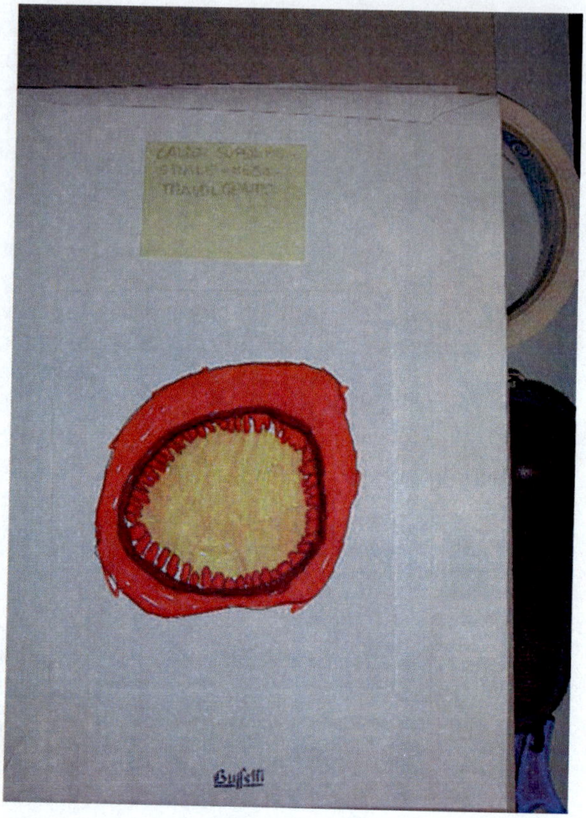

Fig. 9.7 Sun-hot super boiling hot-to the score (+3) "*Very hot*"

9.2.3 Questionnaire Submission (Order of the Questionnaire—The Day)

The experimental work was carried out by architects Francesca Practico and Giulia Goretti during research for their Masters Degree in Architecture.

Questionnaires were given to children at three times of the day:

- The arrival of the children (around 9:30 am), during morning assembly or immediately after;
- Mid-morning (around 11:00 am) after the play hour and before lunch;
- Early in the afternoon (3:00 pm approximately) after the afternoon nap, before the snack and exit of the children (4:00 pm).

In support of the activities two educators were present Laura Rubizzi, for the class of 5-year olds, and Evelina Reverberi, for the class of 4 year-olds.

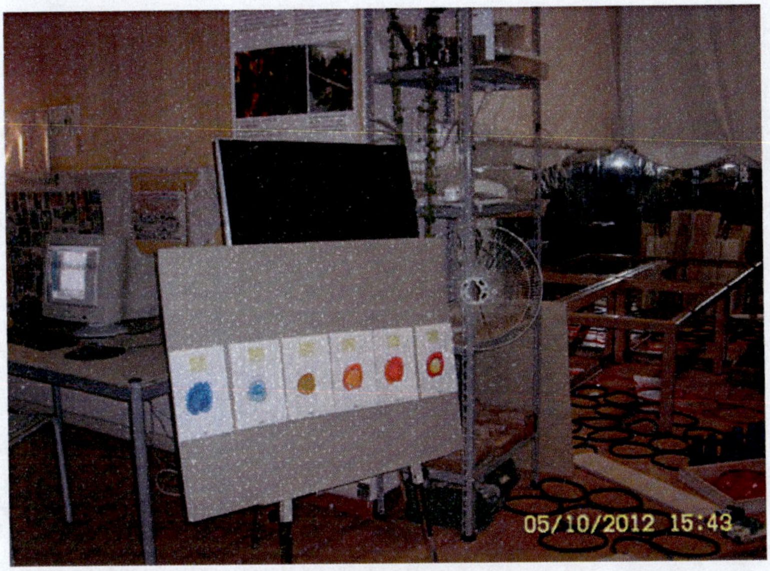

Fig. 9.8 Final questionnaire notice board after morning assembly

Fig. 9.9 Board with other children's designs

Fig. 9.10 One child putting her judgment on the board

Class of 5 year-Old Children (9 October 2012)

The first day of testing concerns the class 5 year-old children. At the beginning of the day, before the entry of children, the two probes for measuring the indoor microclimate and CO_2 concentration were positioned.

The monitor was started after waiting for the arrival of all children within the class and has remained active throughout the day. During the course of the questionnaires, in three distinct moments of the day, the probes were set so as to acquire the data every 30 s, to get more detailed results about the parameters investigated; for the remaining time, the interval was set to 5 min.

Upon arrival the children sat on a piece of furniture in the shape of steps for carrying out the "*Morning Assembly*". The teachers explained to them the kind of activities that would take place during the day along with the two researchers, parallel to the normal activities of the section. The children were placed at tables in groups to draw or work with clay. Some moved into the mini studio to paint or play.

In the classroom there were 24 children, three teachers and two researchers.

The type of activity carried out by children can be described as an average metabolic activity, somewhere between a sedentary activity and a slight walk (1.0–1.2 half of an adult).

Fig. 9.11 The wood stick and the envelopes

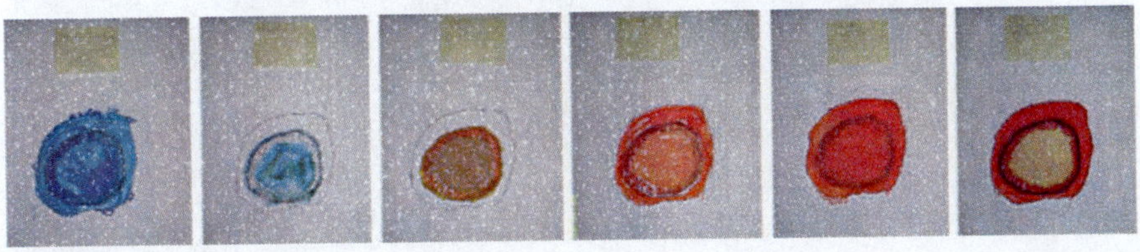

Fig. 9.12 5 year-old questionnaire draws of thermal comfort scale symbols

Fig. 9.13 4 year-old questionnaire draws of thermal comfort scale symbols

Fig. 9.14 Children's symbols for IAQ, from left to right: (0) great air, (+2) very good air, (+3) good air, (+3) bad air and (+4) repugnant air

Fig. 9.15 The morning assembly children of 4 years

The children were dressed in long cotton trousers or jeans, long-sleeved or short-sleeved cotton T-shirt, long-sleeved top and gym shoes (some children had a sweatshirt).

At 9:00 the first Questionnaire was carried out, the indoor conditions were 22.5 °C and the windows slightly open.

In order to know the opinion of individual children, the teachers chosen to call the children to meet in groups of two or three, so that they did not influence each other. The children called continued to play their normal game activities.

The images of the questionnaires, as described in the above paragraphs, had been created in the days before, on the day of measurement, the task of the children was only to insert the stick inside the envelope corresponding to their thermal sensation, the one they preferred and Indoor Air Quality.

The children responded one by one, without condition themselves mutually, in some cases a couple of children, particularly timid, they chose the envelope selected by his friend who had responded before him.

At 10:15, the first questionnaire was completed and the windows were opened for a change of air and the results collected.

At 11:00 the teachers decided to join the children for a snack and discuss the activities.

At 11:40 the children went to lunch in the space adjacent to the class set in the refectory (canteen). Simultaneously an attendant washed the floors and prepared the mattresses on the ground, for the afternoon rest.

While the children ate their meal (pasta, fish and vegetables), they were called in groups of three or four by the educator to again perform the questionnaire. This time, the children participated with more enthusiasm having become more familiar with the two researchers (before they were completely unknown). Others responded in a manner more hasty and were somewhat reluctant to play with their classmates before the afternoon rest. After lunch, the children showed hyperactivity, with an increase in the tone of voice and most 'frenzied' in their movements (running from one side of the classroom to the other).

At 12:40 the children were back in class and lying on their beds for an afternoon rest, some children had changed and wore pajamas, 5 children had returned home.

At 15:00 the children were awakened by the educators, the mattresses were removed and the furniture put back in the classroom with tables and chairs, where children sat and ate a snack (bread and jam, biscuits and juice fruit).

At 15:05 the third questionnaire was administered, children were called in groups of three and responded to the sensation perceived at the time.

The children continued to play in the classroom until the arrival of the parents at 16:00.

To questions like "What is temperature?" a 5 year-old children answered with qualitative judgments about climate (e.g. A. said: "When it rains, it is cold and when the sun appears, it is hot"), or fantastic statements (L.: "If you are near a volcano, you feel sizzling hot"). Someone answered getting closer to the scientific discipline of comfort (F.: "Maybe it is our skin that feels cold or warm").

Class of 4 year-Old Children (16 October 2012)

The second day of experimentation concerns the class of 4 year-old children, also in this case the probes for the microclimatic monitoring and CO_2 concentration were activated upon arrival of the children. Like every morning the children sat on the small staircase inside the classroom for the conduct of the morning assembly, when the teachers explained the type of activities.

In the classroom there were 22 children, three teachers and two researchers.

Again the type of activity carried out by children can be described with an average metabolic activity, somewhere between sedentary activity and a slight walk (1.0–1.2 half of an adult) and the children were dressed in long cotton trousers or jeans, long-sleeved or short-sleeved cotton T-shirt, long-sleeved top or sweatshirt and gym shoes.

At 9:30 the first Questionnaire was carried out, the indoor conditions were 22.0 °C and the windows were closed.

To submit the questionnaires two boards with the symbols drawn by children were used and envelopes to put the wooden stick for voting/rating. The two boards were one with the symbols for the temperature and the other for air quality.

The children were called one by one to the stairs to answer questions, while his companions listened sitting in a circle in front of us, first the questions on the temperature (Thermal Comfort—"how you perceive the environment at the moment?"), and then those on the Indoor Air Quality ("how you perceive the quality of the air at the moment?").

Each child was asked the questions and they, in turn, responded by placing a stick in the envelope corresponding to the chosen option; without mutually conditioning themselves; only one child carried out the questionnaire in part, answering the first question. Some children in answering the question "how would you like the environment at this time?" were affected by the season they prefer. *For example, one child chose 'super cool' because he is a winter enthusiast.*

After answering everything, the children return to their normal activities of daily play, in this case the children are out in the garden to play.

At 11:10 there was the second questionnaire. After returning from the activities in the garden the children were called in groups of two or three to answer questions, so as to prepare for 11:40 for lunch in the refectory.

The door of the mini-workshop was open to circulate air and the classroom has cooled. The children entered classrooms; some had a sweatshirt, some without and some with a jacket. During lunch the attendants prepared the room with mattresses for an afternoon rest. The windows of the classroom remained closed.

At 12:35, after the end of the meal, the children were back in class and lying in their beds, which consisted of a mattress placed on the floor covered with a sheet or "*fleece*" blanket. Some children wore pyjamas, others remained only with the cotton T-Shirt.

At 15:05 the children were awakened, one by one, by the educators and were administered the third questionnaire, the children responded in pairs. Meanwhile the children were seated at the tables and given a snack of bread and jam, biscuits and juice.

At 16:00 the parents came to pick up the children, and ended the whole experimental activity.

Answers of 4 year-old children were strictly related to outside climate (M.: "When it is sunny, it is hot. In Winter, it is cold because the sun is not in sky"; or A.: "In the morning it is hot. At the gym you feel much warmer because you run.") and some of the answers were very funny (A.: "I feel excited when I talk about temperature, because I think about Winter and I really love it.") (Fig. 9.15).

9.3 The Monitoring of Indoor Microclimate and IAQ

Simultaneously with the activities of administration of the questionnaires, the whole day the microclimate was monitored with two probes:

- "*Microclimate datalogger*" for the measurement of the microclimate data indoor: Air-Temperature, Mean Radiant Temperature, Air Velocity and Relative Humidity (RH);
- "*CO_2 measurement Datalogger*", with Instrument HD 37AB13D (Del-taOhm) for the measurement of: Relative Humidity RH (%), Environmental temperature T (°C), Carbon Monoxide CO (ppm) and Carbon Dioxide CO_2 (ppm). The CO_2 measurement was obtained with a special infrared sensor (NDIR technology: Non-Dispersive Infrared Technology) which, thanks to the use of a double filter and special measurement techniques, guaranteed accurate and stable measurements over time.

The Dataloggers were positioned according to the recommendations of teachers in relation to the size and activities carried out in the classroom, so as not to be hit, touched or moved by the children. For this reason we decided not to place the Datalogger neither at the level of children nor near the floor or in the middle of the room.

The probes were placed at 1.30 m above the level of the items of furniture. Before to set the Datalogger we measured air temperature and mean radiant temperature at children level, at 0.60 and 1.30 m level. The difference of temperature was lower of $\pm$ 0.10 °C, so we considered it unimportant. The measuring station was not located in the perfect location, but the classroom space is small and we can suppose "observational error" negligible.

The study used as the sole indicator of air quality (IAQ) the concentration of CO_2 as this pollutant is a continuously measurable value without having to alter the school activity (e.g. use of tracer gases), and because the concentration of CO_2 affects the ability of attention and fatigue by children (the highest concentration of CO_2 coincides with a difficulty concentrating and feeling tired) and the ol-active feeling (air that smells) evaluated empirically.

We decided the Datalogger location following educators recommendation, in order to consider classroom and children play space. We decided to not place the Datalogger at the children level or near the floor because children could have damaged or played with the measuring station, for the same reason we couldn't put it in the center of the classroom because it could be a delay. The measuring station set at 1.30 m level on piece of furniture. Before to set the Datalogger we measured air temperature and mean radiant temperature at children level, at 0.60 and 1.30 m level. The difference of temperature was lower of $\pm$ 0.10 °C, so we considered it unimportant. The measuring station was not located in the perfect location, but the classroom space is small and we can suppose "observational error" negligible.

From Figs. 9.16, 9.17 and 9.18 show datalogger location for each classroom.

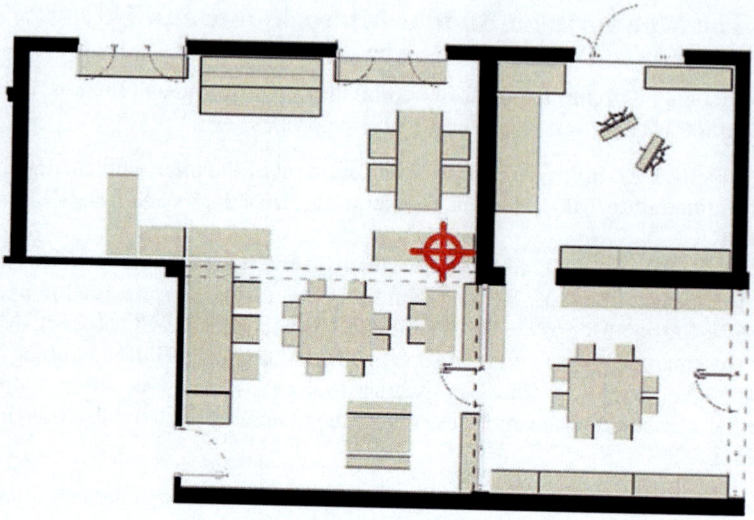

Fig. 9.16 5 year-old classroom and datalogger position

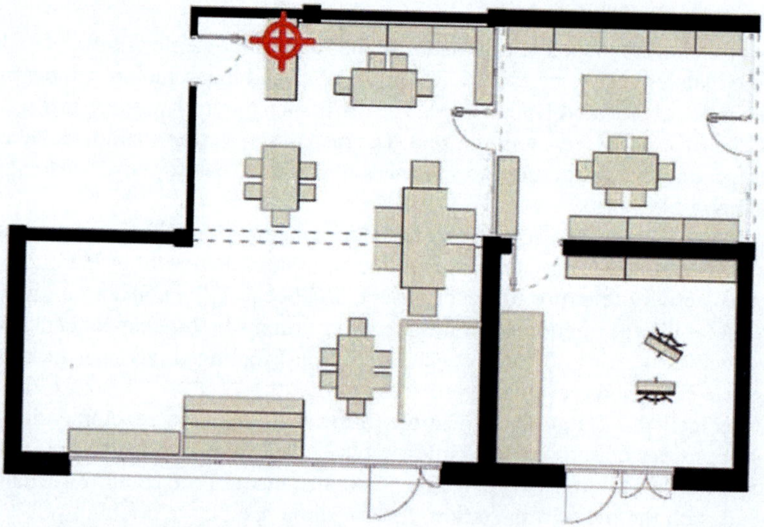

Fig. 9.17 4 year-old classroom and datalogger position

We did not measure the outdoor CO_2 concentration, Temperature and Relative Humidity (RH). The building is located in the centre of city, in a public garden, without car traffic in the range of 700–900 m. Therefore, we cannot consider outdoor pollution as significant for our research.

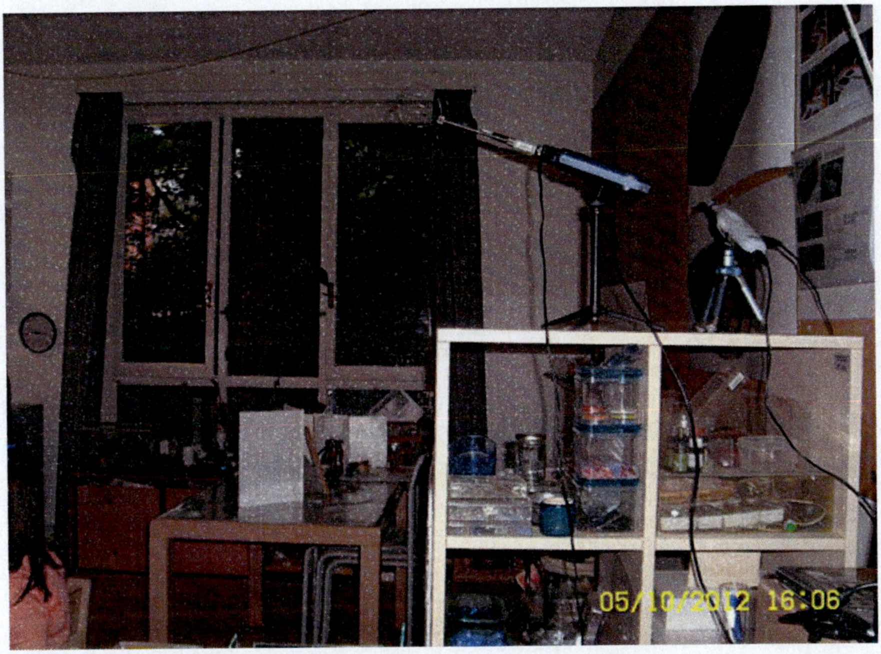

Fig. 9.18 Posizionamento sonde Aula 5 anni

9.4 The Results of the Research

The research activity has seen three steps for the collection of data:

(1) to define the pedagogical approach adopted to present the questionnaire to children, together with children, pedagogues, educators and evaluators. This step was absolutely necessary because children aged 4 and 5 cannot read questions. Questions were defined based on the specific kindergarten pedagogical approach.
(2) to submit the questionnaire in 3 different moments of the day;
(3) to measure, at the same time, the microclimate parameters of the room using a Datalogger.

Since we collected evaluations about thermal comfort with two different tools (Datalogger and children questionnaire), the aim of our study is to analyze the difference between the two.

The research consists of the elaboration of the results of steps (2) and (3) and a comparison.

9.4.1 The Reference Entity

The monitoring is used to calculate the values of PMV and PPD calculated in accordance with ISO 7730 (adults); to do this we needed to locate a "*reference entity*".

The ISO 7730 is applicable to a "reference entity", defined as.

"*healthy men and women exposed to indoor environments where thermal comfort is desirable, but where moderate deviations from thermal comfort occur.*". So the calculation is reported to grown man. In specific cases we need to define the metabolic conditions (meth) and apparel (clo) of the "*child*" and define the "*child equivalent*" ("*standard-kid*"), or the child that for characteristics of met and clo is comparable with characteristics of the adult for the calculation of comfort according to ISO 7730.

Children have different physical structures (weight and height) and clothes. In order to standardize monitoring and questionnaire, we used metabolic activity (met) and clothes (clo) information in accordance with ISO EN 7730, ISO EN 8996, and ISO EN 9920 standards.

We established that a "standard kid" has:

(a) metabolic activity, that depends on the activities performed during the day (similarly to adults):

For the 1st test (after arrival), we considered a medium-intensity activity (play) = 1.6 met (93 W/m^2).

For the 2nd test (in the morning), we considered a moderate-intensity activity (play, draw or similar) = 1.2 met (70 W/m^2). For the 3rd test (in the afternoon after sleeping), we considered a low-intensity activity = 1.0 met (58 W/m^2), where m^2 makes reference to DuBois area.

(b) clothes:

Underwear (0.04 clo), cotton t-shirt (0.06 clo), shirt (0.33 clo), cotton trousers (0.26 clo), socks (0.02 clo) and shoes (0.05 clo), for a total clothing insulation of 0.76 clo.

9.4.2 Results of the Monitoring Campaign and Questionnaire

In this paragraph, we report the results of microclimate measurement and questionnaire. The ***results*** of measurement and investigation have been developed in order to relate the opinions expressed by the children with those detected by the instrumentation, divided into:

- Measuring microclimatic parameters,
- Evaluation perception Thermal Comfort
- Perception Indoor Air Quality Air Quality.

The values obtained from the responses to the questionnaires were then compared with the physical ones, through the PMV and PPD, and triggered a series of considerations about the critical issues of the kindergarten in terms of thermal comfort and indoor air quality.

For Datalogger measurement and evaluation of PMVd and PPDd indexes, we used Deltalog 10 software in accordance with EN ISO 7730 standards. Il pedice d indica I risultati ottenuti dalle misurazioni con I datalogger.

Children Thermal Comfort—Results

Children's response to the questionnaire allowed us to calculate PMVc and PPDc indexes. PMVc is the weighted average obtained with formula (9.1), while PPD was calculated with formula (9.2) following ISO 10551 Annex 2:

$$\mathrm{PMV} = (\Sigma_{k,j} Q_k \cdot n_j)/(\Sigma_j n_j) \tag{9.1}$$

where:

Q is the value representing questionnaire vote (−2 cold, + 2 hot etc.), k
n is the number of children who answered at questionnaire, j

$$\mathrm{PPD} = 100{-}95 \cdot e^{[-(0.03356 \cdot \mathrm{PMV}^4 + 0.21749 \cdot \mathrm{PMV}^2)]} \tag{9.2}$$

where:

PMV is the value of PMV calculated with formula (9.1)

In Tables 9.1 and 9.2, we show PMV and PPD values obtained with Datalogger and questionnaire evaluation for 4 year-old and 5 year-old children.

Tables 9.3 and 9.4 show the results of the questionnaire about subjective judgment and preferred conditions.

Graphs in Fig. 9.9 show PMVd by Datalogger and PMVc by children. Graphs in Fig. 9.10 show the gap between PMVc,sj as subjective judgment and PMVc,p as preferred comfort judgment.

Table 9.1 Section 5 years-old preschool child: PMV measured by datalogger and subjective judgement scales (questionnaire)

Test (hours)	PMV_d Datalogger	PMV_c Children	Difference $PMV_d - PMV_c$	PPD_d Dataloggher (%)	PPD_c Children (%)
9:30	+ 0.47	+ 0.59	+ 0.12	9.57	12.32
11:50	– 0.25	+ 0.35	+ 0.60	6.30	7.52
15:00	– 0.47	+ 0.67	+ 1.14	9.52	14.34
Average			+ 0.62	8.46	11.39

Table 9.2 Section 4 years-old preschool child: PMV measured by datalogger and subjective judgement scales (questionnaire)

Test (hours)	PMV_d Datalogger	PMV_c Children	Difference PMV_d – PMV_c	PPD_d Dataloggher (%)	PPD_c Children (%)
9:30	+ 0.09	+ 0.68	+ 0.59	5.29	14.77
11:00	+ 0.55	+ 0.57	+ 0.02	11.30	11.69
15:00	– 0.42	+ 0.71	+ 1.13	8.88	15.73
Average			+ 0.59	8.49	14.06

Table 9.3 Section 5 years-old preschool child: comparison between $PMV_{c,p}$ preferred and $PMV_{c,sj}$ subjective judgment by questionnaire

Test (hours)	$PMV_{c,p}$ preferred (children)	$PMV_{c,sj}$ subjective judgment	Difference $PMV_{c,p}$ – $PMV_{c,sj}$	PMV_d Datalogger	Difference $PMV_{c,p}$ – PMV_d
9:30	– 0.05	+ 0.59	– 0.64	+ 0.47	– 0.52
11:50	+ 0.00	+ 0.35	– 0.35	– 0.25	0.25
15:00	– 0.11	+ 0.67	– 0.78	– 0.47	0.36
Average	– 0.05	+ 0.53	– 0.59	– 0.08	– 0.03

Table 9.4 Section 4 years-old preschool child: Comparison between $PMV_{c,p}$ preferred and $PMV_{c,sj}$ subjective judgment by questionnaire

Test (hours)	$PMV_{c,p}$ preferred (children)	$PMV_{c,sj}$ subjective judgment	Difference $PMV_{c,p}$ – $PMV_{c,sj}$	PMV_d Datalogger	Difference $PMV_{c,p}$ – PMV_d
9:30	– 0.14	+ 0.68	– 0.82	+ 0.09	– 0.14
11:00	+ 0.18	+ 0.57	– 0.39	+ 0.55	– 0.55
15:00	+ 0.10	+ 0.71	– 0.61	– 0.42	+ 0.31
Average	+ 0.04	+ 0.65	– 0.60	+ 0.07	– 0.12

In Tables 9.1 and 9.2, we show PMV and PPD values obtained with Datalogger and questionnaire evaluation for 4 year-old and 5 year-old children. Tables 9.3 and 9.4 show the results of the questionnaire about subjective judgment and preferred conditions.

We decided to define the metabolic activity value (met in W) following EN ISO 7730 standard; the met value of activity, expressed in W/m^2, is related to the DuBois area of children. As far as children clothing is concerned, we considered the fact that children do not dress themselves on their own; their mother decides what they have to wear, depending on the comfort perception she feels in the morning before taking the kid to kindergarten. Maybe parents usually have a lower metabolic activity than children and tend to overestimate the cold sensation, of course this is a hypothesis or just an opinion.

Tables 9.3 and 9.4 show the results of preferred PMVc,p and subjective PMVc,sj obtained from the questionnaire. In both classrooms, preferred PMVc,p is 0.60 lower than subjective PMVc,sj. Thus, we can state that children would prefer a cooler environment, probably because of their greater metabolic activity and clothing insulation.

So preferred and subjective PMV indexes could have different reasons, in this case we presume they depended on metabolism and clothes, but, similar to adults studies and according to literature about adaptive thermal comfort, the difference might have been various. The research put in evidence a fact: both classrooms prefer a slightly cooler indoor environment.

Another interesting result we obtained is the correspondence between preferred PMVc,p for children and PMVd Datalogger value for adults, evaluated according to EN ISO 7730.

Preferred PMVc,p for children is 0.6 lower than adults' PMV and this means that, with identical microclimate conditions, children feel "slightly warmer" than adults (Tables 9.5 and 9.6).

Tables 9.7, 9.8, 9.9 and 9.10 (Annex) describe the relationship between PMV and PPD results, obtained from questionnaire and Datalogger. For 5 year-old children, the gap between PMV and PPD is 0.12 in the morning, 0.60 at lunch time and 1.14 after sleeping in the afternoon.

For 4 year-old children, the gap in the morning and at lunch time is 0.59 and 1.15 after sleeping in the afternoon. The gap between subjective PPD and Datalogger PPD equals only 3–5%, i.e. the judgment of 1 out of 20 children.

Children thermal comfort slightly exceeds adults' perception, probably due to physiological and/or psychological factors, and/or physical values, like mean radiant temperature (children are nearer to the floor) or relative humidity (children have a greater breathing quotient).

Physiological factors could be investigated analyzing clothes or metabolic activity, whereas psychological factors are very difficult to examine. For psychological and microclimate values, we established that during all day the relative humidity value was greater than 60–65% and that could have had an incidence on children breathing quotient and/or on evapotranspiration.

The average measurement between subjective PMVc,sj and preferred PMVc,p is 0.50 more than Datalogger measurement, especially after sleeping in the afternoon. Children thermal comfort perception is slightly higher (warmer) than standard values. The −0.5 gap between subjective PMVc,sj and preferred PMVc,p equals the gap between PMVc,sj and PMVd Datalogger evaluation following EN ISO 7730.

Children Indoor Air Quality Sensation—Results

To evaluate IAQ comfort perception, we adopted standard EN ISO 10551, that defines the questionnaire requirements to evaluate PMV. Whereas PPD index following the same PPD formula of standard ISO 7730 [26]. PMV and PPD indexes can satisfy and not-satisfy all indoor environmental indexes, as described in a study by d'Ambrosio

Table 9.5 Section 5 years-old: questionnaire results

Hours 9:00	Without odour	With slightly bad odour	With bad odour	With offensive odour	With a very bad offensive odour	PMV_{CO_2}	PPD_{CO_2}	PPD_{CO_2} datalogger
IAQ.child	Delicious air	Very good air	Good air	Bad air	Repugnant air			
Vote value	0	1	2	3	4			
Scoring	13	1	3	1	4			
Results						1.18	34.36%	28.40%
Hours 11:50	Without odour	With slightly bad odour	With bad odour	With offensive odour	With a very bad offensive odour	PMV_{CO_2}	PPD_{CO_2}	PPD_{CO_2} sonda
IAQ.child	Delicious air	Very good air	Good air	Bad air	Repugnant air			
Vote value	0	1	2	3	4			
Scoring	9	3	7	1	2			
Results						1.27	38.87%	38.80%
Hours 15:05	Without odour	With slightly bad odour	With bad odour	With offensive odour	With a very bad offensive odour	PMV_{CO_2}	PPD_{CO_2}	PPD_{CO_2} datalogger
IAQ.child	Delicious air	Very good air	Good air	Bad air	Repugnant air			
Vote value	0	1	2	3	4			
Scoring	5	3	7	1	2			
Results						1.56	53.92%	42.90%

Table 9.6 Section 4 years-old: questionnaire results

Hours 9:00	Without odour	With slightly bad odour	With bad odour	With offensive odour	With a very bad offensive odour	PMV_{CO_2}	PPD_{CO_2}	PPD_{CO_2} datalogger
IAQ.child	Delicious air	Very good air	Good air	Bad air	Repugnant air			
Vote value	0	1	2	3	4			
Scoring	8	2	4	4	3			
Results						1.62	57.38%	32.50
Hours 11:50	Without odour	With slightly bad odour	With bad odour	With offensive odour	With a very bad offensive odour	PMV_{CO_2}	PPD_{CO_2}	PPD_{CO_2} datalogger
IAQ.child	Delicious air	Very good air	Good air	Bad air	Repugnant air			
Vote value	0	1	2	3	4			
Scoring	6	3	5	3	6			
Results						2.00	76.76%	31.00%
Hours 15:05	Without odour	With slightly bad odour	With bad odour	With offensive odour	With a very bad offensive odour	PMV_{CO_2}	PPD_{CO_2}	PPD_{CO_2} datalogger
IAQ.child	Delicious air	Very good air	Good air	Bad air	Repugnant air			
Vote value	0	1	2	3	4			
Scoring	5	2	4	7	3			
Results						2.05	78.87%	35.30%

Alfano et al. (2013). In this case we decided to evaluate the IAQ following the CO_2 marker and the correlate sensation of "foul air". The questionnaire allows to evaluate the children's sensation of CO_2 (Buratti and Ricciardi 2009; Mors et al. 2011).

In Tables 9.5 and 9.6 we reported children judgment values and PMV_{CO_2} and PPD_{CO_2} indexes for both 4 and 5 year-olds classroom. The PMV_{CO_2} index is calculate as the weighted mean of all judgments scores of Tables 9.5 and 9.6, following ISO 10551.

PPD_{CO_2} index is calculate evaluated with Formula 9.1, similar to ISO 7730 formula for thermal comfort.

$$PPD_{CO2} = 100 - 95 \cdot e^{[-0.03353 \cdot PMV_{CO_2}^4 + 0.2179 \cdot PMV_{CO2}^2)]} \tag{9.3}$$

where PMV is PMV_{CO_2} obtained with the questionnaire

Datalogger results (PPDdata) were evaluated with formula (9.4), following ASHRAE 62.1.

$$PPD = 935 \cdot e^{(-15.15C_{CO_2}^{-0.25})} \tag{9.4}$$

where C_{CO_2} is CO_2 concentration expressed in ppm.

The answers to the second question "how do you feel right now?" were not taken into account, because children did not understand the question about the CO_2 level and they preferred to answer with full sentences. In other words, the children did not understand our questions about IAQ, their judgment was linked to the smell to a specific place, for example, the kitchen, or the memory of a bad smell. Their answers are very interesting because show how children do not understand IAQ and CO_2 concentration levels, therefore we need to define physiological breath rate in order to define IAQ level.

9.5 Thermal Comfort Results—Discussion

The results obtained from the monitoring campaign and questionnaires used convey some considerations regarding:

- The "pedagocial point of view",
- The perception of the "*Thermal Comfort by children*" and "*children's metabolism.*"

The design phase of the questionnaires showed that children perceive the indoor microclimate environments, in a synaesthetic manner, or children have clear notions of hot, cold—*and they feel hot or cold inside the rooms*—but in expressing these feelings bind their feelings to a story or image, usually tied to nature or, more often evocative images (freezing cold, boiling hot, sun, ice, volcano, etc.).

Once clarified to the children the request to assign a value, compared to the scale of perception, to their feeling of *Thermal Comfort*, the children were able to express their opinion. This assessment was prepared in such a way to achieve a result comparable to that measured by the probes. It should be emphasized that the contents of the questionnaire are related and can be used only for children who have been involved.

Children have their own world understanding and perception, which are strictly connected to their inner life. Their point of view has never been investigated by scientific literature, neither from the psycho-pedagogical nor from the comfort discipline perspective. Children's thermal comfort perception is strictly related to their cognitive construction of the world, that includes outdoor climate conditions, preferred season or day-time. Cognitive factors modify subjective judgment about indoor climate condition and, as a consequence, they have an impact on the questionnaire.

The results showed a question: why children PMVc,sj is higher than standard PMVd? This difference could depend on a gap between children and adults standard metabolism rate. In this paragraph we would like to debate about children metabolism following ISO 8996. The physiological studies, also reported in references, in order to compare people, several factors should be considered: age, health, etc. but these studies are specific and not standardize for several people. In this way: (a) we define adults and children Metabolism rate, following Standard formula, based on body surface are (Dubois Area). In Metabolism rate (M) we consider standard energetic and breathing value the same for adults and children, since we use Standard values. In Metabolis rate formula, between adults and children, the most important difference is body surface; (b) we compare adults and children metabolism rate; and we (c) suggest a simplified "met-child" in order to standardized PMV index in case of children.

- Metabolism rate.

We can calculate metabolic rate with the following formula (ISO EN 8996):

$$M = EE \cdot V_{O2} \cdot (1/A_{Du}) \tag{9.5}$$

where

M is the metabolic rate, in watts per square meter;
EE is the energetic equivalent, in watt hours per liter of oxygen (Wh/l_{O2});
V_{O2} is the oxygen consumption rate, in liters of oxygen per hour;
A_{Du} is the body surface area, in square meters, obtained with the DuBois formula.

This is the formula to calculate Energetic Equivalent EE:

$$EE = (0.23 \cdot RQ + 0.77) \cdot 5.88 \tag{9.6}$$

where

RQ is the ratio between oxygen consumption rate, in liters of oxygen per hour, and carbon dioxide, in liters of carbon dioxide per hour. Following ISO 8996, RQ value for the case study is 0.85, whereas EE value is 5.88 (Wh/l_{O2}).

The body surface area (ADU), in square meters, is given by the Du Bois formula:

$$ADu = 0.202 \cdot Wb^{0.425} \cdot Hb^{0.725} \tag{9.7}$$

Where

W_b stands for body weight, in kilograms;
H_b stands for body height, in meters.

- Metabolism comparison

If we consider a standard adult body with 70 kg of weight and 1.75 m of height (ISO EN 8996), with 1 met activity for 1.84 m^2 of DuBois surface area, we obtain a metabolic energy (ME) value of 106.64 W.

In case of children for a standard 4 or 5 year-old child body, with 20 kg of weight and 1.20 m of height, the DuBois surface area is 0.82 m^2.

If we calculate this simple proposition

$$ME_{adults} = ADu \cdot met = 1.84(m^2) \cdot 58(W/m^2) = 106.64(W) \tag{9.8}$$

in order to compare the standard evaluation, we must suppose that

$$ME_{adults} = ME_{children} = 106.64(W) \tag{9.9}$$

so

$$met_{children} = (ME_{children})/(A_{Du,children}) = 106.64/0.82 = 129.85(W) \tag{9.10}$$

and then

$$ME_{children}/ME_{adults} = 1.21 \tag{9.11}$$

Suggest Met-Child Standardization

Therefore, children's metabolic energy value is 1.21 time higher than adults' and the standard 1 met for adults corresponds to 1.2 met for children (met-child value). Standard EN ISO 8996 allows us to calculate other metabolic values. One marker about thermal comfort sensation is the oxygen consumption rate (V_{O2}). In the case of children, if we consider a metabolic energy of 129.85 W/m^2 with RQ value of 0.85 and EE value of 5.67, the V_{O2} calculated with formula (9.3) is 18.83 l_{O2}/h (liters of oxygen per hour). In the case of adults, we obtain a nearly doubled value, 34.73 l_{O2}/h.

For the metabolic activities we obtained, we have:

- for 1.2 met, metabolic energy is 156.72 W and V_{O2} is 22.73 (l_{O2}/h) (adults 41.91 l_{O2}/h)

- for 1.6 met, metabolic energy is 208.21 W and V_{O2} is 30.20 (l_{O2}/h) (adults 55.69 l_{O2}/h).

PMV value obtained with EN ISO 7730 formula could be adapted in relation to met value. If in the PMV formula (EN ISO 7730) we adopt the 1.2 met instead of the 1.0 met value, the PMV value undergoes an increase of 0.50, which corresponds to children's perception value.

In brief, we can say that, in the case of children, PMV evaluation can be corrected adopting the "met-child" value obtained with the metabolic energy proposition that contains adults' and children's DuBois surface areas, so "met-child" standardization should then be extended to more subjects with new studies.

9.6 Indoor Air Quality

Children have a different understanding of the world and this aspect has never been investigated in scientific literature, neither from the psycho-pedagogical nor from the comfort point of view. With our experimental activity, we understood the effectiveness of the questionnaire tool to discuss air quality perception with children. If we had used only a data logger for CO_2 evaluation, results would not have explained the influence of IAQ on children activity.

Figure 9.19 shows CO_2 concentration trend, RH and temperature in the 5 year-olds classroom. When we did the first questionnaire in this classroom, the CO_2 level was 1200 ppm (IDA 4), so the results show a PMV_{CO_2} equal to 1.18, which corresponds to the "with slightly bad odour" and "with bad odour" judgment.

The second questionnaire shows a PMV_{CO_2} equal to 1.27, similarly to the first questionnaire, with a level of CO_2 equal to 800 ppm (IDA 3). The third questionnaire shows a PMV_{CO_2} equal to 1.56, which corresponds to the "with offensive odour" judgment, with a level of CO_2 equal to 3.000 ppm (IDA 4).

The kindergarten building is too small for all its functions, so children have to perform several activities in the same room (play, lessons, sleeping). After lunch, the classroom furniture is put in a corner in order to put mattresses on the floor. A higher PPD value for the third questionnaire may depend on children tiredness after sleeping and a high level of CO_2 concentration during the sleeping activity. Some children changed their clothes and wore pyjamas, others preferred not to change clothes for sleeping.

Figure 9.9 shows CO_2 and temperature trends in the 4 year-olds classroom. In this case, the ventilation rate is lower than in the 5 year-olds room and CO_2 concentration is 1800 ppm (IDA 4). In the first questionnaire, children's subjective judgment confirms this fact with a PMV value of 1.62 (between "with slightly bad odour" [+1] and "with bad odour" [+2]).

In the second questionnaire, CO_2 concentration is 800 ppm (IDA 3) and children's judgment generated a PMV value of 2.00 ("with bad odour"[+2]). In the third

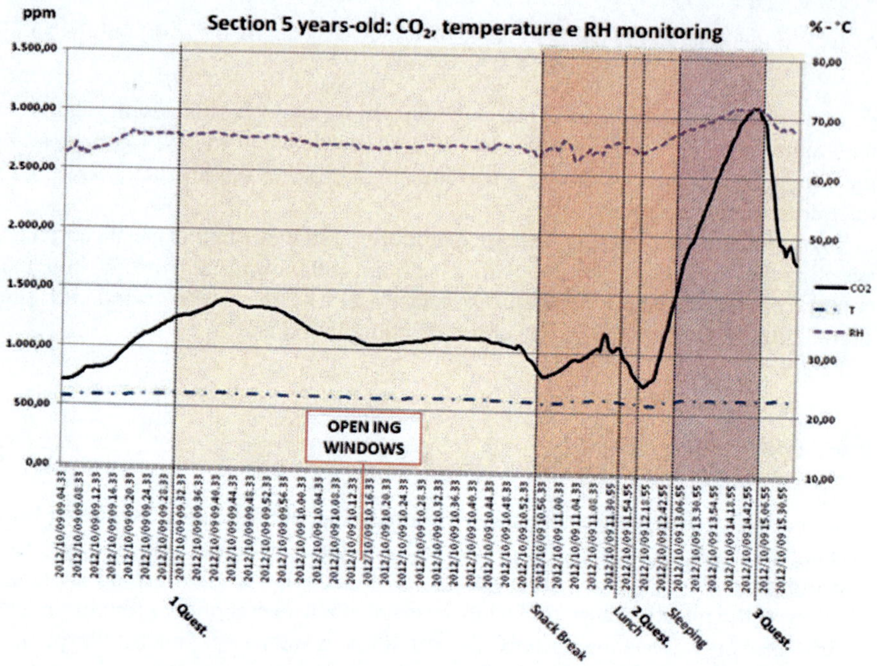

Fig. 9.19 5-year-old classroom: results for temperature (°C), relative humidity (%) and CO_2 (ppm) concentration level

questionnaire, we have the same judgment, PMV 2.05 ("with bad odour" [+2]), but a CO_2 concentration of 1400 ppm (IDA 4).

These results show that children's judgment about air smell is independent of CO_2 concentration measured with the data logger: the increase or decrease of CO_2 does not correspond to an increment or a reduction of the judgment "with bad odour". Children's judgment is independent of CO_2 and IAQ IDA values. A possible cause for this can be the fact that children's opinion about IAQ is not related to CO_2 concentration, but to other kinds of smell: food, people, etc. Figures 9.19 and 9.20 show graphics results.

Another important aspect shown in Tables 9.5 and 9.6 is the relationship between PPD_{CO_2} obtained from the PMV of the questionnaires and PPD_{CO_2} following data logger evaluation. On the one hand, in the case of 5 year-old children, these values are similar; on the other hand, in the case of 4 year-old children, these values are very different. These results demonstrate that children's judgment is not in line with CO_2 concentration. Children's sensation of weariness does not depend on the increase of CO_2 concentration and decrease of the oxygen level.

The 5 year-olds classroom has a volume of 227 m^3 and during the afternoon it hosts 19 children for sleeping, because 5 children leave earlier and go home.

Figure 9.19 shows CO_2 trend during the afternoon nap in the 5 year-olds classroom. It starts from 1500 ppm (at 1.00 pm), with 340.50 l_{CO_2}/h (litres of CO_2 per

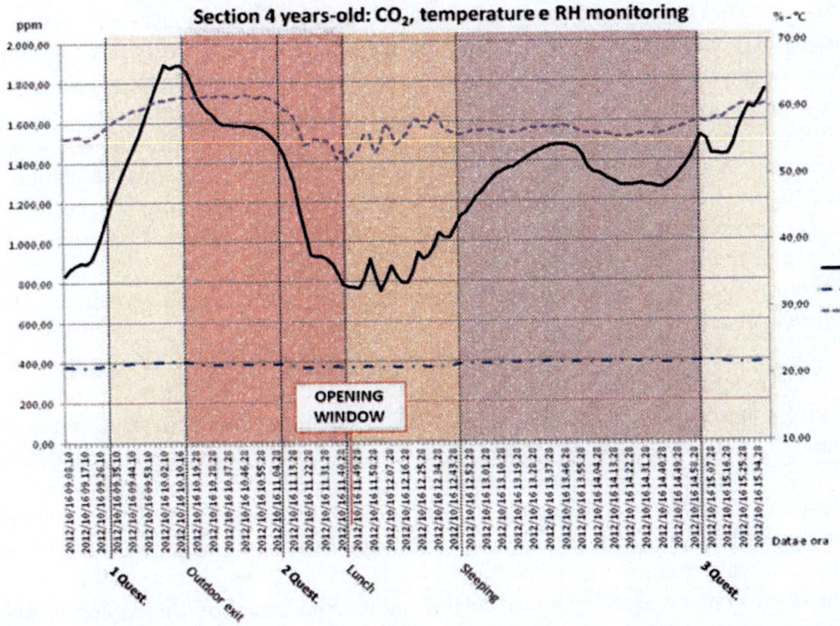

Fig. 9.20 4-year-old classroom: results for temperature (°C), relative humidity (%) and CO_2 (ppm) concentration level

hour), or 17.92 l_{CO_2}/h per child, and reaches 3000 ppm (at 3.00 pm), with 681.00 l_{CO_2}/h, or 35.84 l_{CO_2}/h per child, with an increase of 100%.

Figure 9.20 shows the data logger results in the 4 year-olds classroom. If we compare it with Fig. 9.19, we notice that CO_2 trend is different. It doesn't have the same increase: the CO_2 concentration starts from 1000 ppm (at 1.00 pm) and reaches 1400 ppm (at 3.00 pm), so it goes from 14.17 l_{CO_2}/h to 19.84 l_{CO_2}/h per child, with an increase of 40%. CO_2 ppm concentration is lower than in the 5 year-olds classroom. The difference depends on the volume of the classroom. In fact, the 4 year-olds classroom has a volume of 326 m^3 and during the afternoon nap it hosts 23 children. The 4 year-old children's judgment is not coherent with data logger values (Table 9.4): PPD index calculated with the third questionnaire is 78.87%, whereas PPD measured by data logger is 35.30%.

9.7 Conclusions

We will try to clarify the links between the Children Questionnaire and children response. The children cannot understand the standard thermal comfort questionnaire, follow Standard EN 10551, for this reason we did, with the children in assembly, some specific signs in order to identify each thermal comfort judgment (from cold to hot). The children classroom decided (during morning assembly) these signs, so they understand each sign because they had decided it. It is possible, for another classroom, to do not understand these classroom signs, but each child of a classroom understood the classroom signs. If we compare Fig. 9.5 (5 years-old classroom signs) with Fig. 9.6 (4 years-old signs) the signs are different. When we asked, at single child, to choose a sign, on classroom desk board, in order to explain his judgment about indoor thermal comfort, we are sure his meaning link between sign and response. In this way we are sure about children judgment because they choose their thermal comfort signs.

Therefore we consider acceptable questionnaire results in order to compare with Datalogger results.

The answers of children were correct, plausible and honest. Children demonstrated to understand questions and discussions about comfort and basic concepts like temperature and heating, but they seemed not to understand the difference between them. Educators and pedagogues were very collaborative because of Loris Malaguzzi's pedagogical method characteristics and the strict correlation between comfort and the psycho-pedagogical development of children. In this paper, we do not analyze all psycho-pedagogical issues in depth, but we are well aware that they deserve more attention.

The results can be used for future design and refurbishment of kindergartens and have the potential to be used for improving national and international standards. In conclusion, we were able to understand how to improve thermal comfort studies for children and to identify specific correction factors to evaluate their PMV index. This study can be considered a contribution to improve standards and thermal comfort design for different kind of people and should further promote study execution in this field.

The ***discussion*** of the results is addressed in four lines:

1. the test run and pedagogical aspects;
2. adaptive thermal comfort;
3. air quality related to the activities carried out and the ventilation of rooms;
4. the behaviour of the building over time, made possible thanks to the maintenance of the probes in environments even after the conclusion of the survey with the questionnaire, a choice that has allowed us to highlight the close relationship between technical-architecture conformation of the building and microclimatic parameters such as temperature and relative humidity.

To this were complemented observations regarding the different perception that the child has on the environmental well-being with respect to the adult, for the definition of possible design strategies to improve the comfort conditions of the Diana kindergarten.

The *conclusions* of the experimental relate to:

(a) the insights from the data comparison with the children and the teachers, and the adaptive behavior that they adopted both in terms of metabolic activity, both

for the psychological and perceptual elements on which it is possible to continue with specific studies;
(b) how it is possible to correct the deficiencies of the project for each microclimate parameter;
(c) those which can be further developments in the study and definitions of the parameters of comfort for those specific users, including the educational aspect.

In this study we have analyzed the comfort conditions inside the Diana kindergarten, both by measuring the physical parameters (temperature, relative humidity, CO_2 concentration, etc.), and asking the children attending the nursery to make a judgment about their satisfaction with the environment in which they spend most of their day.

From the point of view of microclimatic parameters, the main problems were encountered on two fronts: the mean radiant temperature and the percentage of relative humidity. In all situations studied, in fact, the surface temperature of the walls was lower than that of air, resulting in the generation of the phenomenon of "cold walls". Moreover, the percentage of relative humidity has proven high in both classrooms (always higher than 55%) and the fact that it is not exclusively dependent from external moisture (as described above) suggests that the problem is due to inadequate characteristics of construction of the building. These problems could be solved by acting on the packages of the building, such as walls, floors and fixtures, to ensure a better seal and at the same time good ventilation, which prevents the accumulation of excessive moisture.

From the point of view of the quality of air, it is noticed that the concentration of CO_2 within both sections in each measurement is above the limits of the law; this is an indication of poor ventilation. In this regard, it was observed that the windows were opened in section only during the break and rarely during the day, although a change of air more frequently would is necessary. In addition, the opening of doors and windows mainly depends on the preferences of educators and serves to facilitate the exchange of air for only a short period of time, after which the concentration of CO_2 readjusts at high levels. Therefore, in order to ensure a good indoor environmental quality it would be appropriate to provide the kindergarten with of a management system of mechanized ventilation.

Another relevant aspect, during the campaign, the measurement was carried out with the Lux meter, at different points of the classrooms and different heights (on the tables, on top of taller furniture) that identified insufficient levels of brightness within the environment. In this case, a possible design strategy might be to open some openings in the roof in correspondence of the classroom and to expand the windows in the facade, on the outer perimeter.

Considering the CO_2 results monitored with the data logger, we suggest to rethink the ventilation system of the building and to take into consideration a mechanical ventilation system. The observed CO_2 increase depends on the inadequate ventilation rate of the classrooms. Children feel a little more tired, but they don't correlate this feeling neither to IAQ odour, nor to CO_2 concentration.

The windows were opened only during lunch-time and snack-time in the morning. This simple operation allowed to significantly reduce CO_2 concentration. The children's opinion about air smell (IAQ) is strictly related to the smell of objects, people and/or activities like cooking, running or playing.

In conclusion, we can state that pre-school children are not able to evaluate IAQ based on CO_2 concentration, so they cannot be considered a good indicator of CO_2 concentration level.

This prompts two questions:

(a) "***Why carry out the study in a nursery school if you cannot get reliable results***?"

We would like to highlight a particular fact: our research results in the nursery school are reliable; but the results show that we cannot consider the children reliable in order to use their sensation as a marker for the CO_2 level of concentration. Children do not understand stale air or foul air. In the case of a building with natural ventilation, it is a human and not a probe that decides to open a window, e.g. the owner or user. In the case of a school building or nursery, it is the teacher and not the user (that is, the children). The study of De Giuli et al. (2012) shows IAQ in the case of school children, and our research has the same aim but with a younger target of school children. Some studies on classroom ventilation, such as Santamouris et al. (2008), Gładyszewska-Fiedoruk (2013), use only monitoring instead of questioning the children. This is one difference between our research and that of others—we studied the IAQ sensitivity of children in a nursery school. Both approaches, however, are correct.

(b) "***Why not using only monitoring instead of questioning the children***?"

Our goals are to understand if children, at a nursery age, understand IAQ, therefore if they could be a good marker to confront with the IAQ standard. If we monitored only the CO_2 trend, we could not know anything about children's sensitivity. Monitoring results could be compared only with the standard, but standard values are defined according to ventilation rate, or oxygen concentration, etc.

In our opinion, the IAQ depends on building users' habits and behaviour, so there could indeed be a marker. Our goals are not only the CO_2 level, the CO_2 trend, or ventilation rate, but the children's sensation. Results show children are not a good markers in understanding the CO_2 level, but they understand smell. Before commencing our research, we were unaware of this fact.

They begin to perceive some sort of discomfort only in the case of high CO_2 levels, only because they feel a bit tired. This can be a problem because it means that children do not adopt any "adaptive action" to contrast CO_2 level, for example they do not decide by themselves to open doors or windows and they do not warn adults about low oxygen levels. Thus, the ventilation rate of a room only depends on adults' actions (in this case, manual window opening). The gap between CO_2 concentration, IAQ level and children's perception needs to be further investigated in order to identify an indicator of inadequate IAQ level for children.

What are our conclusions about building energy research? Building energy performance and energy behaviour depend on IEQ, as reported by the WHO (2000) in

"Addressing the Links between Indoor Air Pollution, Household Energy and Human Health" and Beko et al. (2011) Yu et al. (2011). IEQ, human behaviour and energy consumption are strictly correlated, also IEQ derive for building users knowledge about it. Obviously IEQ follows building design, HVAC systems and user habits. Consequently user habits depend on indoor environmental perception. In the case of specific and "not standard" users, such as children, researchers need to understand their knowledge about indoor environmental.

Finally, the object of our research was children's IAQ cognitive knowledge. We decided to adopt two modalities: a questionnaire asked in three different moments of the day and all-day CO_2 level monitoring.

Research concern 1 kindergarten measurement during 2 days, for questionnaire test, and 2 weeks (1 for classroom) to evaluate indoor CO_2 concentration. Furthermore survey questionnaire concern 22 children 5 years-old and 21 children 4 years-old, at 3 different time by day (morning, lunch, afternoon) for each classroom. The total numbers of answers is 258, if we compare this number with others research with survey questionnaire, that database is enough to do our conclusion, for this specific case study in order to confirm our methods. Of course these research could be done again, and again, in other kindergarten, and/or in other social and pedagogical contest. Following as is usual in research study we decide to report our research results on purpose to suggest to apply and to compare our results in other social and scientific contest.

The numbers of answers was sufficient to draw conclusions: this research confirms our methodology (questionnaire + monitoring). A future goal will be to apply the same methodology in other kindergartens in order to increase case studies, also in other pedagogical contests.

The results of the research show that in the case of children, the questionnaire is a useful tool, improving assessments according to the regulations, in assessing the comfort perceived by children. Always processing and administering the questionnaire so as to be sure of the understanding of the children, as described above.

Acknowledgements The activities described in this chapter are the result of the thesis "*Island in the green. Redevelopment of the Diana kindergarten in Reggio Emilia*"

Undergraduates: Giulia Goretti and Francesca Praticò (YA 2011/2012), rel. prof. Andrea Boeri, co-rel. Professor Kristian Fabbri, Final Synthesis Laboratory "Sustainable Architecture" of the degree course in Architecture, Department of Architecture, University of Bologna.

The Laboratory has its own teaching method in the elaboration of the thesis, whose approach is the integration of the skills of the various disciplines that comprise it: Architecture Technology, Architectural Design and Environmental Physic Technique. The author acknowledges Giulia Goretti, Francesca Praticò, and prof. Andrea Boeri For Their valuable support during experimental activity and the educators Laura Rubizzi and Evelina Reverbero for their helpfulness during questionnaire submission. We thank the authors and collaborators for their availability.

Thanks also to the precious collaboration of the Scuola d'Infanzia Diana of Reggio Emilia.

Annex 1: Transcript of Conversations Before and During the Search of the Symbols

We report the transcript of what was discussed during the meetings, in the framing of the questions. The questions were put to groups of three children at a time and by topic.

Conversation to identify the symbologies for Temperature (13 September 2012) 4-years-old

In the classroom of 4 year-old children the teacher asked the questions:

Italian Version	English Version
"Cosa sarà la temperatura?"	"What is temperature?"
A: La temperatura è quando c'è caldo fuori potete mettere il fresco, quando c'è il fresco fuori potete mettere il caldo;	A: The temperature is when it's hot out, and you can put on the cool, when it is cool outside you can put on the heat;
C: Perché poi se c'è freddo come si fa a riscaldarsi?	C: How then if it's cold, how do you warm up?
F: La temperatura sono le cose che, sono tipo, adesso forse c'è una temperatura che c'è il vento La temperatura che c'è troppo caldo, freddo, delle volte piove viene il sole. Qui l'educatrice interviene per chiarire la relazione tra temperatura e clima, chiedendo "Quando piove o quando torna il sole, la temperatura è sempre uguale o cambia"?	F: The temperature are the things that, they are that kinds, now maybe there is a temperature that there is the wind the temperature that is too hot, cold, sometimes it rains and out comes the sun. Here the teacher intervenes to clarify the relationship between temperature and climate, asking "When it rains or when the sun returns, the temperature is always the same or it changes"?
A: quando piove viene freddo e quando c'è il sole diventa caldo	A: When it rains it is cold when there's the sun it gets hot
Educatrice: "con che cosa sentite il freddo?"	Educator: "with what do you feel the cold?"
F: io non lo sento molto perché mi metto una giacca pesante e ho anche l'ombrello	F: I don't feel it much because I wear a heavy jacket and I also have an umbrella
A: oppure può venire freddo quando c'è la neve	A: or it can be cold when there's snow
F: per è il contrario invece di avere freddo ho caldo, mi metto una giacca troppo pesante	F: for it's the other way round, instead being cold I'm hot, I wear a jacket too heavy
(qui si nota l'egocentrismo dei bambini che esprimono la propria sensazione, in relazione alla propria condizione di vestiario)	(Here we see the self-centeredness of the children who express their feeling, in relation to their own clothing)
Insegnante: "con che cosa sentite il caldo e il freddo?"	Teacher: "with what do you feel the heat and cold?"
F: io riesco a capire	F: I understand
A: ci abbiamo il cervello	A: We have a brain
F: forse è la nostra pelle che sente freddo o caldo	F: perhaps it is our skin that feels cold or heat

(continued)

(continued)

Italian Version	English Version
A: quando la pelle ci ha freddo vengono tutti i peli su e quando c'è caldo la pelle diventa tuta liscia	A: when the skin is cold all the hairs are up and when it's warm, the skin becomes smooth
F: poi anche vengono già delle gocce d'acqua. Sai qual è l'acqua del sudore? Quella che beviamo noi. Quando sudiamo viene fuori. La disegnamo?	F: then there are also water droplets. You know the water of sweat? What we drink. When we sweat comes out. Shall we draw it?
A: aspetta eh! Sento caldino	A: Ah wait! I feel a bit hot
F: io sento un po dei due, perché di colpo freddo poi un po caldo, molta pelle mia mi fa caldo eh!	F: I feel a bit of both, because I suddenly feel cold then a little hot, ah a lot of my skin makes me hot!
C: io mi sento tanto caldo, le mie mani sono fredde	C: I feel very hot, my hands are cold
F: le mie mani sono fredde	F: my hands are cold
A: io sento freddo alle mani e caldo per tutto il corpo tranne la testa e tranne le mani	A: My hands feel cold and warm for all my body except for my head and hands
C: davanti alla cucina sentiremmo caldo, le cuoche sudano. Caldo lo chiamiamo "Caldo bestiale"	C: in the kitchen I would feel hot, the cooks sweat. The heat is called "Boiling Hot"
(da questo momento i bambini incominciano a proporre possibili analogie e risposte per il questionario)	(by this time children begin to propose possible analogies and answers to the questionnaire)
A: poi sarà freddo congelato, ci mettiamo gli scarponi	A: then when it is freezing cold, we put on boots
L: se ti metti vicino un vulcano? Caldo Bestiale, Caldo caldino	L: If you get near a volcano? Boiling Hot, Pretty Hot
A: con la neve sarà freddo congelato, scarpe guanti, scarponi	A: with snow it will be freezing cold, shoes, gloves, boots
L: però anche se ci metti lì un vulcano	L: Even if you put there a volcano
A: è super bestiale	A: it's super beastly (boiling hot)
F: super bestiale mega … il vulcano ti fa morire (dal caldo)	F: super mega boiling hot … the volcano makes you die (from the heat)
(interessante l'analogia con elementi naturali evocativi, come un vulcano, e l'enfasi delle parole)	(Interesting analogy with evocative natural elements, like a volcano, and the emphasis of the words)
L: il fuoco non ti lascia nemmeno le ossa	L: the fire doesn't leave even the bones
F: pure la luce! Però ci rimane la forma	F: pure light! But the form remains
A: lo sapete che il fuoco è una palla gigantesca di fuoco?	A: Do you know that fire is a gigantic ball of fire?
L: e la luna? E' vero che l'acqua trionfa sempre sul fuoco?	L: and the moon? It's true that water always triumphs over fire?
F: il vulcano è travolgente	F: the volcano is overwhelming

(continued)

(continued)

Italian Version	English Version
C: io però lo strumento lo preferisco colorato	C: I though, the instrument I prefer coloured
(a questo punto l'educatrice cerca di mettere ordine)	(At this point the teacher tries to bring order)
Educatrice: Ora io vi rileggo le parole della temperatura che avete detto e voi cercate un modo per segnare le vostre parole. Potete scegliere i pennarelli che volte	Educator: Now will I re-read the words of the temperature that you have said, are looking for a way to mark your words. You can choose the markers that you want
(i bambini scelgono i pennarelli e si mettono a disegnare)	(Children choose markers and begin to draw)
L: io parto dal caldo	L: I start with the heat
A: perché non disegniamo una nuvola?	A: why don't we draw a cloud?
Educatrice: se credi che vada bene per il freddo disegnala	Educator: If you think it's fine for the cold draw it
C: io disegno un sole	C: I'm drawing a sun
F: io voglio il freddo, c'è il bianco? Ah il colore del ghiaccio! (azzurro chiaro)	F: I want the cold, is there a white? Ah the colour of ice! (Light blue)
L: io faccio il vulcano	L: I'm making a volcano
C: questo è un caldo terribile (indicando il disegno)	C: This is a terrible heat (indicating the drawing)
A: un caldo bestiale	A: a beastly hot (boiling hot)
F: è tutto congelato dal freddo	F: it's all frozen from the cold
C: poi se è ancora più caldino bisogna fare tutti i soli	C: then if it's even more hot you have to do it all alone
L: sta uscendo la lava! Un terremoto per uscire	L: the lava is coming out! An earthquake is erupting
F: guarda è così freddo che anche il sole è congelato	F: look it's so cold that even the sun is frozen
A: io comincio ora.. m sento? Oh ghiacciata!	A: I'll start now.. I feel? Oh icy!
L: molto fuoco	L: very fiery
(si fa notare come anche durante la preparazione dei disegni continuino a parlare e disegnare in base alla propria sensazione in quel momento o alla sensazione che immaginano associata al vulcano o al ghiaccio)	(do note that even during the preparation of the designs they continue to talk and draw based on their feelings at that moment or feeling they imagine associated with the volcano or ice)
A: questo è un bambino che suda	A: This is a child who sweats
C: io sto facendo tutti i soli così c'è un caldo terribile. Caldo bestiale mega!	C: I'm doing it all alone so there is a terrible heat. Beastly mega hot!
F: ma è il sole che scioglie il ghiaccio o il ghiaccio che congela il sole?	F: but it is the sun that melts the ice or ice that freezes the sun?
A: se c'è un caldo bestiale si scioglie il sole, quindi!	A: If it's boiling hot, it melts the sun, then!

(continued)

(continued)

Italian Version	English Version
L: guarda pieno di acqua	L: look it's full of water
Educatrice: il caldino come potrebbe essere?	Educator: the heat as it could be?
A: io faccio una palla di fuoco media	A: I'm doing a medium-sized fireball
F: c'è anche il caldo stratosferico	F: There is also the stratospheric heat
C: io faccio le nuvole	C: I'm doing clouds
L: io faccio il ghiaccio	L: I'm making ice
F: io sono tutto congelato	F: I'm all frozen
C: le nuvole vuol dire che piove, poi le palle di neve per il freddo congelato	C: clouds mean rain, then snow balls for the freezing cold
L: io faccio le valanghe	L: I'm doing avalanches
F: il vulcano butta fuori la lava e scioglie tutto il sole	F: the volcano throws out lava and melts all the sun
C: io faccio il caldo normale	C: I'm doing the normal heat
(a questo punto cominciano a giocare e fantasticare)	(at this point they begin to play and fantasize)
L: ti stai riempiendo di ghiaccio e questi sono i lancia-giacchio	L: You're filling with ice and these are the ice-spears
F: il mio galattico lancia-lava! Vediamo se vinco io	F: my galactic lava-spear! Let's see if I win
L: valanga travolgente! Freddo congelato	L: avalanche overwhelming! Freezing cold
Educatrice: stiamo cercando dei segni per far capire le temperature: il freddo e il freddo congelato mancano	Educator: we are trying to understand the signs to understand the temperatures: we're missing the cold and the freezing cold
L: freddo normale	L: normal cold
F: ora facci il caldo normale. Per il caldino un minisole	F: Now let's do warm normal. For a bit hot a mini sun
A: ho cominciato dal congelato	A: I started from frozen
F: vorrei stamparlo e darlo ai miei genitori che lo mettono in quadro	F: I want to print it out and give it to my parents to frame

Conversation to identify the symbologies for Indoor Air Quality (20 September 2012) years-old

Italian Version	English Version
F:. L'aria respiro qua dentro non mi sembra calda	F: The air breathed in here does not seem hot
T: io sento odore di cioccolato	T: I smell chocolate
E: se respiri con la bocca senti un po' caldo	E: And if you breathe with your mouth it feels a bit hot
GL: io sento odore di T. (un altro bambino)	GL: I can smell T. (another child)

(continued)

(continued)

Italian Version	English Version
M: la W. (un altro bambino) profuma di cioccolato. Se chiudo gti occhi sento odore di creta	M: W. (another child) smells of chocolate. If I close my eyes I can smell clay
F: io sento odore di fresco	F: I'm smelling fresh
Educatrice: secondo voi all'ora di pranzo l'aria sarà sempre come ora?	Educator: according to you at lunch-time the air will be like now?
M: sarà più calda perché la pasta e la zuppa è calda ed esce il vapore dal piatto. Quando la cuoca fa pizza sentiamo subito l'odore	M: it will be warmer because the pasta and the soup are hot and the steam comes from the plate. When the cook makes pizza we immediately smell it
A: sarà un caldo bestiale	A: will be a really/boiling hot
Educatrice. E quando vi alzate l'aria sarà come ora o come all'ora di pranzo?	Educator. And when you get up, the air will be like now or like at lunchtime?
A: Sarà freschina, invece quando dormiamo è calda perché abbiamo le coperte	A: It will be a bit fresh; instead when we sleep it's hot because we have blankets
M: nel bagno se lasciamo la pipì c'è odore di pipì o di cacca	M: If we leave the pee in the bathroom it smells like pee or poop
A: c'è un profumino diverso, ogni sacco ha un odore	A: There is a different fragrance, each bag has a smell
An: nell'agriturismo c'è sempre odore di gallina	An: on the farm there is always the smell of chicken
A: Quando l'aria sa di polvere si starnutisce	A: When the air smells of dust you sneeze
An: l'aria la sento con le ossa, con la bocc4 con il naso. Lo scheletro si stringe e io mi muovo invece quando l'aria va bene sono felice e corro. Quando è calda sono triste e cammino; quando c'è molto caldo l'aria puzza	An: the air you smell with the bones, with mouth, with the nose. The skeleton tightens, I move instead when the air is good and I'm happy and I run. When it's hot and I'm sad I walk; when it's very hot the air stinks
Insieme: puzza del passerotto, del grillo, delle ascelle, del collo, dei piedi, anche la pelle L'odore viene anche dalle ossa, sono in tutto il corpo, il profumo può uscire anche dalla bocca come quando mangiamo l'aglio. L'odore è invisibile	Together: smell of sparrow, cricket, armpit, neck, feet, even the smell of the skin comes from the bones, they are all over the body, and the scent can also exit from the mouth like when we eat garlic. The smell is invisible
A: l'odore si può sentire e non vedere	A: the smell you feel and don't see

Conversation to identify the symbologies for Temperature (5 october 2012) 4-years-old

Italian Version	English Version
Educatrice: come vi sembra la temperatura della sezione?	Educator: what do you think about the temperature of the classroom?

(continued)

(continued)

Italian Version	English Version
M: è calda e fredda perché c'è solo un termosifone. Sento un po' di freddo perché sono senza giacca e il mio papà mi ha messo solo una maglia	M: It is hot and cold because there is only one radiator. I feel a little bit cold because I'm without a jacket and my dad put on just one jersey
A: è calda per sentire la temperatura, si sente anche il fiato (alita in maniera tale da far uscire la nuvoletta di fiato) se c'è freddo vedi il fiato	A: It is hot to feel the temperature, you also feel your breath (breathes in such a way to let out the puff of breath) if it's cold you see your breath
(i bambini danno sempre del tu ed usano espressioni come se stessero insegnando agli altri)	(Children always speak informally and use expressions like they are teaching others)
Al: per me c'è freddo, ho bisogno di un'altra felpa =	Al: for me it's cold, I need another sweatshirt =
O: io sento caldo e freddo	O: I feel hot and cold
T: un po' caldo e un po' freddo, perché il caldo e il freddo si possono mescolare, quando c'è molto caldo si sta in maniche corte, quando c'è molto freddo ci si toglie la felpa	T: a bit hot and a bit cold, because the heat and the cold you can mix, when it's really hot you are in short sleeves, when it's very cold you take off your sweatshirt
Educatrice: la temperatura è sempre uguale in aula?	Educator: the temperature is always the same in the classroom?
Al: Può fare freddo, caldo caldissimo, freddissimo, freddo medino, in bagno c'è una temperatura diversa perché in ogni posto c'è una temperatura diversa. Ad esempio qui c'è freddo medio, di là (atelier) c'è freddissimo, in bagno c'è caldino medio, in cortile caldo caldo, se il sole batte, se c'è una brutta giornata c'è freddo sempre	Al: It can get cold, warm, very hot, very cold, a bit cold, in the bathroom there is a different temperature because in every place there is a different temperature. For example, here's a bit cold, over there (studio) it's cold, in the bathroom it's a bit hot, in the courtyard hot hot, if the sun shines, if it's a bad day it's always cold
T: in cucina quando sono andato a salutare la Lucia (la cuoca) c'era molto caldo, caldissimo, poi quando sono tornato in aula c'era un po' caldo e un po' freddo, il freddo si può mescolare con il caldo, quando si mescolano diventa freddino/caldo	T: in the kitchen when I went to say goodbye to Lucy (the cook) it was very hot, very very hot, when I came back in the classroom it was a bit hot and a bit cold, the cold mixes with the heat, when you mix it, it becomes chilly/hot
(è interessante notare come sia chiaro, anche se a livello intuitivo, il concetto di miscelazione delle correnti d'aria a temperatura diversa, come la sensazione dipenda dal passaggio da un luogo ad una temperatura ad un altro a temperatura diversa e come, di conseguenza, il termine per definire la sensazione sia una mescolanza delle diverse sensazioni)	*(It is interesting to note that it is clear, although at an intuitive level, the concept of mixing of air currents at different temperatures, as the sensation depends on the passage from one place to a temperature to another at a different temperature, and consequently, the term for defining the feeling is a mixture of different sensations)*
M: vicino a termosifone c'è più caldo, vicino al muro c'è più freddo, vicino al veto c'è molto più freddo, perché il vetro è sempre freddo!	M: near a radiator is warmer, near the wall is colder, near the window is much colder, because the glass is always cold!

(continued)

(continued)

Italian Version	English Version
Educatrice: avete sempre caldo allo stesso modo durante la mattina?	Educator: are you always warm in the same way in the morning?
M: quando giochiamo tanto abbiamo molto caldo, quando siamo sulla scala [assemblea] caldo medio	M: When we play a lot, we are very hot, when we are on the stairs [assembly] medium hot
E: quando ci alziamo dal letto [dopo il riposo pomeridiano] abbiamo caldo perché siamo sotto le coperte, poi ti viene un po' freddino	E: when we get up from the bed [after the afternoon rest] we're warm because we're under the covers, then you are a bit chilly
Educatrice: come vi sentite in questo momento?	Educator: how do you feel right now?
Al: io sto bene perché c'è freddini caldino, quando c'è freddo freddo non mi piace	Al: I'm fine because it's chilly hot, when it's cold cold I don't like it
T: per me oggi c'è un po' freddino, quando c'è caldo caldo caldo io non sto bene perché sudo e non posso correre	T: For me today is a bit chilly, when it's hot hot hot I'm not good because I sweat and I can't run
M: io sento odore di giochi, quando si entra c'è sempre un buon profumo, cambiano perché cambiano le cose, in cucina c'è un buon odore, in bagno si sente un odore cattivo di più e di cacca, cattivo, molto cattivo, cattivissimo!	M: I smell the toys, when you enter there is always a good smell, they change because things change, there is a good smell in the kitchen, and in the bathroom it smells bad, mostly of poop, bad, very bad, very very bad!
K: io sento profumo di giochi, in sezione gli odori sono buoni, buoni, buonissimi!	K: I smell the toys, in the classroom it smells good, good, very good!

Conversation to identify the symbologies for Indoor Air Quality (5 October 2012) 4-years-old

Italian Version	English Version
Educatrice: Come vi sembra la temperatura dell'aula? Cosa vuol dire per voi temperatura?	Educator: What do you think of the temperature of the classroom? What does it mean for you to temperature?
A: la temperatura è quella che dice agli uomini se c'è caldo o freddo, quando c'è caldo il termometro va sul rosso	A: The temperature is the one that says to the men if it is hot or cold, when it's hot the thermometer goes on red
M: quando c'è il sole c'è tanto caldo, in inverno c'è freddo perché è inverno non c'è il sole, per me c'è freddo	M: When it's sunny it's very hot, in the winter it's cold because there isn't any sun, for me it's cold
A: io sto benissimo perché c'è un po' di caldo e un po' di freddo	A: I am very well because it's a bit hot and a bit cold
M: per me è come essere nel latte tiepido	M: For me it's like being in warm milk
D: io ho caldo, se mi tolgo la felpa forse sto meglio	Q: I'm hot, if I take off my sweatshirt maybe I'm better
L: io mi sento tanto freddo	L: I feel so cold
M: per me oggi c'è tiepido	M: For me today is warm

(continued)

(continued)

Italian Version	English Version
Educatrice: Secondo voi la temperatura della sezione è sempre uguale?	Educator: Do you think the temperature of the classroom is always the same?
M: la temperatura cambia, a volte senti freddo a volte senti caldo	M: the temperature changes, sometimes you feel cold and at times you feel warm
A: al mattino c'è caldo, in palestra ti viene ancora più caldo perché corri	A: this morning it's warm, the gym will be even hotter because we run
(si sottolinea quanto siano esatte le relazioni tra temperatura e sensazione, e tra ambiente, attività e sensazione)	(emphasizes how exact are the relationships between temperature and sensation, and between environment, activities and feeling)
Al: quando andiamo a mangiare c'è più freddo, perché là c'è molto più freddo, in piazza (l'ambiente all'interno dell'edificio) c'è molto freddo, freddissimo, in aula c'è freddo e basta	Al: When we go to eat it's colder, because there it's much colder, in the square (the environment inside the building) it's is very cold, freezing cold, in the classroom it's just cold
An: io il freddo lo sento nella pancia	An: I can feel the cold in my belly
M: io li annuso il caldo e il freddo, li sento dentro al naso	M: I smell the hot and the cold, I feel them inside my nose
An: non si può annusare il caldo e il freddo, si annusano gli odori	An: you cannot smell the heat and cold, you sniff odours
D: adesso che non ho la felpa ho freddo, me la metto	Q: Now I don't have a sweatshirt and I'm cold, I'll put it on
An: in aula c'è caldissimo, nella stanza segreta (la parte a L dell'aula) c'è mezzo caldo e in atelier c'è freddo, freddissimo, se spostiamo il caldo della sezione con un retino invisibile nell'atelier viene caldo. L'atelier è un po' di ghiaccio. In aula è tutto caldo, se apriamo la porta dell'atelier, l'atelier si scada	An: in the classroom it's very hot, in the secret room (the L part of the classroom) it's half hot and in the studio its cold, very cold, if we move the heat of the classroom with an invisible net, the studio will become hot. The studio is a bit of ice. In the classroom it's all hot, if we open the door of the studio, the studio will expire
D: in sezione c'è caldo, in atelier sembra di essere al Polo Nord, è un po' come l'Era Glaciale (il cartone animato)	D: In the classroom it's hot, the studio seems to be at the North Pole, it's a bit like Ice Age (the cartoon)
Educatrice: Come è l'aria in sezione, che odore sentite?	Educator: How is the air in here, what do you smell?
M: c'è odore di pennarelli, a volte c'è profumo di fiori, l'odore cambia con le cose che cambiano	M: There is the smell of markers, sometimes there is the scent of flowers, the smell changes with the things that change
An: ognuno annusa profumi diversi, in sezione io sento profumo di nuvole, di foglie, c' anche profumo di M. A. e D. (gli altri studenti)	An: each perfume smells different; in the classroom I smell the scent of clouds, of leaves, and also the scent of M. A and D. (other students)
D: c'è odore di pennelli e di colore	Q: There is the smell of brushes and colours
Educatrice: e come sono questi odori?	Educator: and how are these odours?
D: sono tutti odore di buono	Q: they all smell good

(continued)

(continued)

Italian Version	English Version
A: in bagno c'è un odore cattivo di cacca e di pipì, se vai nella piazza vicino alla cucina senti degli odori buoni, anzi buonissimi	A: In the bathroom there is a bad smell of poop and pee, if you go in the square near the kitchen it smells good, very good indeed

Comments to the symbols proposed by researchers

The first day the researchers offered to children drawings and symbols to represent the different responses of feeling. The children did not understand the meaning of the drawings. Below are the drawings and the children's responses.

Comments to the feeling of Thermal Comfort (Fig. 9.21).

Italian Version	English Version
C: l'uomo è diverso, gli viene la febbre	C: the man is different, he has a fever
L: il sesto e il quinto non lo so	L: the sixth and fifth I don't know
E: queste persone stanno in piedi, dormendo, gli altri stanno guardando	E: these people are standing, sleeping, the others are watching
F: un signore che ha freddo, uno freddo, no ha caldo, così così, caldino, caldino che sta sudando	F: a man who is cold, a cold one, isn't hot, so-so, a bit hot, hot and sweating

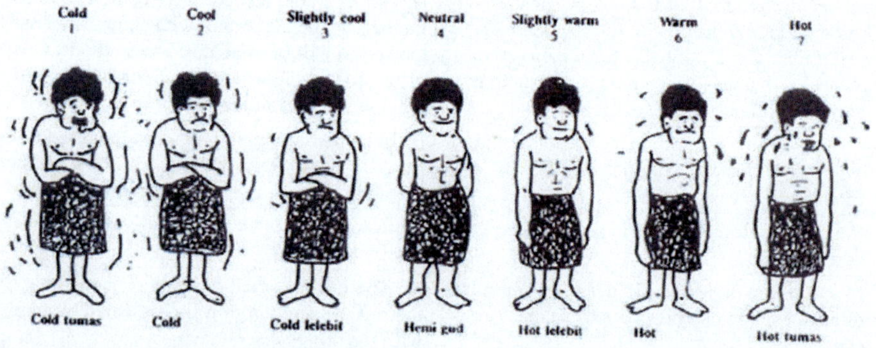

Fig. 9.21 Souls: man and thermal comfort

Comments with respect to the Thermal Comfort Desired (Fig. 9.22)

Italian Version	English Version
C: niente! Ha la varicella, sta bene, sta bene, sta bene,	C: nothing! He has chickenpox, he's fine, he's fine, he's fine,
L: in estate, in autunno, in primavera, in inverno	L: in summer, autumn, spring, winter
E: stanno guardando la neve. Uno è suo padre e poi i figli e gli zii	And: they are watching the snow. One is his father, then the sons and uncles
F: qua è vestito da estate, da ufficio, da uomo, da pioggia, da inverno, da esquimese	F: here is summer dress, office, man, rain, winter, Eskimo

Fig. 9.22 Soul man and clothes

Fig. 9.23 Anime: bad odour, smell

Comments regarding Indoor Air Quality (odour) (Fig. 9.23)

Italian Version	English Version
C: sono i formaggi che il signore vuole mangiare tutti	C: these are the cheeses and the gentleman wants to eat them all
L: forse sono delle casette piccola, media, grande	L: maybe they are small houses, medium, large
E: tanti formaggi, sono tutti uguali, uno è piccolo, medio grande. Le linee dicono che è caduto	E: many cheeses, they are all the same, one is small, medium, large. The lines say they fell
F: vuol dire che qua c'è il naso che sta sentendo un formaggio che non puzza, n pochino, un po' di più, un po' di più	F: to say that here there is the nose that is smelling a cheese that does not smell, a little, a bit more, a little bit more

Comments with respect to the question how do you feel now? (Fig. 9.24).

Italian Version	English Version
C: bisogna ritagliare	C: you must cut this
L: sbadiglia, sembra Dotto	L: yawning, looks like Dotto
E: questo è il nano che vuole guardare suo figlio che vuole dormire, aveva visto un bambino che stava girando per catturarlo	E: this is the dwarf who wants to watch his son who wants to sleep, he saw a child who was trying to capture him
F: dormire, stare alzati, uscire dalla sabbia	F: sleep, stay up, get out of the sand

Fig. 9.24 Soul: tired, normal, shaken/agitated

Annex 2: Prepared Drawings

We report other drawings made by children during discussions

05/10/2012 19:41

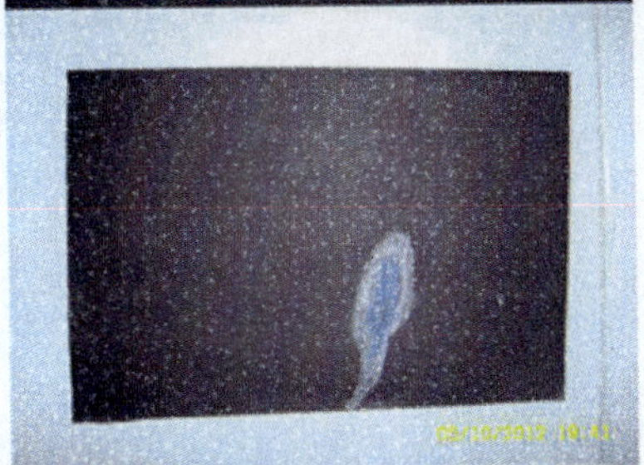

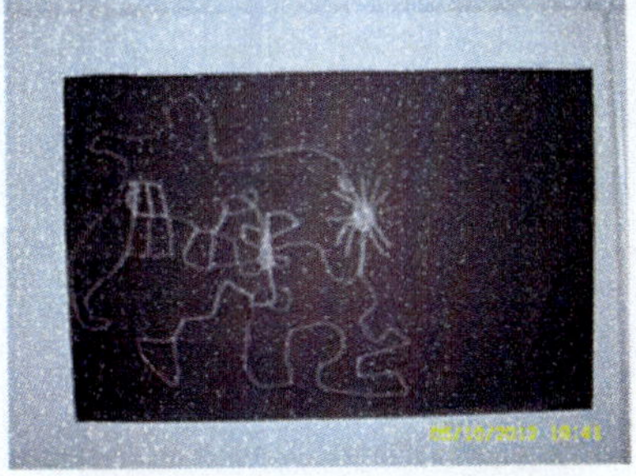

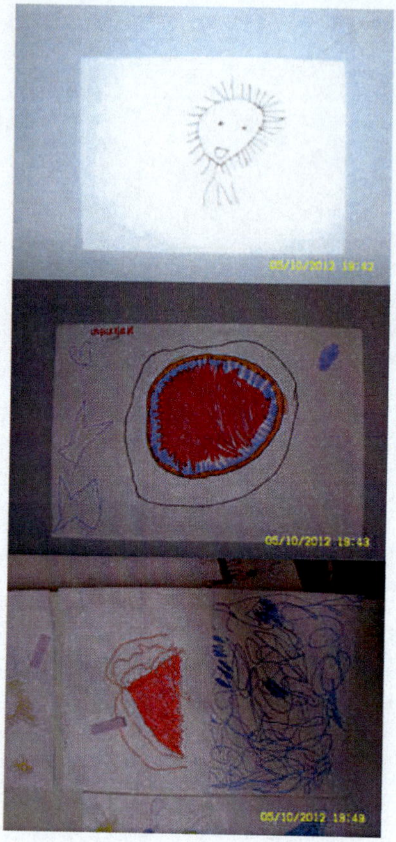

Annex 3: Questionnaire Results

See Tables 9.7, 9.8, 9.9 and 9.10.

Table 9.7 Section 4 years-old preschool child. Thermal comfort perceived Questionnaire subjective judgment results

Hour 9:30	Very cold	Cold	Cool	Neutral	Warm	Hot	Very hot	No reply	TOT	PMV	PPD (%)
Child scale		Freezing cold	Cold	Lukewarm	Warm	Beastly hot	Beastly mega-overwhelm hot				
$PMV_{c,sj}$ subjective judgment	−3	−2	−1	0	1	2	3				
	0	2	1	5	10	2	2		22		
Results										0.68	14.77
Hour 11:00	Very cold	Cold	Cool	Neutral	Warm	Hot	Very hot	No reply	TOT	PMV	PPD (%)
$PMV_{c,sj}$ subjective judgment	−3	−2	−1	0	1	2	3				
	0	2	3	2	7	2	5		23		
Results										0.57	11.69
Hour 15:05	Very cold	Cold	Cool	Neutral	Warm	Hot	Very hot	No reply	TOT	PMV	PPD (%)
$PMV_{c,sj}$ subjective judgment	−3	−2	−1	0	1	2	3				
	0	2	3	1	11	1	3		21		
Results										0.71	15.73

Table 9.8 Section 4 years-old preschool child. Thermal comfort preferred

		Much cooler	Cooler	Neither warmer ner coolre	A little warmer	Much warmer	No reply	Total	PMV
		−2	−1	0	1	2			
Hour 9:30	$PMV_{c,p}$ preferred (children)	2	5	8	6	0		21	
	Results								−0.14
Hour 11:00	$PMV_{c,p}$ preferred (children)	0	3	12	7	0		22	
	Results								0.18
Hour 15:05	$PMV_{c,p}$ preferred (children)	0	4	11	6	0		21	
	Results								0.10

Table 9.9 Section 5 years-old preschool child. Thermal comfort perceived Questionnaire subjective judgment results

Hour 9:30	Very cold	Cold	Cool	Neutral	Warm	Hot	Very hot	No reply	TOT	PMV	PPD (%)
Child scale		Freezing cold	Cold	Lukewarm	Warm	Beastly hot	Beastly mega-overwhelm hot				
$PMV_{c,sj}$ subjective judgment	−3	−2	−1	0	1	2	3				
	0	2	4	8	1	1	6		22		
Results										0.59	12.32%
Hour 11:50	Very cold	Cold	Cool	Neutral	Warm	Hot	Very hot	No reply	TOT		
$PMV_{c,sj}$ subjective judgment	−3	−2	−1	0	1	2	3				
	0	4	5	6	1	1	6		23		
Results										0.35	7.52%
Hour 15:05	Very cold	Cold	Cool	Neutral	Warm	Hot	Very hot	No reply	TOT		
$PMV_{c,sj}$ subjective judgment	−3	−2	−1	0	1	2	3				
	0	2	5	3	1	1	6		18		
Results										0.67	14.34%

Table 9.10 Sect. 9.5 years-old preschool child. Thermal comfort preferred

		Much cooler	Cooler	Neither warmer ner coolre	A little warmer	Much warmer	No reply	Total	PMV
		−2	−1	0	1	2			
Hour 9:30	$PMV_{c,p}$ preferred (children)	2	6	7	5	2		22	
	Results								−0.05
Hour 11:50	$PMV_{c,p}$ preferred (children)	2	2	12	6	0		22	
	Results								0.00
Hour 15:05	$PMV_{c,p}$ preferred (children)	2	4	6	6	0		21	
	Results								0.11

References

ASHRAE Standard 62.1 Ventilation for Acceptable Indoor Air Quality

Bekö G, Toftum J, Clausen G (2011) Modelling ventilation rates in bedrooms based on building characteristics and occupant behavior. Build Environ 46(11):2230–2237

Buratti C, Paola Ricciardi P (2009) Adaptive analysis of thermal comfort in university classrooms: correlation between experimental data and mathematical models. Build Environ 44:674–687

Copple C (2003) Fostering young children's representation, planning, and reflection: a focus in three current early childhood models. J Appl Dev Psychol 23(6):763–771

d'Ambrosio Alfano FR, Ianniello E, Palella BI (2013) PMV–PPD and acceptability in naturally ventilated schools. Build Environ 67:129–137

De Giuli V, Da Pos O, De Carli M (2012) Indoor environmental quality and pupil perception in Italian primary schools. Build Environ 56:335–345

EN 15251:2008. Criteri per la progettazione dell'ambiente interno e per la valutazione della prestazione energetica degli edifici, in relazione alla qualità dell'aria interna, all'ambiente termico, all'illuminazione e all'acustica

Gładyszewska-Fiedoruk K (2013) Correlations of air humidity and carbon dioxide concentration in the kindergarten. Energy Build 62:45–50

ISO EN 7730 Ergonomics of the thermal environment—analytical determination and interpretation of thermal comfort using calculation of the PMV and PPD indices and local thermal comfort criteria (2005)

ISO EN 8996 Ergonomics of the thermal environment. Determination of metabolic rate

ISO EN 9920 Ergonomics of the thermal environment—estimation of thermal insulation and water vapour resistance of a clothing ensemble

MacDonald M (2007) Toward formative assessment: the use of pedagogical documentation in early elementary classrooms. Early Childhood Res Quartely 22(2):232–242

Mors S, Hensen JLM, Loomans MGLC, Boerstra AC (2011) Adaptive thermal comfort in primary school classrooms: creating and validating PMV-based comfort charts. Build Environ 46:2545–2461

Santamouris M, Synnefa A, Asssimakopoulos M, Livada I, Pavlou K, Papaglastra M, Gaitani N, Kolokotsa D, Assimakopoulos V (2008) Experimental investigation of the air flow and indoor carbon dioxide concentration in classrooms with intermittent natural ventilation. Energy Build 40:1833–1843

WHO (2000) Addressing the links between indoor air pollution, household energy and human health based on the WHO-USAID; Global Consultation on the Health Impact of Indoor Air Pollution and Household Energy in Developing Countries (Meeting report); Washington, DC 3–4 May 2000

Yu Z, Fung BMC, Haghighat F, Yoshino H, Morofsky E (2011) A systematic procedure to study the influence of occupant behaviour on building energy consumption. Energy Build 43(6):1409–1417

Chapter 10
Elementary School Case Study

Abstract In this chapter we report a case study of the Elementary School Centro Educativo Italo Svizzero (CEIS) of Rimini. The research concerns the evaluation of Thermal Comfort through questionnaires administered to classes of children between 7 and 10 years and of between 4 and 5 years of age. The questionnaires were carried out with the children and the teachers, according to the CEIS pedagogical model, recalling the experience of Piaget, reviewed by Margerita Zoebli. The questionnaires used images and references related to ideas from school programs.

10.1 The Case Study of the Elementary School Centro Educativo Italio Svizzero C.E.I.S.

This case study concerns the testing held at the Centro Educativo Italio Svizzero C.E.I.S. (Italio-Swiss Educational Centre), located in Rimini, in north central Italy. The CEIS is a childhood village in the historic centre of the city, built on the ruins of World War II, thanks to the reuse of the of military barracks to house the displaced, and consists of a complex of wooden buildings (called "shacks") positioned planimetrically so as to create a small village with a central square. The individual buildings are independent from the functional point of view, and host one or more classes, in relation to the size, and are equipped with an independent heating system.

The building complex is located on the ancient grounds of the Roman amphitheater, and borders the area where there are parts of the amphitheater, and the layout of the buildings is modelled on the shape of a hemicycle.

The complex was born on May 1, 1946 as a result of international solidarity thanks to the intervention of the Swiss rescue workers, who, under the leadership of Margaret Zoebeli, provided the city a centre for the care of social welfare and training of boys.

This information is useful for research because, given the particular history of the buildings, their architectural shape (lightweight wood consisting of a single plane), has effects on the pedagogical model used in the school, on the basis of imprinting

K. Fabbri, *Thermal Comfort Perception*, https://doi.org/10.1007/978-3-031-52610-7_10

given by Margherita Zoebeli, and consequently on the mode of administration of the questionnaires by the teachers, and the perception of space by the children.

The architectural shape of the individual building, on one floor, the relationships between the other buildings, and the particularities given to a small Village in an urban context of denser citizens and larger buildings, units a pedagogical method adopted, determines an "affection", a "belonging" to the places by the children who attended to, specifically, unlike the attendance of more anonymous school complexes.

Rimini is a town located in the north-east of Italy (geographical coordinates 44°03′00″ N–12°34′00″ E), with an average size and with a standard of living of middle high income. Built as a tourist hub, for what concerns mass tourism and bathing, and one of the most important in Europe, so much so that the city's population can increase by 5 times the resident population. The tourist-accommodation destination characterizes the economy and the urban layout of the city.

The city can be described as characterized by two centres: a historical centre in the old town, where normal activities in the city are carried out during the whole year, and where you find CEIS; the other "*marina*" centre or the centre of the promenade where majority of the tourist activities flow on the coast, consisting predominantly of accommodation facilities and hotels. The city has a phenomenon of petty crime, due to immigration and tourism, nevertheless such activities do not affect the social dynamics of towns, where you see a good rate of welfare. It can be said that there are no phenomena of degradation or unemployment likely to characterize the population.

The CEIS is one of the private school complexes opposing public schools, where several generations of native Rimini citizens have been were formed.

10.1.1 The Building

The design model of CEIS can be traced back to lucky spontaneity, as regards the relationship between architecture and pedagogy, where since the relationship is between inside and outside, the child is not an object but the subject of education.

The original plan of the settlement, of 1946, is the architects Felix Schawrz and Margaret Zoebeli, (Director of the centre and active researcher in pedagogy), who organized the kindergarten with an open articulated space, with areas for different activities, thus widening the scope of experiences and children.

The original design (Fig. 10.1) included 13 wooden barracks, assembled on site, with structured frame, connected with joints "swallow nest" so as to reduce the number of nails and screws. The cabins are positioned on a stand, and are spaced by an interspace of approximately 50 cubic meters from the ground by means of pillars and beams at the base. The vertical walls are made of prefabricated panels in wood, in modules of 90 cm, and a thickness of 10 cm. In the same module of 90 cm are inserted wood frames with single panel glass. The wood used is larch, which remains visible both inside and outside. The roof consists of trusses, with planks of wood and roofing tiles.

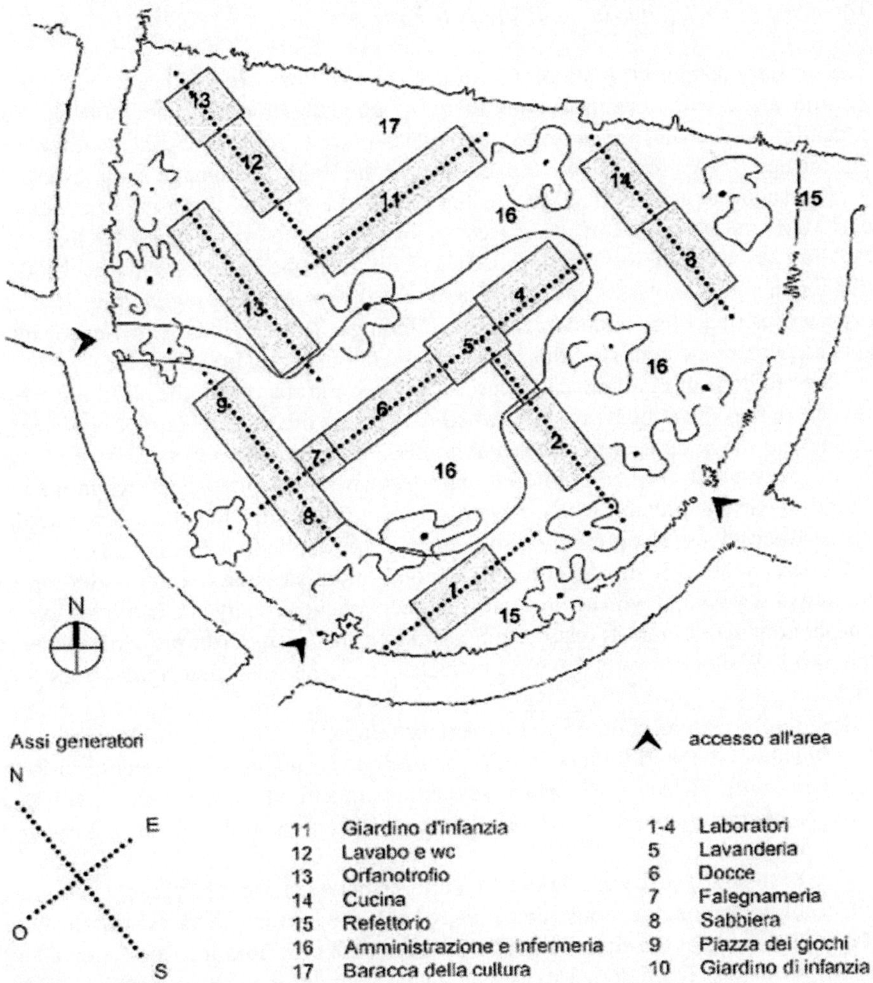

Fig. 10.1 Plan of the school village CEIS

10.1.2 The Children Age 6–10 (Elementary School)

The CEIS is a private secular school, and calls for the payment of registration fees, and is home to the different school levels provided in Italy: Crèche, Kindergarten and Primary School, to which are added a laboratory and after-school care for the treatment of Dyslexia, and a semi-residential community for welcoming young people and the training of teachers, in addition to the M. Zoebeli Foundation and press office.

The research took place in the Primary School.

The primary school consists of 10 classes (two full courses) with a maximum of 22 students, 2 with disabilities, and there are three classroom teachers, one special education teacher and one or two support educators in the case of children with severe disabilities. Added to these are the teachers of the English language and music.

The social dynamics that are created within the groups of children between 6 and 10 years are aimed at the discovery of socializing with others. At this age, children leave the egocentric approach and discover the relations with each other (the companions of classes and adults) and with the world, thanks to their intellectual experiences (teaching) and their empirical findings. Through these experiences they also acquire an awareness of their own judgment (and role) in the world.

The readings of children can be useful tools to interface with the children; these are part of the educational program, whether there are other readings. The books read in this age range depend on the reading skills of children themselves: in the early years of school (6–7 years) children learn to read books independently, unlike the period of early childhood where the stories were told orally, the chance to read the stories themselves (and possibly write) changes the action in learning and curiosity.

The years after 8–10, the concepts become more structured and in addition to reading and grammar, you acquire other concepts in geography, mathematics, physical phenomena, colours, foods, clothing, children call and describe properties objects and facts. Words not only have an evocative role, but also descriptive, they say something.

Reading books of children in elementary tell stories of other children who are learning about the world, Pinocchio, Treasure Island, and books on specific curiosities: Dinosaurs, superheroes, princesses, monsters and other characters or animals strongly characterized, with strong characters and a central role in history and imagination.

One of the most successful books for children between 6 and 10 years is the series of Geronimo Stilton, the mouse detective, who discovers the mysteries with the help of the inhabitants of the city of Topazia. In these books the emotions are followed by a page, some words are drawn so as to coincide with writing the meaning of objects or emotions, so as to learn new things in a funny way.

Kids want to have fun and excitement mainly of two "strong" emotions: fear (monsters) and the laughter, the pure free fun and stupidity of play. In addition, children are learning that words have fun with puns and rhymes, and combine words that interest them for the evocative sound or effects they create. In this sense, in the questionnaires in the description of the sensations "extreme" words are terms invented, combined to maintain their evocative power, even more than descriptive ones.

In the questionnaire the researchers constitute a new element, so the teachers must submit the researchers and the activities of the children, playing an active role, they can be directly managing the entire procedure of administration, provided that this does not lead to an active role, or modification of the responses or the mode of administration of the questionnaire, if not agreed with the researcher. The teachers distribute and collect the questionnaires. It should be noted that, in the Italian primary school, the children remain in the same classroom all day and have the same 2/3 teachers for the whole year and the path of the 5-year study. This involves an "affective-attachment" as much to the teachers as to places, or classroom that becomes their own second "social" bedroom for most of the day and year; that's why they decorate the walls that become also an educational support over the blackboard.

In the specific case of the CEIS, this case study, the classroom is even more marked given that coincides with the cabin and with the report that the cabin has with the surrounding public space, since it is part of the small village of CEIS. This appropriation of the places is strongly felt by the children and adults, on the walls are also placed calendars with daily activities including weekly household tasks.

10.1.3 The Pedagogical Model of C.E.I.S.

The C.E.I.S. adopts an innovative pedagogical model, derived from the will of the founder M. Zoebeli, which was, by waves today, the subject of ongoing research and application.

The school refers to the values, methods, practices and teaching of *active education*, and educational cooperation, scientific psychology, following approaches, respectively, of John Dewey, Jane Piaget, Lev S. Vygotsky and Don L. Mulani, and Raffaele La Porta and Mario Lodi.

Inside the building there is the Foundation "Margherita Zoebeli" that "promotes studies and research of innovative experiences in the pedagogical-didactic area, especially to developing intercultural education, and promotes knowledge of teaching methods and materials for the education of psycho-physical handicapped children" (quote from the site of the CEIS). Multiculturalism and integration through comparison, are at the heart of the pedagogical model of the CEIS.

The school puts the child at the centre so that he/she can participate in and contribute better and with pleasure to the realization of the social experience of learning in the school. Education is aimed at building individual and democratic citizens, the organization reflects the settlement pattern of the village. The educational program reads: *"Objectives of citizenship education are essential building the sense of law and the development of an ethic of responsibility, which are realized in the duty to choose and act consciously and involving the commitment to develop ideas and to promote actions aimed at continuous improvement of their living environment."*

The learning environment (methodology) plans to:

- Enhance the experience and knowledge of the students,
- Implement appropriate actions in regards to diversity, to ensure they do not become inequalities,
- Encourage the exploration and discovery,
- Encourage collaborative learning, since learning is not just an individual process,
- Promote awareness of their way of learning, or "*learning to learn*".

The principles are called the democratic principle, that of diversity, individualization, sociability, error, and the principle of "*open horizons*", or open to the world and to diversity.

For the purpose of research on the perception of comfort by children it is good to know the pedagogical model and the curriculum adopted by the school, as the basis of this model must be based on the testing activities, from those who have to administer the questionnaire, in this case the teachers.

The research methodology involves the administration of the questionnaire directly to the children, and it is crucial that the child is able to understand the question and the answers of the questionnaire and responds according to their personal feeling. The fact that the pedagogical method of CEIS provides for the "*educational cooperation*" has placed doubts on the possibility that the feeling of comfort is attributed to the whole class and not by the individual.

In this regard, in defining the search path and the route of administration of the questionnaire, it was agreed, with the head teacher Giovanni Sapucci, that there would be teachers to administer the questionnaire to the children.

The request of researchers has been to ensure that the response from the child was unique and not influenced by the opinions of other children.

In relation to its pedagogical method and the didactic teachers have decided to build the questionnaire, the graphic translation of the meaning of the questions with the children, collectively, in order to maintain the role of the community in the democratic choices, and then, once they have drafted a common questionnaire, giving it to individual children to have their individual opinion.

The construction activities of the questionnaire were carried out during the normal teaching and programming to complement the concepts studied with the questions of the questionnaire.

10.2 The Research Activity

The methodology adopted for the evaluation of comfort involves the administration of questionnaires to children aged between 8 and 10 years.

From discussions with the Director in charge and some teachers, in relation to the teaching program, it was decided to submit the questionnaire to children in grades third and fifth, so as to have a variety of samples. The first and second classes were excluded from the trial, because in the opinion of the teachers, they would have involved (in their opinion) difficulty in understanding the concepts of the questionnaires consequently leading to unreliable values.

In this regard it should be noted that trainers, teachers, educators, school principals, seeking to adapt the content of research in relation to their own internal training organization. The limit of the survey (all students or only a few) have to be defined and agreed, as far as possible with respect to the organization and to the economy and education, always in relation to the limits and objectives of the research, trying to avoid forcing the point of view of educators and teachers, as this could result in a "bad" administration of the questionnaire.

The scope of investigation of this research was to determine whether the level of comfort of the "shacks" (cabins) of the CEIS centre, measured and perceived, was comparable with that of other school buildings.

For this reason, the limits of the object of study were data from performing as many questionnaires as possible within multiple buildings. The results show that, even though they are made of wood with light structures, the perception of the comfort of the children does not have high levels of dissatisfaction. If the objective of the survey was also educational, and not architectural, or if you wanted to check, for example, the effectiveness of the teaching method with respect to the notions of comfort, the search would have to be extended to all classes.

The contents of the questionnaire were defined according to ISO 10551 and in agreement with the educators, as well as with teachers it was agreed:

- The building and classrooms;
- The duration of the questionnaires, choosing only the morning, because according to the educators, children are more attentive,
- The mode of administration, deciding the two days of the first week of December 2012.

Table 10.1 show contents of questionnaire.

Table 10.1 Table of Contents Questionnaire

1 How do you perceive the environment at this time ?						
[_] Very cold (−3)	[_] Cold (−2)	[_] A slightly cold (−1)	[_] neutral (0)	[_] A slightly hot (+1)	[_] Hot (+2)	[_] Very hot (+3)
2. How would you like the environment to be at this time ?						
[_] hotter	[_] as it is	[_] colder				
3. Do you feel ok in this environment ?						
[_] Yes	[_] No					
4. How do you perceive the quality of the air at this time ?						
[_] tolerable (no odours)	[_] with a slight bad odour	[_] with a bad odour	[_] with a strong bad odour	[_] intolerable		
5. How do you feel at this time ?						
[_] tired	[_] normal	[_] awake				

10.3 Questionnaire Submission (Sequence of the Questionnaire—The Day)

Classes third and fifth classes are divided into two subgroups that are both in the same "cabin", but in two different classrooms.

The fifth classes in the cabin are "Friendship" divided in classrooms "Mimosa" and "Jasmine". The questionnaire was submitted prior to the group of children in a classroom and then the other. Similarly the third class in the cabin "Fountain", is divided into two classrooms "Mantis" and "Paganelli". The subdivision of the sections in groups is part of the educational method, and during the week the classrooms were merged to perform common tasks.

We show the detailed narration of processing and administration of the questionnaires for each classroom in order to show the different approach adopted.

The oral approach (explanation) and the correlation with the images—Fifth Classes

The first day of detection was held with *Fifth Class Mimosa Hall*, December 3, 2012 at 8:30 am in the presence of three teachers and 19 children. At the entrance the children were gathered and arranged in a circle to explain the presence of the researchers within the classroom and activities. The teacher explained, orally, in a clear way, the questions and answers.

In conjunction with the administration of the questionnaire instrumental monitoring was carried out, by the use of probes for monitoring the indoor microclimate.

The children, whenever they did not understand the terms and concepts, showed their hands (democracy and discipline) and asked questions together (cooperation) which were clarified. Then the questionnaires were distributed.

To the question "*How do you perceive the environment at this time*?" The children have understood the meaning, but the teacher said the word "*perceive*" in the question, and so the question referred to a subjective evaluation and the individual child. There was a debate among the children, about the meaning of the word "*neutral*" which was interpreted as an environment neither hot nor cold.

The children have understood the meaning of the numbers (+3, +2, etc.) and the relationship between positive numbers and negative numbers from heat to cold.

In order to translate into a unique language to all of the questions and the meaning of the scores, they were associated, *by the children*, to nations, so

- (+3) Africa
- (+2) Egypt
- (+1) Spain
- (−1) England
- (−2) Russia
- (−3) North Pole

it should be noted that the association with the nations was adopted in a democratic and collaborative approach, not so much (or not only) for understanding the meaning of the question, but especially in order to identify a common set of values, according

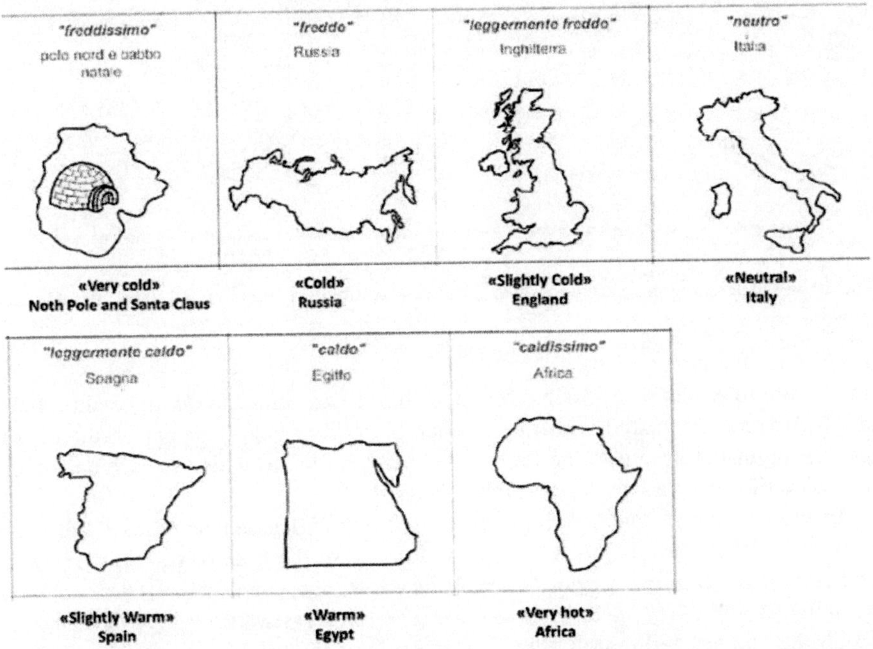

Fig. 10.2 Questionnaires of the classroom fifth Mimosa—attribution judgments to a nation

to a participatory criterion. In this way it is clear for all, the differences between the cold that you can feel in Russia compared to that you may feel at the North Pole (Fig. 10.2).

To the question "*Are you okay in this environment*?" The teacher wanted to clarify that "*feeling good*" should not be confused with the social aspect, the feeling good with his classmates, the emotions, comfort etc., but only to the environment, to air and temperature.

To the question "*how you perceive the quality of the air at this time*?" The doubt was the meaning of the word "quality" a term that describes the characteristics of objects and at the same time it gives it a value. The responses of the children referred to different attributes related to air: the quality is "*if the air moves too fast*", "*if it is hot or cold*" "*if it is strong or slow from the windows*", "*if there is pollution*", "*if there is the smell of smoke coming from the outside, smog*", "*if you the air doesn't change and circulate*", "*if it smells*". The children have understood that the term quality was reported to not only smell but to ventilation or temperature, so they could attribute to the sense of smell as an example with attributes related to the level of hygiene.

The next question, "*How would you like the environment to be now*?" was clear to the children and did not require an explanation.

The tolerable answer "*without odour*" (0) is equivalent to a situation without odor, like odourless soap; "*with slight odour*" (+1) if you forget to wash your feet in the evening (allocation to something that you can feel with your senses, or rather to a

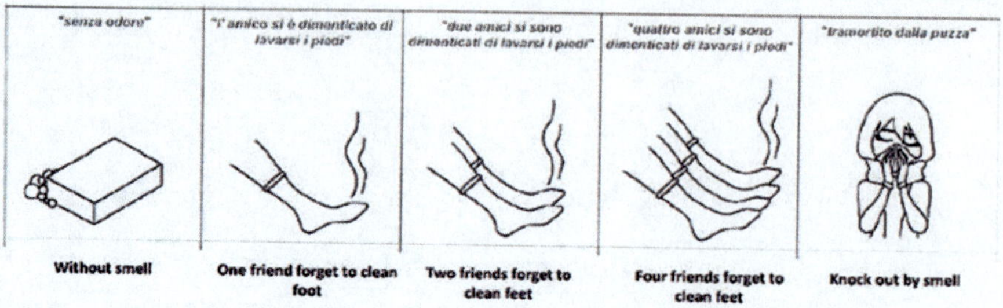

Fig. 10.3 Questionnaires classroom fifth Mimosa—attribution judgments Indoor Air Quality (smells)

transaction to which it is possible for each child to associate an odour); values (+2) and (+3) were associated with the number of children who had not washed their feet the night before, up to the level (+4) "*intolerable*" to which is assigned, even playfully, the term "a knock-out stench" (Fig. 10.3).

To the question "*How are you feeling right now*?" The teacher stressed the fact that being tired or awake must not be conditioned to the time of day, but "*it's that feeling that get in the very crowded environments (shopping centres) where you feel the sensation of a lack the air and you feel more tired (greater need for oxygen).*" It should be emphasized that, in this case, the role of the teacher, who plays the part of reasoning for the children, is active, or proposes an interpretation of the sentence, while it is the children who translate the interpretation in a common language.

In this question the teacher hypothesizes that children are able to distinguish and separate their opinion about the boredom or the pleasure of the school day, which affects mood, compared to a physiological phenomenon tied to breathing such as the lack of oxygen.

At 9:10, the combination of explanations and questions was completed and answered; the questionnaire was distributed to each child who answered individually. Questions concerning the age, weight and height were added so as to develop a "child standard" for the classroom in order to calculate the PMV measured by monitoring equipment for the site, and simultaneous administration of the questionnaire.

In the class there were two girls with disabilities serious, for which the teachers recommended not to submit the questionnaire, as they could give answers unreliable that would detract from the validity of the questionnaire. This was an error in the research.

In this regard, it is stressed that the judgment on the validity of the questionnaire *exclusively the responsibility of the researcher* and not the teachers, school administrators or other parties that supported and cooperated with the research, because, although they provide useful information for the specific context, they do not have knowledge in the aims of the research and the thinking in relation to measurement of comfort. In particular excluded children, such as handicapped, establishes their inability to feel about the environment, it is a value judgment that does not affect the search or data collected, and is likely to result in unjust discrimination towards these children. Where possible everyone needs to be involved in the testing, the feedback about the social aspect or errors and uncertainty of the research belong to the researcher, during the processing of results.

On the same morning starting at 10:30 the experiment was carried out in the *Fifth Class Jasmine Hall*, where there were three teachers and 18 children. Also in this case we proceeded to the explanation of the questionnaire by the teachers.

Regarding the first question, we focused on the meaning of "*perceive*", to which the children attributed the meaning: "*how do you see the environment*", "*climate*", "*understand and feel the environment*", "*perceive if it's cold or if it's hot,* " and "*air quality* ". The teacher has correctly pointed out that the term perceive refers to the fact that the evaluation is subjective, it is an individual valuation and therefore there are no wrong answers, but each answer is correct if it is perceived. In the administration of the questionnaires of the schools there is a risk that the child sees in the questionnaire a tool for self-assessment, as if it was a test at school, or school homework, and then by his answer may depend on an impact on the school, on building and the teacher. Children tend to be "self-centered" with respect to the effects of their actions. This aspect has to be clarified, several times.

In this case the children associated with each level of sensation the application of a type of clothing and the relative month, or the clothing to respond to the thermal sensation, so.

- (+3) Bathing costume, month of August (baking, sweaty)
- (+2) Vest, month of June (heat of early summer)
- (+1) T-Shirt and trousers, month of April (spring heat)
- (0) Neutral (neither too hot nor too cold)
- (−1) Trousers and long-sleeved top, month of September (a bit cold)
- (−2) Jacket and boots, month of November (cold)
- (−3) Heavy coat, scarf, gloves, freezing cold (snow and freezing cold)

it should be noted that the association of a season threatens to give the child the judgment of the season in which he/she carried out the questionnaire. In this case it was not so.

An interesting aspect to both questionnaires is that, when they can, the children prefer to attribute to "extreme" sensations (very hot/very cold) or extremely playful phrases and expressions, such as "*I'm baking from the heat*" (Fig. 10.4).

Again to the question "*do you feel well in this environment*?" the teacher found it useful to specify that it was not a judgment relating to social relationships with other children and/or with the teachers, but that if there is a good temperature or good air.

Asked about the perception of air quality, the children explained that according to their explanations if the air is "*clean or dirty*", polluted, in every case something you can "*perceive through smell*", a smell that can be "*bad, fragrant, good, pleasant or unpleasant*" (compared to younger children the smells are not exclusively associated to activities, but are designed with their own meaning and quality). For each degree of the responses of the rating scale children have associated a stage of occupation of the classroom.

Fig. 10.4 Questionnaires classroom fifth Jasmine—attribution judgments to a clothing

So the air is "*tolerable*" (0) in the early morning, when it's just been cleaned and there is no-one: while it is a "*Light bad smell*" (+1) in mid-morning, and it is a sustainable thing; values (+2) and (+3) correspond to the level of pollution of the classroom at lunch and in the afternoon. The assessment of the air quality is linear based on the increase of the permanence of the people in the environment, with the consequent increase of pollutants due to people, no account is taken of ventilation. The value (+4) intolerable is an unbearable smell, that of having to leave the environment, like when you come back from gym class (gymnastics) (Fig. 10.5).

Unlike the classroom to the question "*how do you feel right now*?" the case was reported to the feeling of physical exhaustion, so "*Tired*" is the end of the tasks, "*Normal*" at the beginning of the lessons and "*Awake*" at the time of recreation. The judgment is associated with the feeling of mood and level of boredom of children, which is correlated with the judgment on the environmental parameters.

Fig. 10.5 Questionnaires classroom fifth Jasmine—attribution judgments Indoor Air Quality (presence people and time)

When asked "*how would you like the environment at this time*?" The teacher made it clear that this is not desired (for example, "*I wish there was snow*", "*I would like to be at the seaside*") but the feeling inside the classroom, giving a reason (for example: "*I wish it was warmer because my hands are cold*").

Figures 10.6 and 10.10 show delivery of questionnaire in fifth classes.

Fig. 10.6 Delivery of the questionnaire fifth classes

10.4 The Chromatic Approach/Sensitive—Third Classes

The second day of the survey took place in the *Fountain Cabin of the Third Class, classroom Canocchie*, December 5, 2012 at 8:30 am in the presence of three teachers and 20 children. Earlier in the day the teachers sat children at their own benches, inviting them to "*listen to your body*". Then the questionnaires were distributed, and the teacher asked the children to read the questions and answers, and the children began to read and respond, without waiting for an explanation of the meaning of the questions. The questionnaire was interrupted and the teacher explained that they had to tick only one answer and did not have to look at the responses of the companions.

The teacher asked a child to read the first question aloud to the whole class, asking her what she did not understand the sentence. The girl replied that she did not understand the word "perceive" the teacher asked the other children what, according to them, that word meant. The responses were varied, but after reading the options of the answers of the question, the children understood that it was referring to how they about felt the temperature of the environment at that time.

Children associated each level of feeling a colour and a definition (Fig. 10.7), so

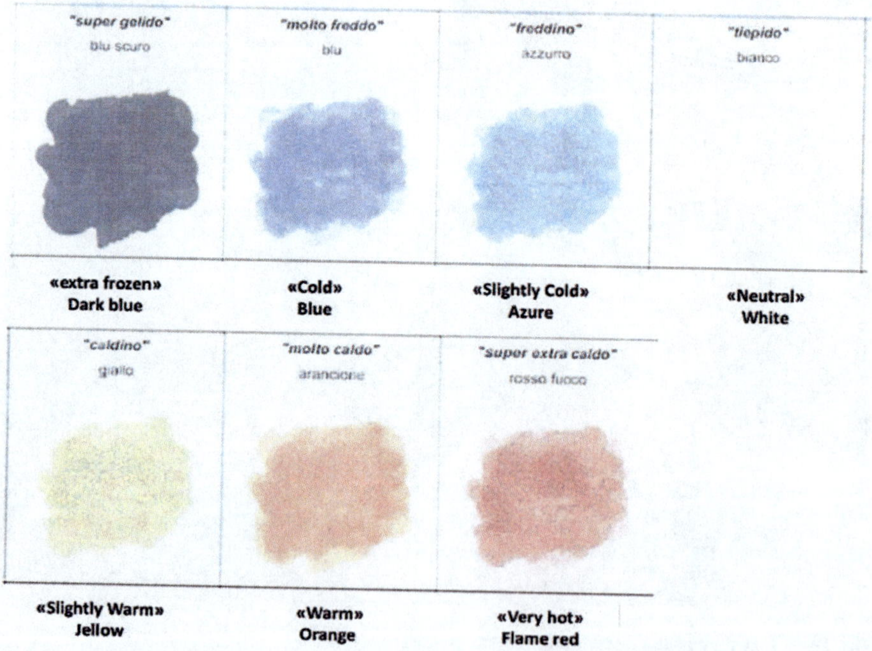

Fig. 10.7 Questionnaires third class Fontana—attribution judgments to a colour

Italian version	English version
(+3) super extra caldo, colore rosso fuoco	(+3) very hot, colour fire red
(+2) caldo, colore arancione	(+2) hot, colour orange
(+1) caldino, colore giallo	(+1) slightly hot, colour yellow
(0) tiepido, colore bianco	(0) warm, colour white
(−1) freddino, colore azzurro	(−1) slightly cold, colour light blue
(−2) molto freddo, colore blu	(−2) very cold, colour blue
(−3) Super gelido, colore blu scuro	(−3) freezing cold, colour dark blue

The question "*Do you feel well in this environment*?" did not result in difficulties in understanding. The children have understood that it was the feeling with respect to temperature.

In the question "*How do you perceive the quality of the air at this time*?" the children did not understand the concept of "*perceived quality*", from the discussion and the associated responses, there revealed a common understanding on the smell of the room, but it was not clear whether they have understood the terms "*sense*" and "*quality*". The smell was associated with the opening of the windows, so the value (0) "*Tolerable*" was associated in a condition in which there is no need to open the windows; (+1) is associated with a condition in which it is necessary to open the windows for 10 min, (+2), the air is smelly and there is a need to open the windows for half an hour, while with (+3) the windows must be open for at least 1 h; Finally (+4) is when the air is so suffocating that you have to open the windows for many hours (Fig. 10.8).

It should be pointed out that in this case the perceived odour is associated with the definition of an action required to return the room to the normal condition. In this case children think, cognitively, in an operational way, association to the feeling action; therefore they can be considered as "sensors" active to the level of indoor pollution, although it is not clear what the smell (the pollutant) is that makes them decide to open the windows.

On the same morning starting at 10:30 the experiment in the *Fountain cabin of the Third Class Paganelli* was carried out, where there were three teachers and 21

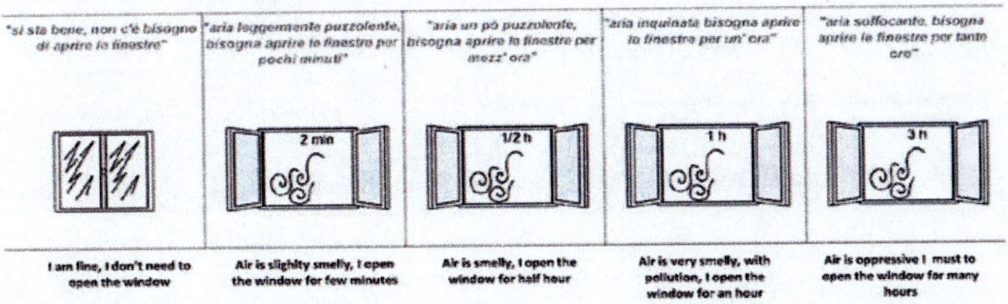

Fig. 10.8 Questionnaires third class Paganelli—attribution judgments Indoor Air Quality (opening time windows)

children. Again the children sat at their desks and the questionnaire was distributed, and a child read out the questions. The children had no clear meaning of the word "perceive", which was given the meaning "*to feel a thing*". We proceeded to the reading of the responses and the association of mental image matching.

Children associated each level of feeling with an "*object*" and the water, so

Italian version	English version
(+3) super mega caldissimo, il vulcano	(+3) super mega hot, a vulcano
(+2) molto molto caldo, l'acqua che bolle	(+2) very very hot, boiling water
(+1) un pochino caldo, l'acqua per la camomilla (non per il thè)	(+1) a bit hot, water for the camomile (not for the tea)
(0) neutro, l'acqua che bevo	(0) neutral, water to drink
(−1) freschino, l'acqua della sorgente	(−1) fresh, water from a spring
(−2) molto freddo, associato al frigorifero	(−2) very cold, associated with a fridge
(−3) freddo bubolante, associato al freezer	(−3) super freezing cold, associated with a freezer

One child said that for him, in the classroom, he sometimes felt hot and sometimes cold, it was explained that the answer was to refer to that time, the precise moment when they were responding to the questionnaire.

In this way it was easier to explain that to the next question, "*Do you feel well in this environment*?" was referred to the feeling that they wanted to perceive.

To the question "*How do you perceive the quality of the air at this time*?" The teacher immediately specified that the question was related to the smell of the air, so the children had for each answer an object with its smell: (0) "*Tolerable*" is a normal condition, such as odourless soap; (+1) is a little smelly, like a sneaker; (+2) is stinking like a pipe; (+3) is very smelly like car fumes; and (+4) "*Intolerable*" is mega super smelly, like a bag of rubbish (Figs. 10.9 and 10.10).

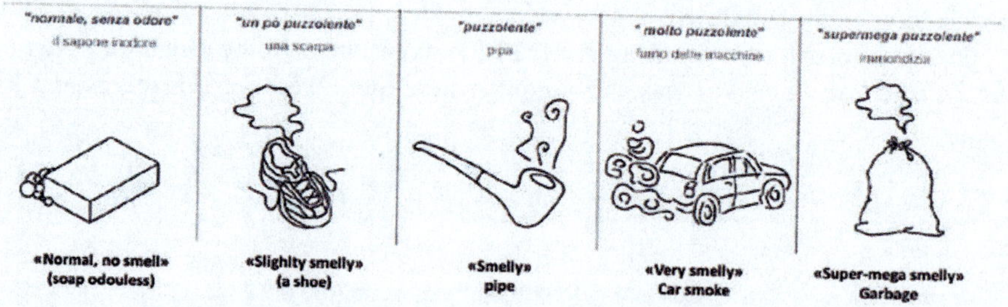

Fig. 10.9 Questionnaires Aula third Canocchie—attribution judgments Indoor Air Quality (smell of objects)

Fig. 10.10 Delivery of questionnaires third class

10.5 Monitoring with Microclimate Datalogger and Probe

In conjunction with the administration of questionnaires to assess the perceived comfort of the children the microclimate of the indoor environment was monitored.

The monitoring was carried out with the probe HDR 32.3 Delta Ohm which involves the measurement of air temperature (T-AIR) and the Relative Humidity (RH) of the air speed, with a hot wire anemometer, and mean radiant temperature (TMR) with a globe thermometer.

The probes were positioned in the classroom for the duration of the administration of the questionnaire, for each classroom. The graphs of the figures show a different trend of the temperature and Relative Humidity: in the classroom Mimosa the T-AIR TMR have a similar trend, but the TMR is about 0.5° C lower, so the walls are slightly "cooler", and HR is always less than 50%, therefore the lower the risk of feeling humidity. The analysis of the PMV (Table 10.2) according to ISO 7730, in the case of an adult, is slightly slightly cold (−0.40) and PPD 9.26%.

The same distribution in the Jasmine classroom, in this case the difference between T and TMR-AIR is about 0.4 °C, and varies by about 2 °C during the entire period of the measurement, while HR is always less than 45% with a higher risk of "dry air", the value of PMV is equal to −0.34 (PPD 7.66%).

In Canocchie Classroom T-AIR and TMR almost coincide but have values between 16 and 18 °C, with HR around 45%, which should correspond to a greater feeling of

Table 10.2 Question results: how do you perceive the environment at this time?

Question score	+3 (very hot)	+ 2 (hot)	+ 1 (slightly hot)	0 (neutral)	−1 (slightly cold)	−2 (cold)	−3 (very cold)		PMV	PPD
Mimosa Classroom	0	3	4	7	5	0	0	19		
	0	6	4	0	−5	0	0	5	**0.26**	**6.44**
Gelsomino Classroom	1	3	8	2	4	0	0	18		
	3	6	8	0	−4	0	0	13	**0.72**	**15.98**
Canocchie Classroom	0	1	2	6	8	3	0	20		
	0	2	2	0	−8	−6	0	−10	**−0.50**	**10.23**
Paganelli Classroom	1	3	7	8	1	0	1	21		
	3	6	7	0	−1	0	−3	12	**0.57**	**11.84**

"cold", in fact the PMV is equal to—1.71 (cold) and PPD 62%. Finally Paganello Classroom T-AIR and TMR vary between 18 and 20 °C with HR to below 40–42%, and the corresponding value of PMV is 0.93 (cold) and PPD 23.32%.

Figures 10.11, 10.12, 10.13 and 10.14 show the results of IAQ monitoring.

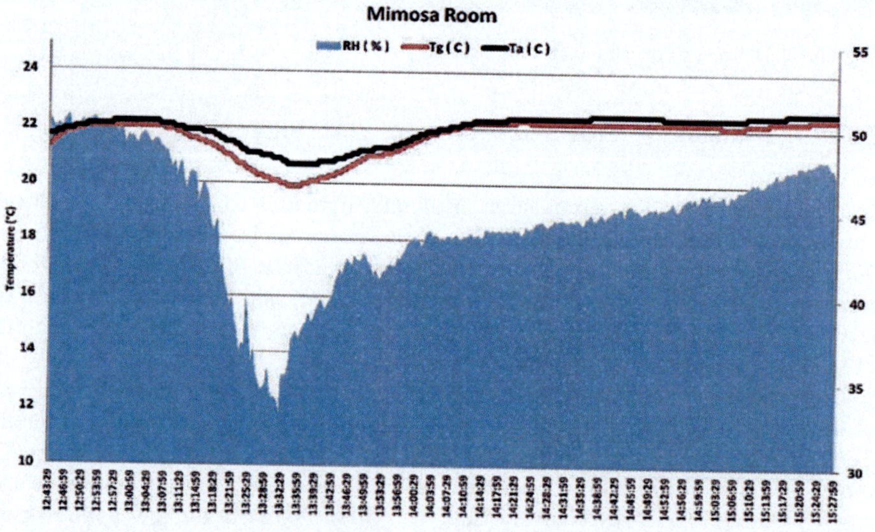

Fig. 10.11 Monitoring indoor microclimatic parameters Mimosa classroom

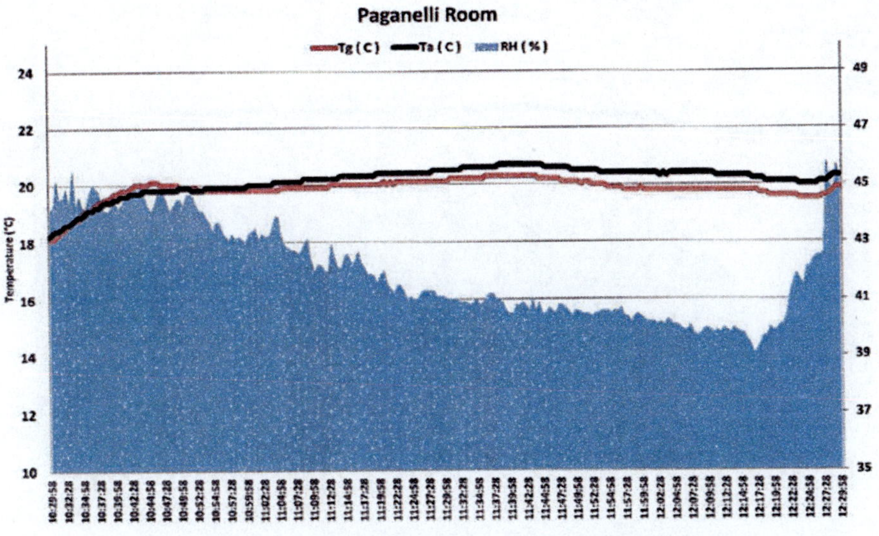

Fig. 10.12 Monitoring indoor microclimatic parameters Gelsomino classroom

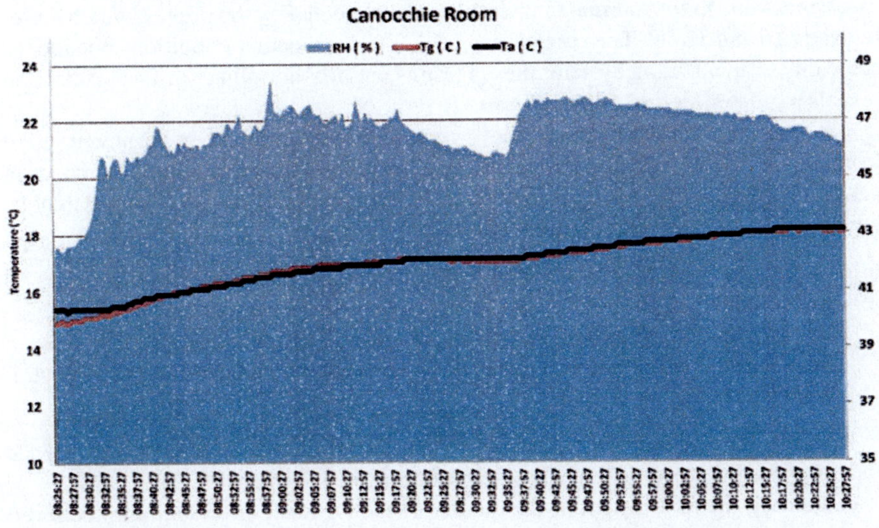

Fig. 10.13 Monitoring indoor microclimatic parameters Canocchine classroom

10.6 Questionnaire Results

The results of the questionnaire show an understanding of the questions from the students, the judgment is "*slightly slightly warm*" for all classrooms (Table 10.2) except the classroom Canocchie showing a judgment "*slightly slightly cold*" (−0.50)

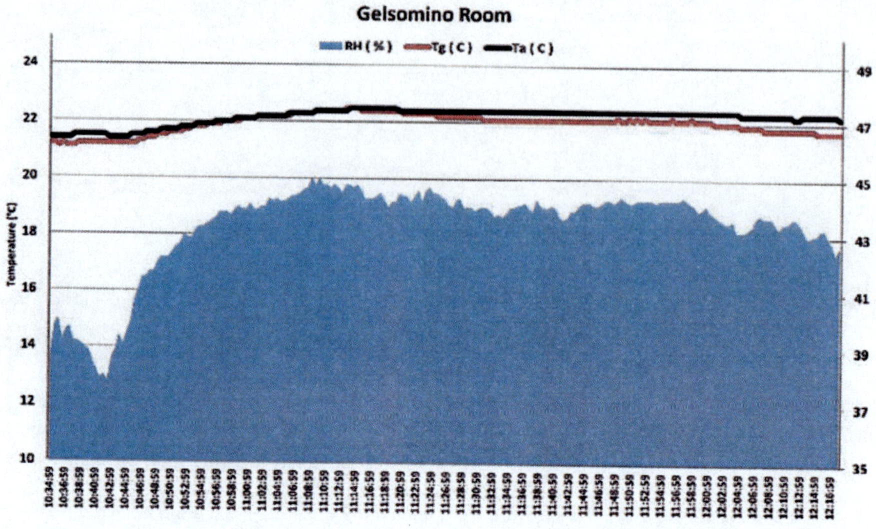

Fig. 10.14 Monitoring indoor microclimatic parameters Paganelli classroom

assessment due to the internal temperature, also detected by the probes, which varies between 16 and 18 °C. The reason may be due to the weather conditions outside the same day or the heating system; these reasons are not the subject of this research, it reveals a greater feeling of discomfort.

Most of the children believed they were comfortable in the environment, apart from two students in the Mimosa classroom and one in the Gelsomino classroom of students of the fifth year. In this regard it should be noted that the response may be affected:

(a) of the assessments related to the social dynamics internal to the classroom;
(b) by habit and "*affection*" to the classroom and the classroom;
(c) the attitude of the children, and people in general, to adopt adaptive behaviors to the desired thermal comfort, especially there where the case is close to neutral conditions.

Finally it is noted that in the fifth class, where the children have more independence and individuality, records the expression of negative judgments.

Regarding the assessment of the quality of the air most of the children believed that the air is "tolerable", that there are no unpleasant odors, smells or anything. The fact that children's positive judgment is confirmed by the responses to the question "*How would you like the environment at this time*?" where for the most part the answers required no adjustments, except for the fifth classroom, while for the third classes they required an environment "slightly slightly warmer" in particular in the Canocchie classroom (+0.35) which recorded the only feeling of discomfort towards the feeling of cold.

Tables 10.2, 10.3, 10.4, 10.5 and 10.6 show questionnaire results by classroom.

Table 10.3 Question results: do you feel well in the environment?

Question score	Yes	No
Mimosa classroom	17	2
Gelsomino classroom	17	1
Canocchie classroom	20	0
Paganelli classroom	21	0

Table 10.4 Question results: how do you perceive the quality of the air at this time?

Question score	0 (Tolerable)	+ 1 (slightly bad odour)	+ 2 (bad odour)	+ 3 (strong bad odour)	+ 4 (Intolerable)		PMV	PPD
Mimosa classroom	18	1	0	0	0	19		
	0	1	0	0	0	1	**0.05**	**5.06**
Gelsomino classroom	17	1	0	0	0	18		
	0	1	0	0	0	1	**0.06**	**5.06**
Canocchie classroom	17	3	0	0	0	20		
	0	3	0	0	0	3	**0.15**	**5.47**
Paganelli classroom	19	1	0	0	1	21		
	0	1	0	0	4	5	**0.24**	**6.18**

Table 10.5 Question results: how do you feel right now?

Question score	−1 (Tired)	0 (Normale)	+ 1 (Awake)		PMV	PPD
Mimosa classroom	2	14	3	19		
	−2	0	3	1	**0.05**	**5.06**
Gelsomino classroom	4	11	3	18		
	−4	0	3	1	**−0.06**	**5.06**
Canocchie classroom	4	12	4	20		
	−4	0	4	0	**0.00**	**5.00**
Paganelli classroom	2	16	3	21		
	−2	0	3	1	**0.05**	**5.05**

The second level of analysis of results concerns the comparison between the measured data from the probes and the results of the questionnaires, in particular with respect to the PMV and PPD.

The first thing to do to be able to compare the two values is to build the figure of the "*Standard child*" or "*Child equivalent*" from the weight (kg) and the average height (m) average, for each class on the basis of which to calculate the area of each Dubois "*Child equivalent*". This area is compared with the area of Dubois of the

Table 10.6 Question results: how would you like the environment to be at this time?

Question score	+ 1 (hotter)	0 (as it is)	−1 (colder)		PMV	PPD
Mimosa classroom	7	6	6	19		
	+ 7	0	−6	1	**+ 0.05**	**5.06**
Gelsomino classroom	5	7	6	18		
	+ 5	0	−6	−1	**−0.06**	**5.06**
Canocchie classroom	8	11	1	20		
	+ 8	0	−1	7	**+ 0.35**	**7.55**
Paganelli classroom	4	16	1	21		
	+ 4	0	−1	3	**+ 0.14**	**5.42**

Adult Standard as defined by ISO Standard 7730, the purpose of calculating the W emitted by the adult equivalent to the metabolism of the child.

EN ISO 7730 was prepared by referring to an adult, "*average*", suggested as an adult of 80 kg and 1.80 m in height who has an area of 1.99 m^2 Dubois and a metabolic activity of 115.52 W for sedentary activity. In order to calculate the index PMV and PPD children you need to re-proportion the metabolism of the child as if it were that of an adult, so to calculate the equivalent half of the child corresponding to an equivalent area of Dubois of an adult the following proportion is used:

$$(\text{Child metabolic activity W})/(1\ \text{met}) = (\text{Adult metabolic activity W})/(\text{X met})$$

so

$$70.53/1 = 115.52/X$$
$$X = (115.52 * 1)/70.53 = 1.64\ \text{met}$$

Table 10.7 shows:

Table 10.7 Average data results related to the metabolism of children

Average values	Age	Weight (kg)	Height (m)	Area Dubois (m^2)	(W/ m^2)	W/child	W/ equivalent person	Met equivalent
Mimosa classroom	9.95	35.61	1.46	1.22	58	70.53	95.00	1.64
Gelsomino classroom	10.00	37.11	1.44	1.22	58	71.02	94.35	1.63
Canocchie classroom	8.00	30.34	1.32	1.05	58	31.01	109.82	1.89
Paganelli classroom	8.10	30.14	1.32	1.05	58	60.95	109.93	1.90

- For each classroom: age, weight and height of children, as given in the results of the questionnaires;
- The area of Dubois's "average child" for the classroom;
- The W/m^2 for the metabolic activity of 1 met;
- The W/child, or the power (W) emitted by the child in relation to its physical dimensions;
- The W/equivalent person, or the power (W) that the child would have if it were an adult, or better, compared to the area of Dubois of an adult;
- The met/person equivalent (met equivalent), representing half of the child if he were an adult, the value used for the simulation of the value of PMV according to ISO 7730.

Identifying the metabolism equivalent to the Child it is possible to calculate the PMV according to ISO 7730, from the measured indoor microclimate for each classroom. The calculation was made based on the data measured by the sensors and calculated with the software Deltalog 10 of DeltaOhm.

The subject, both adult and child equivalent was assumed with clothing to equal to 1 Clo: pants, socks, t-shirt, shirt, trousers and shoes.

Table 10.8 shows the results of the experiments, for each classroom and tests are given the values:

- For children:

  the PMV and PPD from the questionnaire;

  the PMV and PPD obtained from the probe using microclimatic data measured during the administration of the questionnaire and the metabolism of the "Child equivalent" according to ISO 7730;
- For adults:

  The PMV and PPD obtained from the probe related to the metabolism of an adult according to the standard ISO 7730;

Table 10.8 Comparison of the results of the questionnaire and the probes

	Child				Adult		Indoor data (average)		
Average values	PMV Questionnaire	PPD (%) Questionnaire	PMV Probes	PPD (%) Probes	PMV Probes	PPD (%) Probes	T-AIR °C	TMR °C	RH %
Mimosa Classroom	0.26	6.44	0.85	20.49	−0.40	9.26	21.99	21.54	44.62
Gelsomino Classroom	0.72	15.98	0.66	14.14	−0.34	7.66	22.19	21.85	43.54
Canocchie Classroom	−0.50	10.23	0.15	5.77	−1.71	62.00	16.97	16.91	46.60
Paganelli Classroom	0.57	11.84	0.57	11.78	−0.93	23.32	20.14	20.14	42.00

The data microclimatic average recorded during the measurement: Air Temperature (T-AIR in °C), Mean Radiant Temperature (TMR in °C) and the Relative Humidity (RH in %).

The results show different values between the PMV of children, real and equivalent and between children and adults simulated. In particular, it is clear that adults have a lower thermal sensation of about 0.5–1 point of PMV, demonstrating that the metabolism of children is detected by the questionnaires.

Table 10.9 shows the differences between the various questionnaires of the PMV. In particular:

- The comparison between PMV and PMV of the Questionnaire of adults shows a difference between +0.66 and +1.50, for younger children, demonstrating that the metabolism of children is affected significantly compared to the perception of the comfort of the adults. Common sense that tells us that children are less sensitive to cold is confirmed by the data, and this is related to increased metabolism of the lower area of Dubois. Even the Canocchie classroom, where children have a feeling of "*slightly cold*" (−0.50) adults have a feeling of "*cold*" or "*very cold*", with respect to which the teachers adopt adaptive systems related to clothing;
- The comparison between PMV and PMV of the Questionnaire of "*Child equivalent*" shows that in two cases, Jasmine Classroom and Paganelli Classroom values coincide, and the gap is zero or only +0.06;
- In the other two classrooms the two PMV have a difference of −0.59 and −0.65, that the equivalent child receives a thermal comfort warmer than the actual child as measured by the questionnaire.

Figures 10.15 and 10.16 show a graphics with questionnaire results.

This may be due, in the case of Canocchie Classroom to an internal temperature below 20 °C, between 16 and 18 °C, a factor which may have an influence in the perception of children, in this case of 8 years old, the metabolism is subjected to the effort of having to make up for the lower temperatures which can affect the perception of comfort.

In the case of the Mimosa Classroom a deviation may be due to the phenomena of adaptation of children to a temperature slightly warmer (22 °C instead of 20 °C) compared to the standard value by legislation (20 °C).

Table 10.9 Difference between childrend thermal comfort sensation by questionnaire and adult thermal comfort by indoor measurement

Average values	Difference between children (PMV Questionnaire) and adults (PMV Probes)	Difference between children (PMV Questionnaire) and child equivalent (PMV Probes)
Mimosa Classroom	+0.66	−0.59
Gelsomino Classroom	+1.06	+0.06
Canocchie Classroom	+1.21	−0.65
Paganelli Classroom	+1.50	+0.00

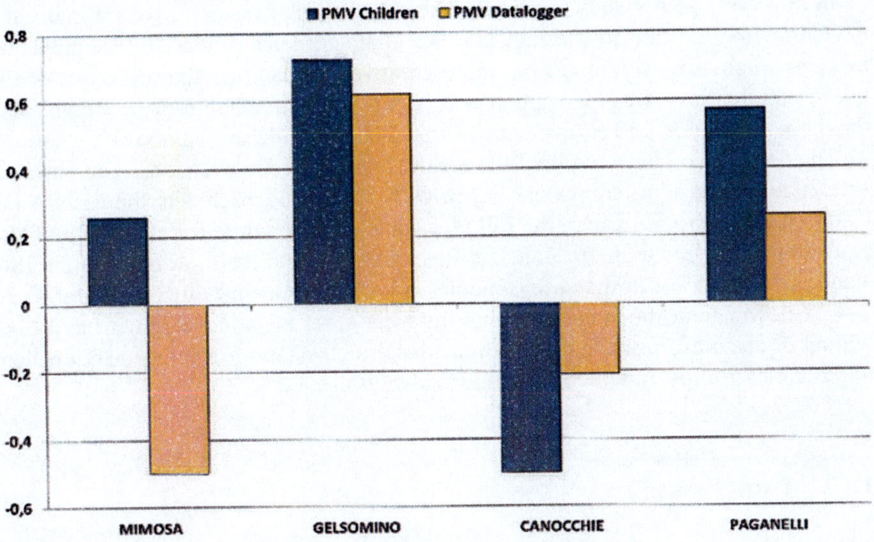

Fig. 10.15 Comparison between the PMV sub-divided by classroom

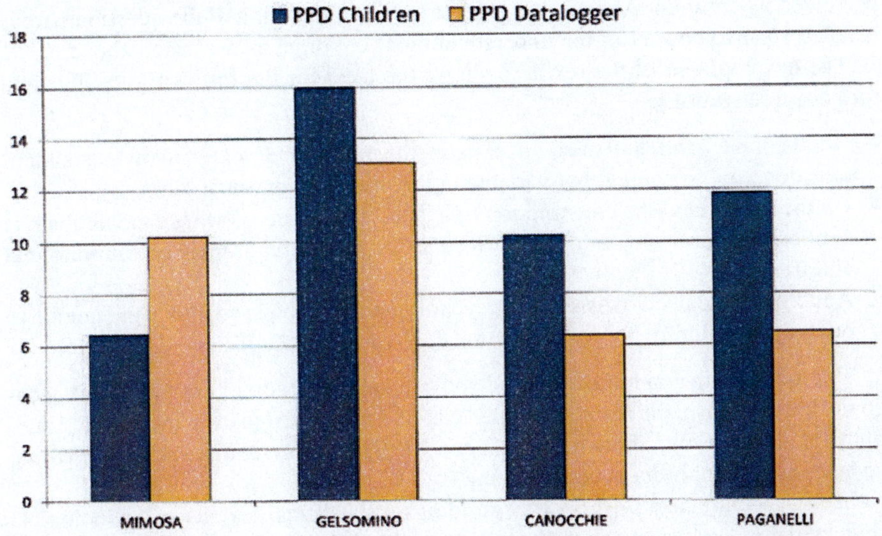

Fig. 10.16 PMV comparison by typology of valuation

In any case, both values can be the subject of further investigations. Finally the deviation may be due to the fact that the characteristics of the child equivalent, average height and weight are not representative compared to the responses of the questionnaires involving perceptual and cognitive phenomena more complex than the mere translation and equivalence of metabolism, as described above.

The comparison between the PPD for the classroom shows that the gap between actual child, and adult equivalent is below 20% (1 in 5), except in the case of the Canocchie Classroom, where the PPD for adults is 62% while for children real and equivalent, it is less than 10% and for the probe equal to the minimum by the ISO standard 7730 (5%). In the case of adults in an environment with a temperature of 2–4 °C lower than the standard value involves a net negative judgment and high feeling of discomfort, while for children adaptation systems are activated equivalent to other classrooms, in which the value is always around 10%.

10.7 Conclusions

The research results show that in the case of children, the questionnaire is a useful tool for better assessments according to the regulations, in assessing the comfort perceived by children. Always, the process and administering of the questionnaire is handled so as to be sure of the understanding of children, as described above.

The development of the research shows the need for the following insights and further investigations:

- On the one hand the need to expand the number of cases with a standard questionnaire in content but variable in the mode of administration;
- On the other hand the construction of a "*Child equivalent*" whose metabolism is applicable even if you adopt the ISO 7730, a model that report or calibrate real children;
- And finally the need to assess the psychological variables, adaptive and cognitive relating to children, because of the random variable test.

The monitoring conducted focused above all on the study of cognitive and pedagogical, in any case, the results obtained can be used in designating the Virtual Environment, as the results show that the Canocchie Classroom requires action regarding its heating system, in its power or sizing.

In the specific case of the CIES buildings, "shacks/cabins" have technological characteristics such as to constitute a unique and historical testimony. The results of the questionnaire and of the monitoring showed that, although they were built and designed with another purpose, and even though they are made of wood and therefore lightweight and not well insulated, they allow the performance of school activities with comfort conditions equal to that of other buildings.

This may also be due to the architectural conformation and to reduced heating volumes, which increases the incidence of free contributions due to solar and internal heat loads (including children) and the distribution and number of radiators.

To these may be added insulation interventions of the cover and the floor, which fortunately is not in contact with the ground, and the replacement of fixtures, where possible.

The different stages of analysis, the study of Human Behaviour and questionnaires of the occupants to the Building Energy Performance Simulation, are not intended to verify the design choices, but they are part of the design process. Calculations and questionnaires are tools to guide, and verify the design choices, like the technological solutions and formal planning.

Acknowledgements The sustainable customary care. Studies and proposals for protocols conserved by the "Educational Center Italo Swiss CEIS" Undergraduates: Chiara Mei, Manuela Mignani, Selina Morri AA 2012/2013, Relatore Prof.Arch. Andrea Ugolini, Correlatori Arch. Kristian Fabbri, Arch.Tessa Matteini.

The author acknowledges Chiara Mei, Manuela Mignani, Selina Morri, and prof. Andrea Ugolini for their valuable support during this experimental activity.

Thanks also for the valuable collaboration of the Educational Center Italo Swiss CEIS.

Chapter 11
Middle School Case Study

Abstract In this chapter we report the case study of the Middle School Plautus Cesena (Italy). The research concerns the evaluation of Thermal Comfort through questionnaires administered to approximately 230 children between 10 and 13 years of age. The questionnaires were developed by researchers and then given to children in two campaigns; the first campaign of surveys was not accompanied by monitoring and reported some errors that we consider useful to report in this book. Monitoring was done for one representative classroom. In the same classroom was made the monitoring and measurement of the levels of lighting and Perceived Lighting Comfort.

11.1 The Case Study Middle School Tito Maccio Plauto

This case study covers the testing which took place at the school complex Ex-Malatesta Novello, located in Cesena, in particular at the Middle School Plautus, already part of the "School of the Future" for the energy upgrading of school buildings.

> School of the future. *Towards Zero Emission with High Performance Indoor Environment* http://www.school-of-the-future.eu/ Program of the 7th Framework Program of the European Union - Energy. The goal of the project is to study examples effective for improving building energy performance of school building. The object of the research is the energy performance and the energy retrofitting, while aspects of the Human behaviour were not taken into consideration. In fact, the City of Cesena took the opportunity of the project to improve the system of remote management and control of all school buildings by adapting the use of heating systems to the effective use of school environments.

Cesena is a municipality located in the north-east of Italy (Geographic coordinates 44°08′78″ N–12°15′03″ E), with an average size and with a standard of living and medium–high income. It can be said that there is no degradation or unemployment likely to characterize the population.

In facing experimentation activities it is useful to understand and describe the building in which the activities take place, in terms of spatial dimensions, the size and the number of classrooms, the provision of more levels, the green space.

K. Fabbri, *Thermal Comfort Perception*, https://doi.org/10.1007/978-3-031-52610-7_11

The spaces are inhabited by the children during the entire span of school, their understandings of space, as well as finishing, the windows, the instrumental equipment of the classroom, the level of maintenance of the school, has an influence on their perception environment, including the condition of comfort. If the building and classrooms are in a state of disrepair, with plaster that is detached or broken parts and elements, the bad opinion that the child has of the school, is reflected also on the opinion they will have on the building.

In an unattractive building it is easier to have feelings of repulsion towards the building itself. This repulsion may increase degradation phenomena, bullying, and therefore the subjective environment of the boy cannot be distinguished by the view that he has on the social relationship and on the characteristics of the building.

Conversely, if the building is new or well-maintained, if there are no blind and dark spaces, the child will pay more care to the building, reducing vandalism and bullying.

There is a direct correlation between the building and the social dynamics of a group of children, but it is good to keep in mind that after 7 years, children begin to interact independently with other children, defining their individuality. This process continues until the age of adolescence, and the Middle School, where the boys are aged between 11 and 14 years, is the first official stage, where they are not (only) children, but start to become adults and their view on their world, formed from the building is less influenced by the opinions of others.

11.2 The Building

The building in which we held the experiment is part of a school of about 2685 m^2 arranged around a courtyard, where there is a gym for educational activities. The building consists of two blocks placed on a hilly slope, and green spaces and trees, outside the perimeter and inside the courtyard, because of the slope cannot be used. Most of the life of the students, during school hours, takes place within the school. The interior space consists of classrooms and wide hallways.

The building was built in 1965, and has undergone changes, an entire block, that of the gym, was added in 1982.

Compared to the standard required by Italian law (Ministerial Decree 18 December 1975) the building is slightly overcrowded, there is 1 person for every 1.74 m^2 instead of 1.8 m^2, and there are architectural barriers.

The building is on three levels for each block and the two blocks have different altitudes, so much so that there is a ladder to join them.

Figures 11.1 and 11.2 report building plan and section.

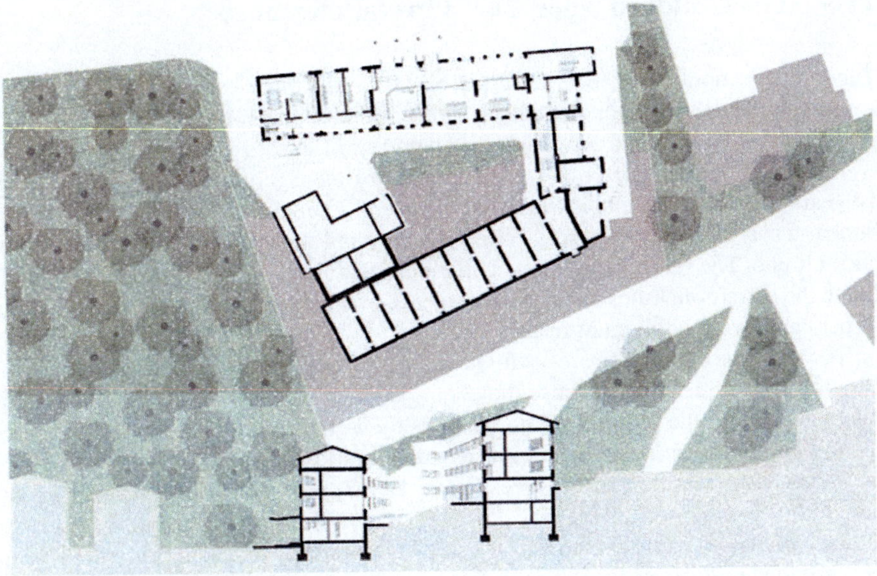

Fig. 11.1 Plan and section of the building—ground floor

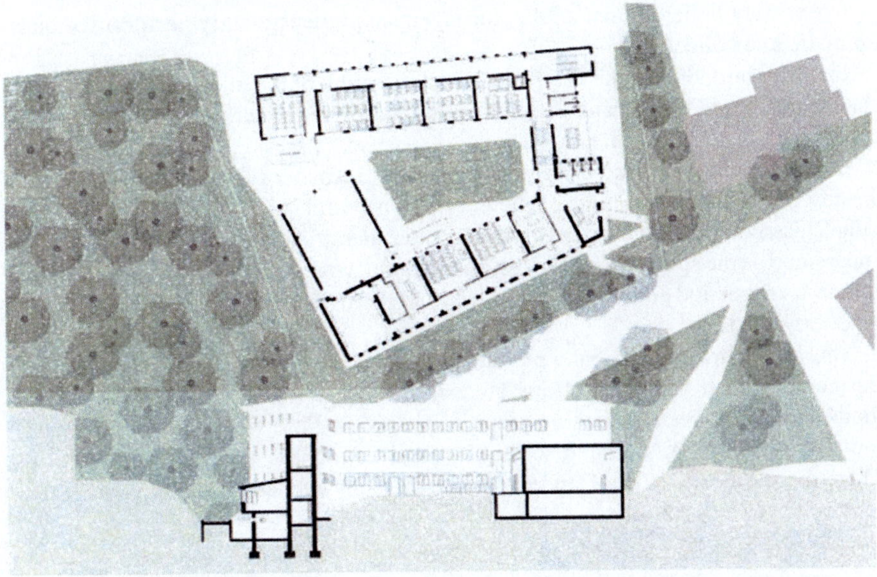

Fig. 11.2 Plan and section of the building—first floor—classrooms

11.3 The Children Aged 11–14 Pre-adolescence

The middle school has 4 sections (A, B, C, D) for the three years of the course, for a total of 12 classrooms, and a total of 236 children (registered year 2011/2012), for an average of 24 students per class, aged between 11 and 14 years.

Regarding the social dynamics of interest to teenagers aged between 11 and 14 years, a useful example is the book "*Diary of a Wimpy Kid*" (Kiney 2007). The book is a sort of diary of Gregory "Greg" Heffley and shows his life in the classroom and at home. The book shows Greg's opinions, but mostly there are anecdotes about life in the classroom, rules and classmates, each of which is identified by a nickname, or pet name with a dominant feature.

The readings of children are useful indicators for approaching the mode of administration of the questions because they help you understand what may be the social dynamics within the group, and the relationship with the authorities, which may be the teacher, or the person who administers the questionnaires.

In the administration of the questionnaires there can always be difficulties in attracting the attention of the boys to the activity. The researcher can be seen as another adult, and while at elementary school the authority of teachers is still present because the boy/girl has not yet completed the process of definition of individuality, in the case of middle school, you run the risk that children are not careful in the explanation of the questionnaire.

Conversely being almost adults, teenagers can independently perform the questionnaire and collect it.

In this regard, during the activity of administration of the questionnaires, in some classrooms, there was greater difficulty in attracting the attention and describing the content of the questionnaires.

In middle school Italian classes are taught by a teacher for the subject, so during the day, every one or two hours, teachers alternate, while boys/girls are always in the same classroom. This aspect involves a greater identification between "*classroom*", understood as the space where the lessons are held, with its geometric characteristics, colour, instrumental equipment, windows, and thermal sensation, and "*class*", as the set of boys/girls.

This aspect involves a greater permanence and identification with the environment classroom, which would constitute a crasis classroom = class. The questionnaires showed envy between classes not equipped with the IWB (interactive whiteboard) and those with IWB, so much so that in the questionnaires carried out by classes where there was the IWB all students wrote that they wanted one, complaining their condition was less fortunate.

11.4 The Questionnaire

11.4.1 First Survey Campaign Following 'School of the Future' Questionnaire

The study of the perception of the comfort of the children of the Middle School was held in two phases. The first experimental campaign was carried out by translating questionnaires from the program "*School of the future*". Table 11.1 report first test questionnaire.

The questionnaire consists of questions in writing, without images, and with the possibility to tick the boxes of multiple responses. In the case of middle-school pupils can administer the questionnaire without using image support. If misunderstandings arise in reading a word, or the meaning of a sentence, it is useful to find a synonym or an example, the "didactic" explanation is likely to be ineffective.

For example, the term "stale", can be translated as "*with a bad smell*" or, "*the smell of the air when you open the windows for a long time*".

The first error in posing questions is that the question refers to memories that children have about their perception in a moment of time different from that in which we conduct the test.

The questionnaire was administered in the spring, asking how you sense the temperature of the classroom in the summer or winter means projecting judgment on an event of the past or the future, for which the association has a greater role mnemonic temperature associated with the season rather than the real feeling.

The questions must be clear about their purpose, if the goal is to study the general opinion of the people about the comfort, without specifying where, then it is possible to refer to different moments. On the other hand, if the objective is to verify and validate the comfort of the environment, and then compare it with monitoring, the question is too specific to be able to give valid results; it is equivalent to the polls on voting intentions, or exit polls that collide with the real data of the polls.

The questions that are repeated are very similar, this can cause boredom in reading and the tendency to speed up the process of compiling skipping the reading of the question.

The questions related to the Visual and Lighting Comfort present errors in the content of the application and in the options. The first aspect to highlight is that the answers contend similar options, but slightly different (*"easily acceptable", "clearly unacceptable", "very acceptable"*). The speed of reading does not allow a person to notice these differences, and in any case they are differences that give doubts in reading, basically all expressing the same thing.

The question 2.(a) is too sophisticated syntactic formulation, it assumes that the boy knows the concept of "intensity" of the light, and in any case it is not clear what meaning can be attributed. Since the question is designed to compare the responses with the measurement of luminance, using a light meter, the wording of the application may contain the expression "amount of light" that is not correct from the

Table 11.1 Questionnnaire. First test

(1) Room Air Temperature

(a) How do you perceive the temperature in the classroom in the summer?

1. Very Hot	2. Hot	3. Warm	4. Neutral	5. Cool	6. Cold	7. Very Cold

(b) The temperature in the classroom is acceptable in summer?

1. Clearly acceptable	2. Just acceptable	3. Not quite acceptable	4. Clearly unacceptable

(c) How do you perceive the temperature in the classroom in the winter?

1. Very Hot	2. Hot	3. Warm	4. Neutral	5. Cool	6. Cold	7. Very Cold

(d) The temperature of the classroom is acceptable in winter?

1. Clearly acceptable	2. Just acceptable	3. Not quite acceptable	4. Clearly unacceptable

(f) The indoor air quality is acceptable in summer ?

1. Clearly acceptable	2. Just acceptable	3. Not quite acceptable	4. Clearly unacceptable

(g) How do you perceive the Indoor Air Quality in winter ?

1. Very stale	2. Stale	3. Slightly stale	4. Not known change	5. Slightly fresh	6. Fresh	7. Very fresh

(h) The Indoor Air Quality is acceptable in winter ?

1. Clearly acceptable	2. Just acceptable	3. Not quite acceptable	4. Clearly unacceptable

(2) Room Lighting

(a) the intensity of the light of day is appropriate to read a text?

1. Too bright	2. Really bright	3. Slightly bright	4. Adequate	5. Slightly low	6. Really low	7. Too low

(b) the intensity of the artificial light (neon) is appropriate to read a text ?

1. Too bright	2. Really bright	3. Slightly bright	4. Adequate	5. Slightly low	6. Really low	7. Too low

(c) Colours are clearly visible in artificial light?

1. easily identifiable	2. Just acceptable	3. Not quite acceptable	4. Clearly unacceptable

(d) The curtains/blinds are acceptable?

1. Clearly acceptable	2. Just acceptable	3. Not quite acceptable	4. Clearly unacceptable

(e) The connections with the outside of the building are acceptable?

1. Clearly acceptable	2. Just acceptable	3. Not quite acceptable	4. Clearly unacceptable

(continued)

Table 11.1 (continued)

(f) The glow of the sunlight shielded from the curtains/blinds, is acceptable?			
1. Clearly acceptable	2. Just acceptable	3. Not quite acceptable	4. Clearly unacceptable

physical point, but it is in the perception of its meaning. Also there are too many options in multiple responses.

The question 2.(b), related to the previous one, is too specific, why repeat the question by varying only between day lighting (natural) and artificial? If it is clear to the evaluator, because it measures the level of illumination with and without artificial light, it is not for those who respond to the question that is too similar to the previous one.

Similarly questions 2.(c) and 2.(d) are too "*Case-Specific*", too detailed to be able to understand the contents, for example, it is unclear what is meant by "visible" and "shadowing". Also question 2.(d), together with the 2.(e) have sophisticated notions for a boy, and even for an adult who does not have knowledge of a technical architecture, like the term connection or the expression "outer-context."

The results showed the gaps of information, such as not allowing the use of data collected, to obtain useful information for the project and for the evaluation of the comfort.

The test consists of 15 multiple choice questions. The first part includes questions on the perception of Thermal Comfort and Indoor Air Quality; the second part concerns the Visual and Lighting Comfort, with questions relating to clothing and the colour of light and environments.

The administration of the questionnaires was carried out with the consent of the head teacher, who requested some clarification on the research activity, including in relation to privacy laws. The questionnaires are all anonymous. The questionnaire was given to all the 263 students present, through self-self compilation.

The results of the first campaign showed the prevalence of some questions, highlighting the gaps in the questionnaire itself that has led to problems on the accuracy of the results (and the possibility of using them to guide the project).

In the administration of the questionnaires in the classroom boys posed some frequently asked questions, in particular clarification on how to determine the Indoor Air Quality. Their perception was closely correlated with the air temperature.

The experiments have shown, both for the nursery and the Middle School, that the Indoor Air Quality, is not perceived by children and young people in a clear manner, the presence of pollutants, dust or odor, is not perceived as such, unless there are no unpleasant odours or specific related to certain activities. Children and young people are not good indicators for IAQ, and this is correlated with temperature, in fact the boys/girls connect the internal IAQ the effect due to their greater permanence in the classroom, a phenomenon that involves increasing the concentration of CO_2.

The results of the questions on the Visual and Lighting Comfort are less effective because of the difficulty of understanding the questions. In particular the application 2.(f) in which reference is made to the curtains, while curtains are not present in the building, but shutters. This little detail puts those who complete the questionnaire in the situation of doubt and delay, and they then choose not to give the answer.

The results of the questionnaire, in the light of the doubts described, are too subjective to be used as a tool for comparison of monitoring activities, or to guide design choices. It was therefore decided to review the contents of the questionnaire, implementing the knowledge gained, and flanked by monitoring activities.

11.5 Second Survey Campaign Following Our Questionnaire

The campaign noted was repeated with variations on the contents of the questionnaire, adding the measurement site of microclimatic parameters and illumination levels. The monitoring operation was carried out in a single classroom representative for the entire building. The questionnaire was also submitted to teach the writer.

The second questionnaire includes 24 multiple choice questions arranged on both sides of an A4 sheet, divided by topics:

- General information, age, height, weight and clothing, in order to have an order of magnitude metabolic activity (met) and on the resistance of clothing (clo)
- Thermal Comfort, with questions related to the perception of indoor comfort, monitored with instrumentation in the site;
- Ventilation, for information on Indoor Air Quality;
- Visual Comfort, with questions relating to the quality of artificial light or nature, compared with the measurement of the levels of luminance (lux);
- Free questions related to the preferences of the boys during the moments of leisure and lesson.

Table 11.2 report second test questionnaire.

The questions in Section "1—*Thermal Comfort*" follow the provisions of the Standard ISO 10551, and refer to the thermal sensation perceived when the questionnaire was administered. The boys understood the content, but we had to make it clear that the questions 1.2 and 1.5 were related solely to the classroom from the physical point of view and not to the class of pupils.

The questions of the Section "2—*Lighting Comfort*" allows us to compare the level of illumination perceived by users with that measured in situ by a light meter.

The first two sections were also present in the first questionnaire, but were reviewed as required by ISO 10551, and the sensation perceived correlated with the level of improvement required or expected.

Section "3—*Ventilation*" relates to questions about how the windows are used, and then the air changes, with the perception of odours, equating "*bad odour*" with the presence of indoor pollutants, in particular with the CO_2.

Section "4—*General Questions*" is to support the first three sections and includes questions that allow us to hypothesize an "*average pupil*" for height and weight so as

Table 11.2 Second test questions (the format is similar to the presented test)

1—Thermal Comfort (following ISO 10551)	
1.1—Hw do you feel at this precise moment?	
☐ I'm too cold (PMV = - 4) ☐ I'm very cold (PMV = - 3) ☐ I'm cold (PMV = - 2) ☐ I'm slightly cold (PMV = - 1) ☐ Neither hot nor cold (PMV = - 0)	☐ I'm slightly hot (PMV = + 1) ☐ I'm hot (PMV = + 2) ☐ I'm very hot (PMV = + 3) ☐ I'm too hot (PMV = + 4)
1.2—Therefore, how are you inside the classroom*?*	
☐ 1—I'm good ☐ 2—I'm OK ☐ 3—I don't feel good	☐ 4—I feel bad ☐ 5—I feel really bad
1.3—At this very moment, how would you like to feel?	
☐ Much colder (PMV = -3) ☐ Colder (PMV = -2) ☐ A bit colder (PMV = -1) ☐ No change (PMV = 0)	☐ A bit hotter (PMV = + 1) ☐ Hotter (PMV = + 2) ☐ Much hotter (PMV = + 3)
1.4—Taking account of your feelings, do you consider acceptable the current conditions of the temperature inside the classroom?	
☐ Yes	☐ No
1.5—In your opinion, this environment?	
☐ 1—Is very acceptable ☐ 2—Not quite acceptable ☐ 3—Quite unacceptable	☐ 4—Difficult to accept ☐ 5—Unacceptable
2—Lighting Comfort	
2.1—Is there enough light in the classroom to read and write?	
☐ 1—It's very dark ☐ 2—It's dark	☐ 3—It's light ☐ 4—It's too light
2.2—When artificial light is off is there enough light in the classroom to read and write?	
☐ 1—It's very dark ☐ 2—It's dark	☐ 3—It's light ☐ 4—It's too light
2.3—Are you near or far from the window?	
☐ 1—Next to the window ☐ 2—In the middle of the classroom	☐ 3—Far from the window
2.4—The light of the classroom lets you read the board well?	
☐ 1—Yes ☐ 2—No, because it's too dark	☐ 3—No, because it's too far ☐ 4—No, because it reflects the light (it's too light)
2.5—Do the sun bother your eyes?	
☐ 1—Yes	☐ 2—No
3—Ventilation	
3.1—How often are the windows opened to change the air?	

(continued)

Table 11.2 (continued)

☐ Never ☐ Once a day	☐ At the end of every lesson ☐ Often during the same lesson
3.2—Who decides to open the window?	
☐ Students ☐ The teacher	☐ When a student asks the teacher
3.3—Are there odours inside the room?	
☐ No odours ☐ There are bad odours by the end of the morning	☐ There's always a bad odour
3.4—What odours would you like in the class?	
☐ No odours ☐ Odours of cleaning	☐ OK as it is
3.5 Where do the bad odours come from?	
☐ From outside ☐ By the sweat of the people	☐ From detergents ☐ Form chalk
4.—General questions	
4.1—*How tall are you?*	
1 ☐ < 120 2 ☐ 120–140	3 ☐ 140–160 4 ☐ > 160
4.2—How much do you weight?	
1 ☐ < 40 2 ☐ 40–60	3 ☐ 60–80 4 ☐ > 80
4.3—Do you have difficulty staying alert, regardless of the subject?	
☐ Yes	☐ No
4.4—In which moments do you feel particularly tired?	
☐ At the end of the lesson ☐ At the start ☐ When the lesson is very difficult	☐ When the lesson is very boring ☐ When the lesson is very long
4.5—Where do you spend the interval?	
☐ Macchinette ☐ Aula ☐ Giardino	☐ Corridoio ☐ Un altro spazio all'interno della scuola
4.6—Where would you like to spend it?	
☐ Vending machine ☐ Classroom ☐ Garden	☐ Corridor ☐ Another internal space
5.—Colours	
5.1—Your school is colourful*?*	
☐ Yes	☐ No
5.2—How would you like it?	
☐ More colourful than now	☐ Less colourful than now

(continued)

Table 11.2 (continued)

6.—Acoustics	
6.1—When are you in the classroom, which are the noises?	
☐ No noises, silent ☐ Outdoor noises	☐ Noises from other classrooms ☐ Internal noises from other classrooms
7.—Write in three words what you like most in your classroom:	
....	
8.—Write in three words what you like least in your classroom:	
....	

to calculate the Dubois Area, to be used to determine the indices of indoor comfort PMV and PPD. The questions were written for the range of values, so as to make the questionnaire anonymous and to avoid doubts or customization in the compilation. Questions 4.3 and 4.4 assess the effects of the amount of CO_2 (in ppm) contained in the classroom, compared with the level of attention.

Question 4.3 allows us to untie the difficulty of the subject with regards to the concentration ability of the pupil, while question 4.4 allows us to evaluate the effects of increased CO_2 concentration in the environment with respect to the ventilation of the classroom. The increased amount of answers like "*When the lesson is very long*" or "T*he end of a lesson*" allows us to make a direct correlation with the ventilation of the amount of CO_2, in fact, the more CO_2 will reduce the rate of inspired oxygen and thus a reduction in the ability of the brain to process information with a consequent increase in the feeling of tiredness and fatigue. Conversely answers as "A*t the start*" or "*When the lesson is very difficult*", are disconnected from the conditions on the Indoor Air Quality.

Questions 4.4 and 4.5 allow us to know the ways the spaces of the school environment are used, compared to how the children would like to use the environment. The question is not directly related to monitoring, but, according to the design, this allows you to check the weakness of the building, as children tend to spend time in environments where they feel the best, the most enlightened where conditions of discomfort, thermal, visual and acoustic are less present.

The last two questions 5 and 6 on the colours and acoustics of classrooms are to assess, in terms of quality and not quantity because they were not provided for in the monitoring activities on the perception of other conditions of discomfort that can impact than the previous answers. In a noisy environment, although well lit or thermally neutral, the feeling of noise discomfort can be a negative affect.

At the end of the questionnaire were added two open questions in order to elicit more information from pupils who were not already covered in the multiple-choice questions. To avoid making these questions too demanding, leading to the boys not responding, it was thought to request key word answers, in order to obtain additional information.

The research: questionnaire submit (administration)

The research took place on November 6, 2012, and was coordinated with the school's Management Institute, to advise all teachers of different subjects. On the same day there were the monitoring activities using probes and a datalogger.

The activity started at the end of the first lesson, so as to be sure of the presence in the courtroom of all students.

The first questionnaire was administered to a second year class placed on the ground floor on the north side near the entrance. The classroom has the entrance corridor from the north side and the window overlooking the courtyard to the south side.

In this chamber the two probes were positioned: the Datalogger for measuring the indoor microclimate and the probe for measuring the concentration of CO_2. Since the probes had to stay for the whole morning we tried to place them in a location not easily accessible by the children, avoiding accidental bumps.

The shape of the building, the shape and size of the classrooms are similar, all with the entrance from the hallway with a window overlooking the north and south. The only differences are given by the presence of shadows cast by the court or other shielding elements, such as trees and other buildings, according to which the classrooms are located on the first or third. In this case, the students and the micro-climate of the classroom in which have been installed are representative of all classrooms and are the reference case to compare the measurement results with that of the questionnaire.

The entrance in the classroom led to the discontinuation of the lesson of history and Italian, this, in turn, caused a distraction and the boys began to talk to each other. The first difficulty was to attract attention and ask the students to complete the questionnaire, individually, so that each expressed their opinion. At the same time we proceeded with laying the probes in the classroom. The boys showed interest in the probes asking for clarification on their operation and use.

Before distributing the questionnaires the purpose of the research was explained with a phrase like this:

> *The questionnaire will deliver all we need to know about your personal feeling of comfort in this classroom. You will respond individually, and you have to make judgments only on the room, not on other classmates or the teachers or the school. We leave the questionnaires, cross through one answer, we will collect them in 10 or 15 minute.*

After the explanation the questionnaires were distributed. The same operation was carried out in all other classrooms at all levels: the questionnaires were delivered and explained, interrupting the lesson, after which we proceeded to withdraw them.

According to the lesson we could find ourselves in front of teachers more or less available to perform the activity, every teacher proved helpful, and compiled their own questionnaire, trying to reduce the time of interruption of the lesson.

The distribution and collection was made for all classes and all 230 students present. The whole operation took place over about two hours after which it was terminated by measuring the indoor microclimate with the probes.

Monitoring activities: thermal comfort and Indoor Air Quality

The monitoring was carried out with two probes:

- Datalogger (model HD 37AB of DeltaOhm) for measuring the concentration of carbon dioxide (CO_2 in ppm) and carbon monoxide (CO in ppm), in addition to Air Temperature (ta in °C) and Relative Humidity (Rhin %);
- Datalogger (model HD 32 of DeltaOhm) for measuring the indoor microclimate and processing of PMV and PPD, by detecting the velocity (va in m/s) measured with a hot wire anemometer, measuring the Wet-bulbe temperature (Tw in °C), air temperature (Ta in °C), Mean Radiant Temperature Tmr in °C) measured with Globe thermometer, and relative Humidity (HR in %).

In the classroom combustion processes do not take place therefore the presence of Carbon Monoxide (CO) was not verified, while the trend of increase and change in CO_2 concentration allows you to check the level of concentration and dilution following the opening of windows or doors. During the two hours of administration of the questionnaire the windows and the front door were closed, the door was opened at the end of the lesson with the beginning of the interval and the exit of the students.

Figure 11.3 shows the results of the monitoring, the opening of the door and window at the change of the hour leads to a fast reduction of the concentration of CO_2, which increases rapidly with the closing of doors and windows from 400 to 900 ppm. The natural ventilation keeps the CO_2 level within acceptable limits. The trend of temperature and humidity are constant, without substantial changes, to show a stable inside microclimate. The measurement of the data assumes highly variable values due to the end of the lessons with consequent opening of doors and windows and exit of pupils.

The results of the Datalogger for measuring the indoor microclimate were processed using the dedicated software DeltaLog 10, inputting data of the metabolic rate of the boys, taken from the questions of Sect. 11.4 of the questionnaire.

Most of the responses to the question 4.1 were the no. 3, between 1.40 and 1.60 m in height, therefore we decided to use an average height of 1.50 m, while for 4.2 the most responded between 40 and 60 kg, we decided to use a value of 50 kg. The area of Dubois for an individual to 1:50 m and 50 kg is equal to 1.43 m^2. The EN ISO 7730 identifies for sedentary activity study the value of one half, equal to 58 W/m2, and the metabolic activity of the child is equal to 82.89 W.

The EN ISO 7730 has been prepared by referring to an "*average*" adult, suggested as an adult of 80 kg and 1.80 m in height has an area of 1.99 m^2 Dubois and a metabolic activity of 115.52 W for sedentary activity. In order to calculate the PMV and PPD index of children it is necessary to re-proportion the metabolism of the child to that of an adult, so if 115.52 W are equal to 1 half, in the case of a child 82.89 W are equal to half of 0.72 an adult.

To calculate the equivalent half of the child corresponding to an equivalent of Dubois of an adult the following proportion is used:

$$\text{(Metabolic activity child W)}/(1\ \text{met}) = \text{(Metabolic activity adult W)}/(X\ \text{met})$$

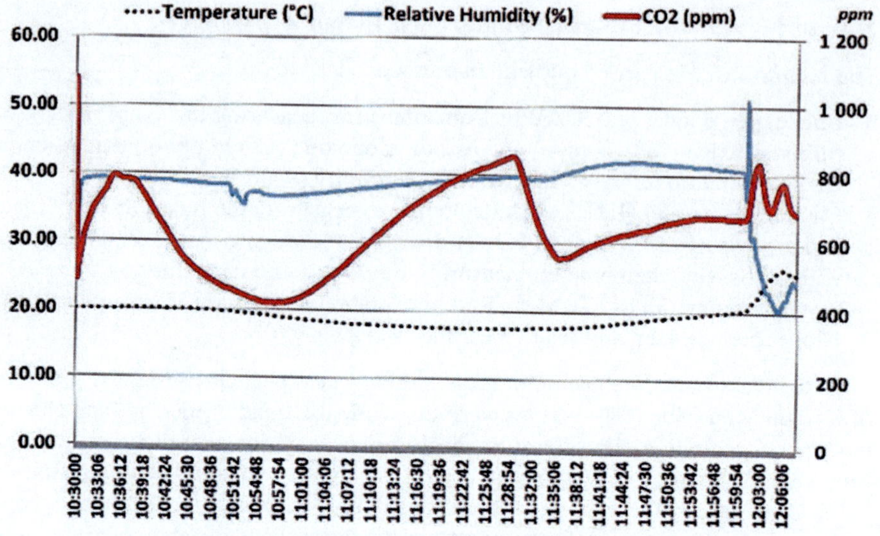

Fig. 11.3 Results of monitoring in the sample probe CO_2 Air Temperature and HR

so

$$82.89/1 = 115.52/X$$
$$X = (115.52 * 1)/82.89 = 1.39 \text{ met}$$

The metabolic activity equivalent of the child amounts to 80.83 W/m^2, then. it is constructed as a subject for simulations of the PMV Index with 1:39 met and a clothing Clo equal to 0.7 (0.1085 m^{2}C/W—pants, tops, trousers, socks, shoes).

Figure 11.4 shows the results of the monitoring in the classroom sample, also in this case the opening of the door and windows at the end of time leads to a fast reduction of the concentration of the temperature, and especially the relative Humidity at the end of the lesson, when the door is opened. The graph also shows the change in the PMV calculated from microclimate data.

The values of PMV are all below 0 corresponding to the value of Thermal Neutral, this may be due to the operation of the heating, the simultaneous presence of more students in the classroom and at a value of metabolism of the boys/girls different from that that it would have had with adults. The value of HR and PMV reduce rapidly with the opening of the doors, the HR reduction depends on the dilution of the water vapour with ventilation due to the opening of doors and windows, and consequently the value of the PMV measured also varies.

The value of PMV and of PPD calculated with the software Deltalog 10, according to the Standard EN ISO 7730:2008 is equal to:

PMV = −0.31 (slightly slightly cold)
PPD = 7.50% (less than one child in 10).

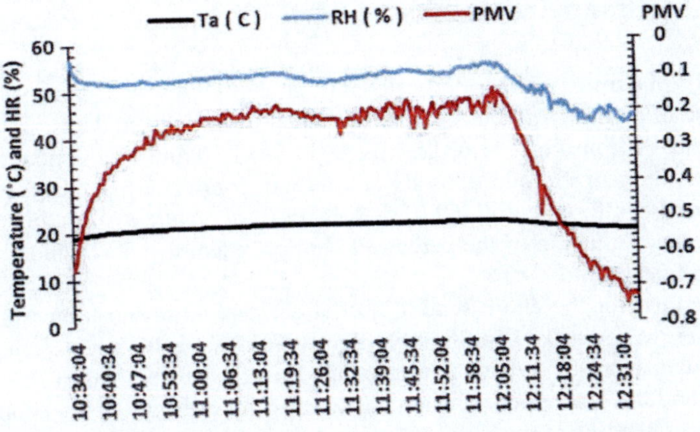

Fig. 11.4 Results monitoring activities in the classroom sample probe HD.32 PMV index (right x-axis), Air Temperature and HR (left y-axis)

11.6 The Measurement Activity: Natural and Artificial Lighiting

The other measurement activities carried out in the same classroom are regards the measurement of the distribution of the level of illumination within the classroom, with only natural light, only artificial illumination and the simultaneous presence of natural and artificial lighting, which is the situation more common during the winter.

The classroom was divided according to a grid identifying six points of measurement of luminance. The results shown in Fig. 11.5, show the distribution of only natural or only artificial light, are not homogeneous, or otherwise does not allow even distribution made with the dual illumination.

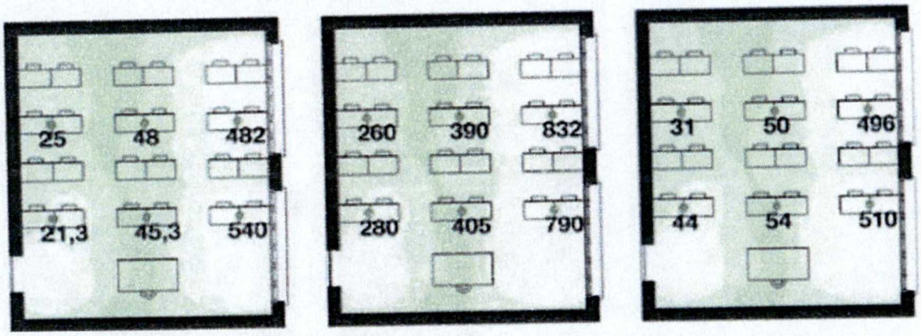

Fig. 11.5 Measuring luminance result activity, right only natural light, left: only artificial light, center: both natural and artificial light

11.7 Questionnaire Results

The results of questionnaires administered by students allow firm conclusions on the perception of comfort and Indoor Environmental Quality by students.

The images from Figs. 11.6, 11.7, 11.8, 11.9, 11.10 and 11.11 show the results of the questionnaires to questions about Thermal Comfort. Most of the boys are in a position of comfort Neutral (62.01% PMV = 0) or Slightly Cold (25.76 PMV = −1), with an evaluation of the perceived comfort equal to a PMV index = −0.32 Slightly Slightly Cold.

The results, for the entire school, do not correspond to those measured in the classroom (PMV = −0.33 55 Slightly Slightly Cold). This highlights a discrepancy between data input of the simulation, related to the metabolism of the boys and clothing, and the perception of Thermal Comfort by the boys. It also confirms the proper understanding of the boys of the contents of the questionnaire, it is even more

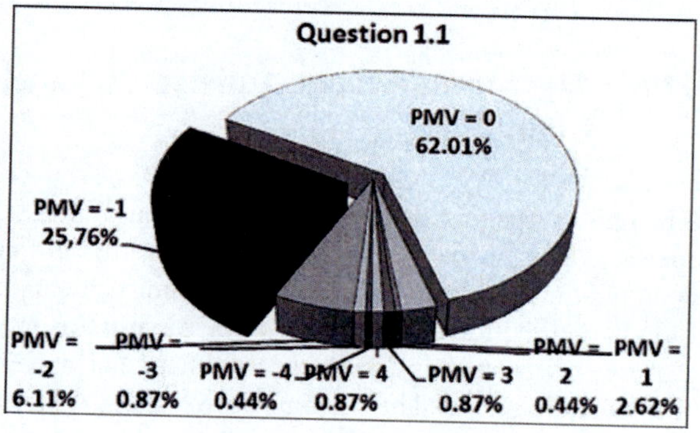

Fig. 11.6 Question 1.1 results

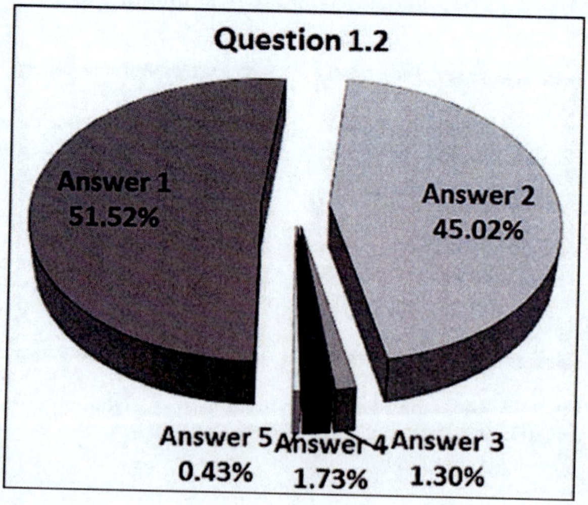

Fig. 11.7 Question 1.2 results

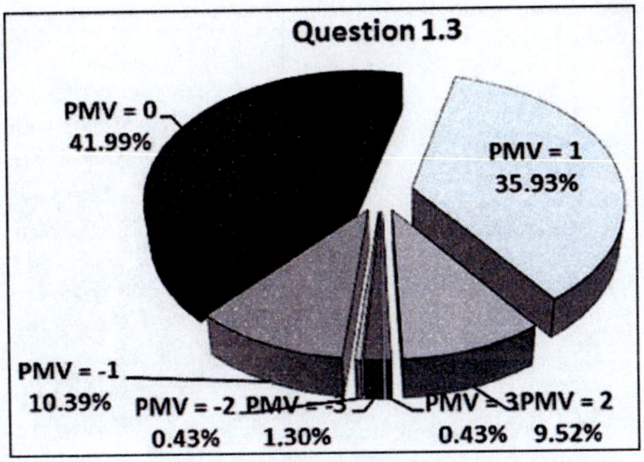

Fig. 11.8 Question 1.3 results

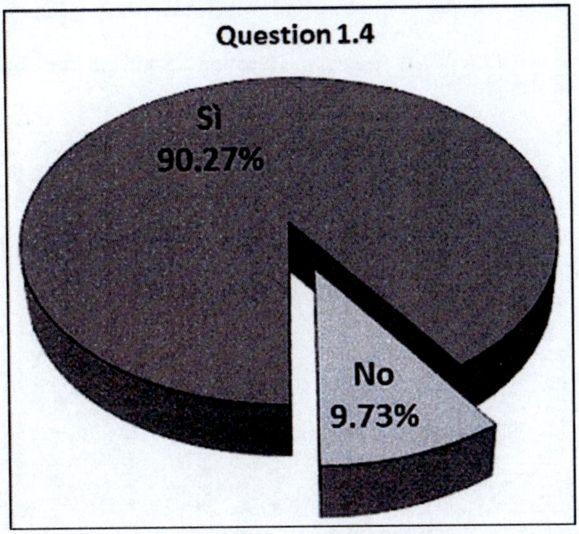

Fig. 11.9 Question 1.4 results

important to compare these results with those of the Thermal Comfort expected from the boys. Most answered "*I'm fine*" inside the classroom (Questionnaire 1.2) and considered the condition as "*Definitely acceptable*" (55.26% Questionnaire 1.5).

The comparison between Thermal Comfort perceived and expected (Questionnaire 1.2 and 1.3) confirms the understanding of the questionnaire, so much so that most of the boys/girls responded that they would like an environment "*Slightly warmer*" (PMV + 1) or "*Hotter*" (PMV + 2) and the need to act on the heating system and on the building (better insulation and/or replacement fixtures). Nevertheless, most of the boys/girls believed "*Definitely acceptable*" (55.26%) the classroom environment, while the remainder believes the environment "*Not quite acceptable*" or "*difficult to acceptable*". Negative feedback is related to the conditions of Thermal

Fig. 11.10 Question 1.5 results

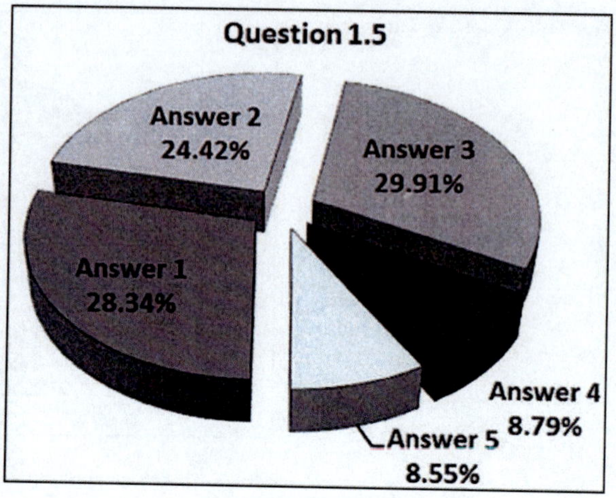

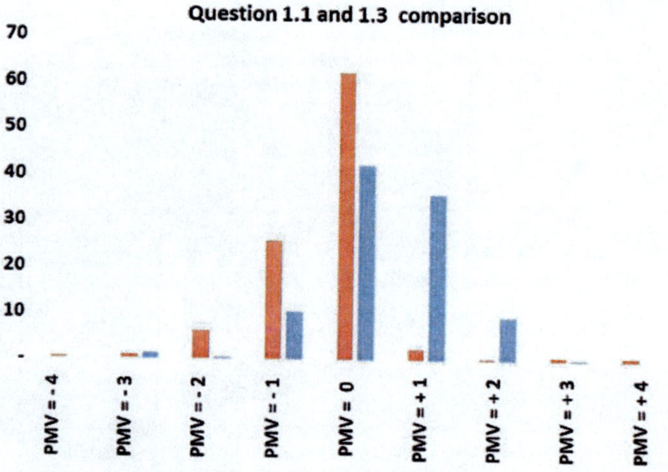

Fig. 11.11 Questionnaire 1.1 and 1.3 results comparison

Comfort, nevertheless they can be influenced, in part, from assessments of the social climate of the classroom or school environment.

The section of the questionnaire on the Visual and Lighting Comfort shows a positive assessment of the level of lighting, both natural and artificial, especially natural light which ensures a more uniform distribution of light in the environment. On this aspect activity of measuring the level of indoor lighting was carried out. The boys/girls are equally distributed in the environment and everyone can read the blackboard (Fig. 11.12).

Figures 11.13, 11.14, 11.15 and 11.16 show the results regarding ventilation and perception of Indoor Air Quality. The boys/girls feel that there are no unpleasant

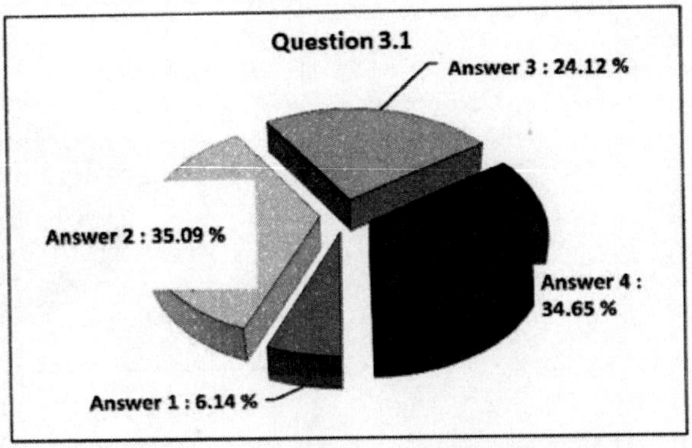

Fig. 11.12 Questionnaire 3.1

odours in the classroom (Questionnaire 3.3 = 74.78%), so the increase of CO_2 concentration that occurs during the morning is not detected as bad odour, while a small percentage (20.43%) believe that there is bad smell at the end of the morning, presumably due to the increase of the concentration of CO_2. Nevertheless, the boys put in relation to the Indoor Air Quality of the classroom the presence of odours or pollutants that come from outside of the classroom or school, or from inside due to the sweat of the people due to the presence of these people. Other phenomena pollutants such as the presence of detergents, chalk or other correlated substances, is not thought possible, and most considered that there is no odour present.

Unlike the Thermal Comfort, Indoor Air Quality is not found by people within the environment a useful indicator to take action to improve the same IAQ. The boys/girls are not good indicators, so much so that almost half would like to smell some odour related to the feeling of a "*clean smell*". In this regard it is noted that the smell of cleanliness, in general, corresponds not to an absence of odours, but the presence of a pleasant smell, which may also be harmful.

Question 4.4 on the feeling of tiredness in the boys/girls, can be correlated, in an indirect way, to the concentration of CO_2 with its perception. On the one hand most

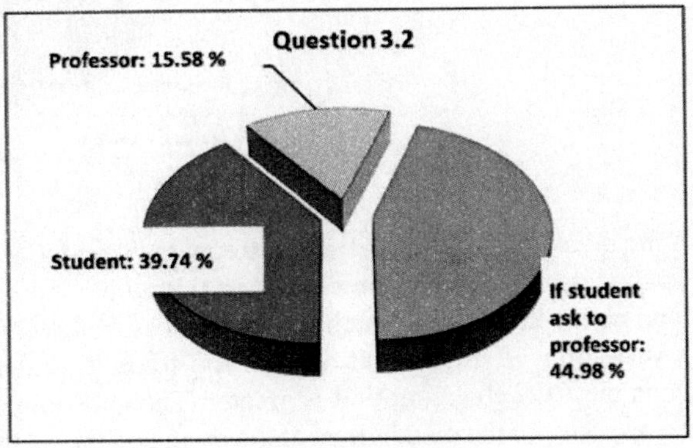

Fig. 11.13 Questionnaire 3.2

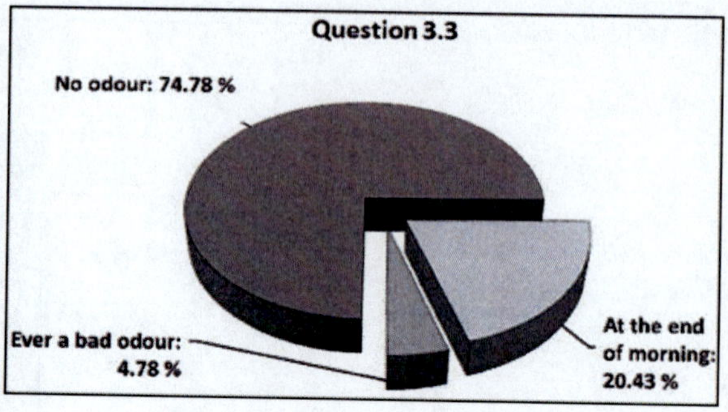

Fig. 11.14 Questionnaire 3.3

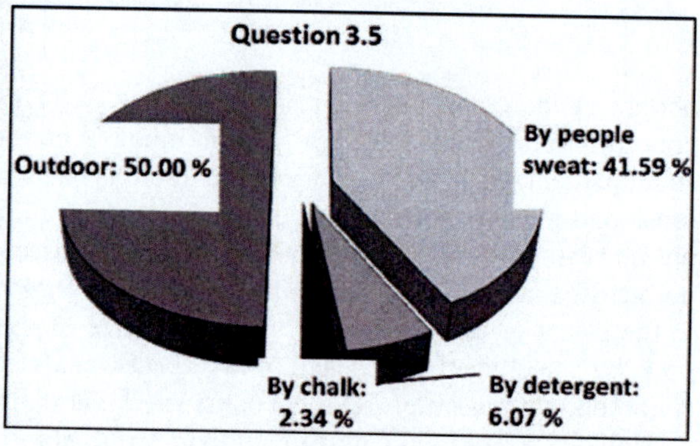

Fig. 11.15 Questionnaire 3.5

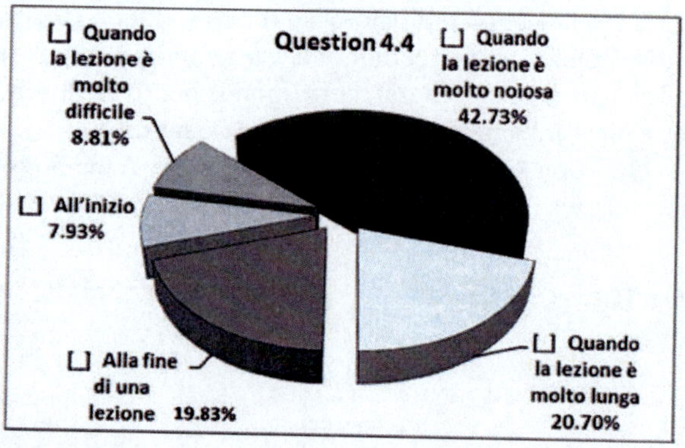

Fig. 11.16 Questionnaire 4.4

of the boys/girls have a feeling of tiredness when the lesson is very boring (42.73%) a parameter that cannot be correlated with the concentration of CO_2. On the other hand about 20% have a feeling of tiredness at the end of the lesson, or when this is very long, in total about 40% of the students, have a greater feeling of fatigue when the CO_2 concentration is greater. The questionnaires, so formulated, confirm the response of the metabolism of the brain and the relationship between the decrease

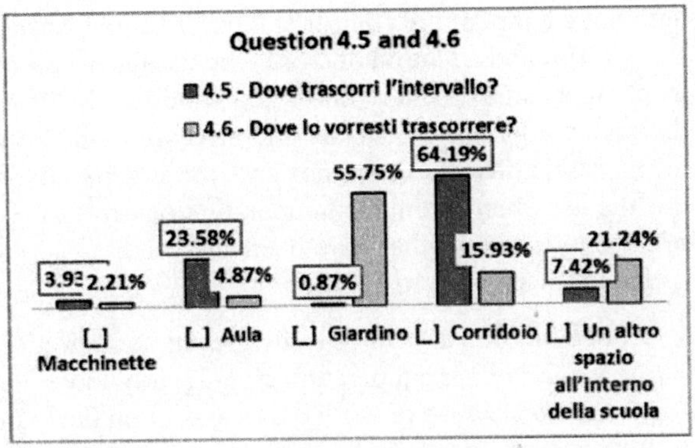

Fig. 11.17 Questionnaire 4.5 and 4.6, comparison. 4.5—Where do you spend the interval? 4.6—Where would you like to spend it? Macchinette = Vending machine, Aula = Classroom, Giardino = Garden, Corridoio = Corridor, Un altro spazio all'interno della scuiola = another internal space

in the concentration of oxygen and the perception of that decrease with the increase of the feeling of tiredness. Fatigue can be an indicator to assess the rate of ventilation and IAQ of an environment, and the relative method of management.

The use of crossed questions is useful for indirectly verifying the physical phenomena related to the microclimate and the indoor comfort.

Other questions let you know how they lived in environments, the Virtual Environmental, on the part of children. For example, most boys/girls believed that the school is colorful and that would like it a lot more colorful, and that the major source of noise is inside the classroom, due, most likely, the vivacity of the children themselves.

Figure 11.17 shows the comparison between the method of use of the spaces of the school and what the boys would like. The majority (64.19%) spend their time during the break in the large hall, or in the classroom or in another school setting (entrance, stairs etc.). Their desire is to spend most of the time in the garden (55.75%) or at least in a space different from the classroom or hallway. This combined with the previous questions can provide valuable information to guide the architectural design.

The section of the questionnaire with free questions was not filled in by all students, and reported comments on the social activity inside the classroom or the school. In particular, there emerged a request and rivalry between the students, all children of the classrooms without the IWB (interactive whiteboard) demanded that the whiteboard should also be installed in their classroom, confirming that schools are a small social community, with its own the social dynamics of small rivalry and comparison between classrooms, student classes and age. The presence of spaces (the garden) allow for socialization among all students.

11.8 Conclusions

The monitoring activity allows us to obtain some useful information for the design of the Virtual Environmental. By analyzing the data obtained with the instrumentation described, and based on the visits it was noticed that there completely lacks an automatic temperature control of the environments and the parameters that influence the comfort and attention of those who use these spaces. The consequence of this is a mechanical adjustment of doors and windows that leads to a strong dispersion of the thermal energy and a considerable increase in the level of noise, both factors which lower the quality of classrooms and, more generally, of the educational spaces. For this the use of an automatic temperature control is needed to improve the conditions of the environments that have been analyzed.

In summary, the project should:

- reduce thermal asymmetry, the greater sensation of cold is also due to a value of the mean radiant temperature significantly lower than that of air, by acting on the thermal resistance of the fixtures and/or on the walls;
- improve uniformity in the temperature distribution within the environment, with low temperature radiators, radiant panels or plates or similar;
- enable temperature control for each environment,
- provide for the common spaces of socialization with a greater variety of colours, space and lighting.

Finally the results show that the use of the questionnaire can be a valuable tool replacing or supporting the monitoring of the indoor microclimate, performed by instrumentation.

Like the difference that can be seen between photography and portraiture, the questionnaire captures results with a different sensibility, which includes social and psychological cognitive-aspects. This last point highlights how important it is to reduce the complexity and number of questions asked to the boys/girls, and above all, to clarify all aspects during the administration of the questionnaire and not during compilation. This activity may be accompanied and supported by other educational activities related to the program lessons of physics, mathematics, science, technology etc.

Acknowledgements The activities described in this chapter is the result of the thesis "A school for tomorrow" Undergraduates: Michele Maraldi and Valentina Keep AA" 2011/2012 rel. prof. Andrea Boeri, co-rel. Professor Kristian Fabbri, Final Synthesis Laboratory "Sustainable Architecture" of the degree course in Architecture, Department of Architecture, University of Bologna.

The Laboratory has its own teaching method in the elaboration of the thesis, whose approach is the integration of the skills of the various disciplines that comprise it: Architecture Technology, Architectural Design and Environmental Physic Technique.

The author acknowledge Michele Maraldi, Valentina Tieni, and prof. Andrea Boeri for their valuable support during experimental activity

Thanks also for the valuable collaboration of the Middle School Tito Maccio Plauto Cesena.

Annex: Results

See Table 11.3

Table 11.3 Second test questionnaire results

1—Thermal Comfort		
1.1—How do you feel at this precise moment?		
☐ I'm too cold (PMV = −4)	1.00	0.44%
☐ I'm very cold (PMV = −3)	2.00	0.87%
☐ I'm cold (PMV = −2)	14.00	6.11%
☐ I'm slightly cold (PMV = −1)	59.00	25.76%
☐ Neither hot nor cold (PMV = −0)	142.00	62.01%
☐ I'm slightly hot (PMV = + 1)	6.00	2.62%
☐ I'm hot (PMV = + 2)	1.00	0.44%
☐ I'm very hot (PMV = + 3)	2.00	0.87%
☐ I'm too hot (PMV = + 4)	2.00	0.87%
Results weighted average PMV I −0.32 (Slightly slightly cold)		
1.2—Consequently then, how do you feel inside the classroom?		
☐ 1—I'm well	119.00	51.52%
☐ 2—I'm quite well	104.00	45.02%
☐ 3—I don't feel well	3.00	1.30%
☐ 4—I feel bad	4.00	1.73%
☐ 5—I feel really bad	1.00	0.43%
1.3—At this very moment, how would you like to feel?		
☐ Much colder (PMV = −3)	0	0.00%
☐ Colder (PMV = −2)	1.30	1.30%
☐ Slightly colder (PMV = −1)	0.43	0.43%
☐ No change (PMV = 0)	10.39	10.43%
☐ Slightly hotter (PMV = + 1)	41.99	42.17%
☐ Hotter (PMV = + 2)	35.93	36.09%
☐ Much hotter (PMV = + 3)	9.52	9.57%
1.4—Taking account of your feelings, do you consider acceptable the current conditions of the temperature inside the classroom?		
☐ Yes	204.00	90.27%
☐ No	22.00	9.73%
1.5—In your opinion, this environment?		
☐ 1—Is very acceptable	126.00	55.26%
☐ 2—Not quite acceptable	60.00	26.32%
☐ 3—Quite unacceptable	35.00	15.35%
☐ 4—Difficult to accept	6.00	2.63%
☐ 5—Unacceptable	1.00	0.44%
2—Lighting Comfort		
2.1—Is there enough light in the classroom to read and write?		

(continued)

Table 11.3 (continued)

☐ 1—It's very dark	6.00	2.61%
☐ 2—It's dark	9.00	3.91%
☐ 3—It's light	204.00	88.70%
☐ 4—It's too light	11.00	4.78%
2.2—When the artificial light is off is there enough light in the classroom to read and write?		
☐ 1—It's very dark	10.00	4.42%
☐ 2—It's dark	62.00	27.43%
☐ 3—It's light	147.00	65.04%
☐ 4—It's too light	7.00	3.10%
2.3—Are you close to or far from the window?		
☐ 1—Next to the window	87.00	37.99%
☐ 2—In the middle of the classroom	72.00	31.44%
☐ 3—Far from the window	70.00	30.57%
2.4—The light in the classroom lets you read the board well?		
☐ 1—Si	191.00	84.14%
☐ 2—No, because it's too dark	6.00	2.64%
☐ 3—No, because it's too far	5.00	2.20%
☐ 4—No, because it reflect the light (it's too light)	25.00	11.01%
2.5 - Do the sun bother your eyes?		
☐ 1—Yes	105.00	47.30%
☐ 2—No	117.00	52.70%
3—Ventilation		
3.1—How often are the windows opened to change the air?		
☐ Never	14.00	6.14%
☐ Once a day	80.00	35.09%
☐ At the end of every lesson	55.00	24.12%
☐ Often during the lessons	79.00	34.65%
3.2—Who decides to open the window?		
☐ Students	91.00	39.74%
☐ Teacher	35.00	15.28%
☐ When a student asks the teacher	103.00	44.98%
3.3—Are there odours inside the room?		
☐ No odours	172.00	74.78%
☐ There is bad odour, towards the end of the morning	47.00	20.43%
☐ There is always bad smell	11.00	4.78%
3.4—What odors would you like in the class?		
☐ No odour	31.00	13.48%

(continued)

Table 11.3 (continued)

☐ Odour of cleaning	114.00	49.57%
☐ It's OK like this	85.00	36.96%
3.5 Where do the bad odours come from?		
☐ From outside	107.00	50.00%
☐ From people's sweat	89.00	41.59%
☐ From detergents	13.00	6.07%
☐ From chalk	5.00	2.34%
4.—General questions		
4.1—How tall are you?		
1 ☐ < 120	2.00	0.88%
2 ☐ 120–140	16.00	7.05%
3 ☐ 140–160	147.00	64.76%
4 ☐ > 160	62.00	27.31%
4.2—How much do you weigh?		
1 ☐ < 40	72.00	32.43%
2 ☐ 40–60	141.00	63.51%
3 ☐ 60–80	8.00	3.60%
4 ☐ > 80	1.00	0.45%
4.3—Do you have difficulty staying alert, regardless of the subject?		
☐ Yes	60.00	26.79%
☐ No	164.00	73.21%
4.4—In which moments do you feel particularly tired?		
☐ At the end of the lesson	45.00	19.82%
☐ At the start	18.00	7.93%
☐ When the lesson is very difficult	20.00	8.81%
☐ When the lesson is very boring	97.00	42.73%
☐ When the lesson is very long	47.00	20.70%
4.5—Where do you spend the break/interval?		
☐ Vending machine	9.00	3.93%
☐ Classroom	54.00	23.58%
☐ Garden	2.00	0.87%
☐ Corridor	147.00	64.19%
☐ Another internal space	17.00	7.42%
4.6—Where would you like to spend it?		
☐ Vending machine	5.00	2.21%
☐ Classroom	11.00	4.87%
☐ Garden	126.00	55.75%
☐ Corridor	36.00	15.93%

(continued)

Table 11.3 (continued)

☐ Another internal space	48.00	21.24%
5.—Colours		
5.1—Your school is colourful?		
☐ Yes	145.00	63.04%
☐ No	85.00	36.96%
5.2—How would you like it?		
☐ More colourful than now	199.00	90.45%
☐ Less colourful than now	21.00	9.55%
6.—Acoustics		
6.1—When you're in the classroom are there noises?		
☐ No noise, silent	8.00	3.48%
☐ Noises outside	44.00	19.13%
☐ Noises in other classrooms	77.00	33.48%
☐ Noise inside the classroom	101.00	43.91%

Reference

Kinney J (2007) Diary of a Wimpy Kid. Amulet Books. www.wimpykid.com/

Part III
New Deal

Classroom environment or Learning environment? Environment is a polysemantic word, that should refer to architecture, as the space between walls, pavement and ceiling, or the relationship between students and teacher during moments of learning or between students as pedagogical and social experiences. In this way, the environment can be good, bad or neutral. Considered in this sense, architecture is a part of the learning experience, it is not just a space. Furthermore, in the space outside the school, such as a playground or green area, the word "environment" will have additional meanings as climate change, climate impact, nature, biology, etc., all of these should be included in the term 'Outdoor education'.

The new part III of the book concerns a New Deal: *Architecture and sensitivity*, in the classroom and how to learn through a space, and the *Outdoor education and thermal comfort*, both are a results of qualitative research in order to promote future studies about these issues.

Chapter 12
Architecture, Comfort and Emotion in the Classroom (Air, Smell, Color)

Abstract Thermal comfort is part of the child's experience in the classroom and at school. The senses, sensations and, consequently, the emotions he or she feels during the day at school, are related to the architecture and to the pedagogical model. This chapter deals with the relationship between the senses and thermal sensation, between emotions and alliesthesia, the role of the senses in the perception of architecture, overcoming the approach to design that sees the predominance of the eye, as illustrated by J. Pallasmaa, towards architecture that considers all the senses.

12.1 Impact of Thermal Comfort Perception During a Lesson: Sense as a Part/element of Pedagogy

The human body comes into contact with the environment that surrounds it in two ways: through the *senses* of its body (sight, hearing, proprioception, touch, etc.) thanks to which it is able to experience *emotion* and *sensation*; or through reason or rationality, through the understanding and '*interpretation of the world*'. The *world* is not to be understood as the *earth* but rather, as what we are able to know. This is true even more so for children, who experience the transition from the world to the universe. To clarify the concept, I borrow two works: Alexandre Koyré's book, *From the Closed World to the Infinite Universe* (Koyré 1957), and the last two lines of Louise Glück's poem *Nostos*:

We look at the world once, in childhood.
The rest is memory.

(Glück 1996)

The environment in which the child is immersed constitutes his or her entire experience: from home to family, from friends to school, everything is part of the world. Experience, growth, interaction with others, children and adults, the school, etc., all bring the child from his or her own world, made up of known things and 'immediate' explanations, to the universe, made up of the immense, infinite and continuously to be discovered. Referring mainly to the role of science, Koyré provides

K. Fabbri, *Thermal Comfort Perception*, https://doi.org/10.1007/978-3-031-52610-7_12

two environments, two ways of understanding the world: one is the *Closed World* that we can assimilate, for the child and for us, as what we know directly that constitutes our experience, the other is the *Infinite Universe*, which includes everything we know thanks to study, interpretations and notions learned.

Children, in defining distances, take themselves as the center of the world, so their own home is the center against which the distance is measured to school, to the friend's house, to the square, etc., places that don't have an effective geographical location. In Italy, for example, the city of Milan or Rome, for a child who does not live there, is a distant place: Milan or Rome or New York or Iceland are the same thing: a place very far from home, unreachable.

The adult, on the other hand, knows the cities even though he or she has never been there and knows where Milan, Rome, New York or Iceland are, and knows the real distance from where he or she is.

Children, at least up to the age of six, before schooling, in defining time use expressions such as '*today*', to indicate the time they are living, '*tomorrow*', to indicate the future, '*yesterday*' to indicate the recent past, or '*the other day*' to indicate the remote past or 'old' experiences of a few weeks or a few months. They sometimes use the expression '*when I was little*' to refer to experiences or events from previous months or years. After schooling and as adults, time becomes objective: the expressions today, yesterday or tomorrow remain, but they refer to moments defined in time with respect to past or future years, the day of the week or the time of day. Adults place events, even personal ones, in a time external to their own experience in objective times. The same reasoning is true for spaces and places: for the adult they are related to each other, there is a path from one point to another, an inside and an outside, the child, on the other hand, may not have a clear path or following the path from one place to another, from home to school, for example, the distances and dimensions of the environments are part of their learning path.

The verses of Glück's poem "say" that the child (all of us, as children) has a complete and total knowledge of the world, he or she knows butterflies, colors, nature, his or her room, the heat, the sun, the cold, the snow, the wind, etc., with all of the child's own body and soul. This knowledge in the adult life is lost. It is lost, perhaps, because of *reason*, interpretation, rationality, or experience—and it is possible to return to it only through memory.

The gaze of a child is not only the sense of sight, but includes all the senses, all the emotions, experiences, feelings that are lived in the environment.

The child perceives the classroom, the external environment, the school through his or her *senses* that determine his or her *emotions* and *sensations*. Thermal comfort, for the child, is not only the condition of balance or the predicted mean vote (see Chap. 4) but it is part of the educational experience, part of the day, and can be a factor of attention or inattention (distraction), and can be (also) a factor of learning.

Thermal Comfort, for children, is not just a thermal or thermo-hygrometric data linked to the environment and their physiology, but rather, *it is part of their world*.

The *senses* are the first link between our body and the world, between the world and architecture. Sensations, emotions and feelings '*pass*' through the senses, as well as rationality, interpretation, knowledge and thought and/or action.

Before continuing, some clarifications on the terms:

a *'sense'* defined as "*one of the five powers (sight, hearing, smell, taste and touch) that your body uses to get information about the world around you*" (Oxford on-line dictionary) thanks to your *'feeling'* as "*something that you feel through the mind or through the senses*".

A *'sensation'* is "*a feeling that you get when something affects your body*" thanks to both, an *'emotion'* "*a strong feeling such as love, fear or anger; the part of a person's character that consists of feelings*" or a *'reason'* "*a cause or an explanation for something that has happened or that somebody has done*". Finally, *'rationality'* "*the fact of being based on reason rather than emotions*".

Following this reasoning, is thermal comfort a sensation, an emotion, or a rational choice deriving from a specific reason?

Thermal Comfort is a judgement on the environment in which the body is located and which it perceives through the senses and with respect to which it feels both emotions (or feelings) and a rational interpretation, thanks to which it issues the judgement: comfortable, warm, cool, and so forth.

Senses and thermal sensation

Being in a given environment, space or place involves a relationship with it determined by one's body, first and foremost, the position of the body. Proprioception or *kinaesthesia* is the sense of "*knowledge of the position and movement of parts of your body, which comes from sense organs in the muscles and joints*" (definition by Oxford Dictionary). The senses each play a role in the relationship with the environment, Thermal Comfort depends on the "sense of warm and cold" due to the thermoregulation systems present in the skin (where the sense of touch is also present), including sweat and perspiration, and in breathing, in particular, the nose and throat (pharynx and larynx), where the sense of taste and smell are present, as well as the internal endocrine and metabolic activity.

Sight (Lighting/Visual comfort) and hearing (Acoustic comfort) are important senses in a classroom for reading, writing, listening or speaking. Smell and taste are influenced by the 'taste' of the air both by the presence of good or bad smells, and by the relative humidity content (water vapor) of the air which determines mass exchanges with the water vapor contained in the breath, skin and saliva. The foul odor or mold can also be felt on the breath and on the tongue.

In this paragraph, we focus on some words related to comfort, starting with the definition of Comfort defined as "*the state of being physically relaxed and free from pain; the state of having a pleasant life, with everything that you need*" (Oxford Dictionary Online).

The senses of the human body are those that allow us to perceive and respond to physical stimuli, to feel physical, emotional or mood stimuli. Through the senses we are able to perceive, passively, and to feel, through sensitivity, actively, attributing a sense to what is perceived of the environment in which our body is located. The word 'sensitive' comes from the Old French *sensitif* *'capable of feeling'* which, in turn, derives from the medieval Latin *sensitivus* from the Latin *sensus* past participle of sentire *'feel, perceive'* (www.etymonline.com/word/sensitive).

The sensory aspect given by the biology of the sense organs is followed by two short-term and long-term reactions: emotion and sentiment.

Sentiment is "*a feeling or an opinion, especially one based on emotions*" (Oxford Dictionary) and it is a long-term response, happiness is a condition that is based on an emotion that lasts or repeats itself over time, such as when a person is together with friends or meets them the next day at school or even after a long time.

Emotion, from the Latin *emovere* '*move out, remove, agitate*', is an immediate response to the situation of relationship with the classroom environment, for example, or with the class of other students. Anger or joy can be related to relationships with other students, to the reward or punishment given by the teacher, but also to the light that enters through the windows, to the warmth or heat. Boredom can be due as much to the lesson as to the CO_2 concentration rate which alters the level of oxygen in the blood and brain by acting on the ability to concentrate.

In any case, the emotion occurs in a momentary datum linked to the experience and, in our case, to the environmental and architectural context in which they live. Emotion allows one to imprint the notions that appear.

The relationships between '*thermal sensation*', '*comfort*' and '*physiological response*' have been at the heart of thermal comfort studies.

P. Gagge himself (Gagge et al. 1967) will consider them to refer to the body's ability to anticipate and respond to changes in the physical environment through a sort of hysteresis effect. The thermal sensation, Gagge argues, can anticipate the body's actions to adapt to the microclimate by means of sweating, vasodilatation, vasoconstriction and shivering.

The response of our body, whether conscious (clothing adaptation) or non-conscious (physiological response) with respect to the climatic conditions of the environment, determines our thermal sensation. Our goal is to ensure a condition of balance and homeothermy between the human body and the environment, when this is achieved the 'neutral' thermal sensation can be considered pleasant or, at least, not annoying, but if this is not achieved and one remains in a condition of discomfort, the thermal sensation is, at least, annoying. The feeling of discomfort can correspond to different emotional states.

The emotional state, in turn, can affect the thermal sensation, for example "*it's cold but I'm having fun playing in the snow with the other children and therefore I can stand the feeling of cold*", or on the contrary, the thermal sensation can affect the emotional state, for example, "*it's too hot in the classroom, this makes me so tired that the lesson is boring*".

The thermal sensation, explains Cabanaìc (1969),depends on two elements: the perception of the characteristics of the physical stimulus and the affectivity which corresponds to the binary condition: pleasure/displeasure, pleasant/unpleasant.

During the school years, learning takes place both through *reason* and *rationality* as well as to *emotion* and *sentiment*. Some notions are learned through the rational way, such as listening to the lesson in the classroom, studying, storytelling, playing, all activities that set in motion the child's conscious states. Depending on the age, in fact, these activities can be more or less complex. Other notions are learned through the emotional way and through the experience lived at that moment. Emotion plays a decisive role in firmly establishing the didactic notions, whatever the topic of the lesson, if the child experiences that moment with annoyance, he or she will reject or associate that notion with a condition of suffering.

Learning takes place *experientially*, for example: immobility, such as sitting for four, six or eight hours in the same desk is an experience, even if it is negative (a condition that reduces the desire and ability to learn). The child actively or passively experiences an emotion related to the lesson and his or her presence in the classroom, an emotion that can be positive (joy) or negative (boredom, anger).

Thermal sensation and the other senses of the human body influence the emotional state and, therefore, have an influence on learning.

Emotion and Alliesthesia

In the first part of the book, we saw the relationship between thermal sensation, indoor thermal comfort and children's perception, and the case studies of the second part confirm that, in relation to the age of the child, the understanding of thermal sensation and comfort depends on different factors than those of adults. Children perceive the world through their own internal narrative of how they know the world, a narrative in which the relationship between thermal sensation and emotional state is '*closer*' than it is in adults. Even more so in an environmental and educational context such as the school where the consensus affects learning.

The adults themselves were once children and each one, with their own personal story, learned notions about thermal comfort. During adulthood, the definitions of thermal sensations conform to common and defined terms and sensations such as hot/cold, wet/dry, notions so common and shared that it is possible to measure Thermal Comfort, based on people's judgements, as shown by Fanger's studies, the study of adaptive thermal comfort and the related ISO and ASHRAE standards. We are talking about a well-established discipline.

The latest innovative research on thermal comfort also addresses the diurnal changes in thermal perception (Vellei et al. 2023), the role of the Circadian Rhythm on Core Body Temperature, known since 1842 (Gierse 1842) as mentioned in Vellei et al. (2021), and *Thermal Alliesthesia*, a term used by Cabanac (1979) to refer to

> *the property of a given stimulus to arouse pleasure or displeasure according to the internal state of the subject.*

The feeling of pleasure or displeasure is influenced by the microclimatic conditions of the environment, whether indoor or outdoor. In an article, Fanger et al. (1977) shows that Thermal Comfort is not influenced by visual stimuli or noise stimuli produced by the environment. The alliesthesia is able to assess how these stimuli affect the internal emotional state and the feeling of pleasure/displeasure. Thermal Comfort affects the emotional state and learning, so it is possible to learn how to design school environments, classrooms, furniture and teaching methods, able to take into account *Emotion* and *Thermal Alliesthesia*.

Alliesthesia describes how the quality of environmental stimuli (environmental as both built environmental, thermal, visual, etc., and social environmental, people, relationships, etc.) affect the internal emotional state, and Cabanac defines it as follows:

> *Alliesthesia refers to a modified outcome or rise in the hedonic dimension of any mental process, from sensation to mental performance, and decision making, leading to physiological comfort and happiness. Positive alliesthesia is the rise in pleasure or joy, and a decrease of displeasure or pain. Negative alliesthesia is the opposite* (Cabanac 2020).

Research in psychology has dedicated much effort in trying to define what an emotion is and what a feeling is (Kleinginna and Kleinginna 1981). The emotion (anger, disgust, fear, joy, sadness, surprise, etc.) consists of a transient psychological phenomenon, corresponding to a mental-state in relation to a mental-experience. To clarify the transitory role of emotion, Cabanac (2002) clarifies that

> *joy is a dynamic, transient and pleasant experience, whereas happiness is a stable indifference.*

The intensity of the emotional experience (internal emotional state) determines the stability of the memory and, I would add, the association of the context, of the place in which the emotion takes place. Pleasure is a mental signal of usefulness, in other words, if I feel pleasure, the mental signal memorizes and learns through the experience. Thermal Comfort can be a condition of pleasure or non-discomfort, Discomfort is always a condition of displeasure.

The place, indoor and outdoor, acts in avoiding situations of non-pleasure or displeasure, or it can promote sensory pleasure and motivate behaviors of curiosity and pleasure (Perlovsky et al. 2010). Outdoor Education, described in the next chapter, makes curiosity and discovery its passe-partout for learning.

12.2 Architecture, Culture and Thermal Comfort Habit of the Children in Space

In the school environment, particularly in the classroom, proprioception or *kinaesthesia* is related to the position of the child's body in space in relation to himself or herself, other children and the teacher.

The child can undergo the school and teaching experience having the perception of his or her body always seated, with little mobility, mostly linked to the upper part of the body—torso, arms, hand, head—while the legs constitute a useless appendage of the body for the duration of the lesson. Educational activities that involve getting up, moving around in the space allow one to stimulate the sense of proprioception and see the classroom in a different way, from several points of view, with a beneficial effect on the educational experience.

In the same way, the arrangement of the desks configures a hierarchy during learning, due to the sense of proprioception (the place of one's body in the 'classroom-world') and sight (based on the position of one's desk, the hierarchy of what I see, who or what is close to me and what is far away, an aspect that determines one's 'role' in the class).

Platooned classrooms, i.e., with desks arranged in an equal way and oriented towards the teacher's desk, involve a hierarchy between the teacher and the students, who are spatially in a condition of spatial inferiority or subordination. The position of the teacher, whether sitting, standing motionless, or moving around the classroom, can change this hierarchical spatial relationship.

This configuration of school furniture involves a hierarchy between peers, among the students themselves, because those who are close to the desk or blackboard will have a different relationship with the teacher and with the other students who are, for example, at the back of the classroom. It is no coincidence that in the configuration of the space with the platooned classroom, students who sit or prefer to sit at the back of the classroom have a 'rebellious' or distracted school behavior, and/or a lower academic performance than others. To be at the bottom is to be at the end of a hierarchy, and it allows one to use the bodies of others to protect or distract oneself.

Peter Weir's *Dead Poets Society* (1989) and Laurent Cantet's *Entre les murs* (2008), just to name two of the best-known films set in a classroom, show this layout of the platooned classroom and its social dynamics. The scene of Prof. John Keating's (Robin Williams) outdoor lecture inviting students to walk, tries to unhinge the hierarchical version of the classroom, the lesson and the pedagogical model of the prestigious College. The film ends with a subversion of the classroom: in homage to Professor Keating who was kicked out of school, the boys, not all of them, climb onto their desks reciting the verses 'O Captain! My Captain!' by Walt Whitman. The shot rises to the level of the students, above the desk: now they see the classroom from a different point of view, the ceiling, the door, the floor, the blackboard, every relationship with architecture has changed.

In *Entre les murs*, young people with difficult family and social situations choose to sit at the back of the classroom.

The arrangement of the desks depends on how the child perceives—with all the senses, sight, hearing, thermal sensation—his or her relationship with the classroom and the class. In the event that he or she stays in the same place for the whole year or for several years, if the arrangement is never changed, the experience the student will have to experience every day will be more and more impoverished, repetitive, without novelty, in other words, boring. Boredom will be the emotion the student will attribute to the lessons, whatever the topic may be.

The position of the desk determines the distance between the teacher and the other children, as we have said, the light from the windows will always be entering from the same direction, and can be a factor of interest or disturbance, there will always be the same lighting and vision of colors, the same view of the outside, the same ventilation, the same smells for all the school years. The position of the windows, of the system terminals (radiators, vents for ventilation, etc.) will affect the local discomfort and the Predicted Percentage Dissatisfied (PPD, see Chap. 4).

During the learning phase, there should be a relationship with the space and the object of the lesson and/or course, because the relationship with the space and the environment influences the senses, the (thermal) sensation, the emotions and, therefore, the emotional component of learning.

The classroom is a device capable of stimulating the senses or, on the contrary, if poorly designed, capable of anesthetizing them. Legislation and standards define levels of temperature and relative humidity, Indoor Air Quality and thermal comfort that have to be guaranteed. The legislation provides that inside the classroom there must not be conditions of too hot or too cold, thermal asymmetries, local discomfort, bad smells, substances, smells or noises coming from outside, noise coming from other environments or reverberation in the same classroom. Standards and literature on the impact of indoor environmental quality in students' comfort (Amoatey et al. 2023) are able to define the levels for measuring, designing and evaluating average standard conditions.

Nothing in the standards prohibits the design of the classroom as a teaching device capable of stimulating interest, even providing elements of distraction or interruption of routine. The furniture should *always* be mobile, able to orient the desks towards the teacher's desk, or to arrange them in a U-shape like the stalls of a theater or in an O-shape, in a circle, like an assembly. The teacher can also stand in the center and not just on one side of the classroom. Windows should be used to look out, and not just to let in light and sunshine; even to the point of deciding to open them in winter to feel a little cold or in spring to be kissed by the sun, smells or insects.

The school is the institution responsible for training as a place of learning and the classroom has been, until now, the main environment within which teaching activities take place. The twenty-first century requires a new perspective, as recalled in the OECD document '*Back to the Future of Education. Four OECD Scenarios for Schooling*' (OECD 2020) and UNESCO, *Learning to become* (UNESCO 2020).

The school space is part of the educational setting. Not just the space, but the entire '*classroom environment*', including the child's Thermal Comfort. In some cases, for example in conditions of local, momentary or chronic discomfort that is repeated over time, thermal comfort can affect school learning, concentration, relationships with others. Added to this is the fact that the condition of perfect neutrality (e.g., PMV = 0) does not occur in reality since the body is looking for new stimuli and constant adaptation to the environment, even more so in children and young people who are constantly looking for sensory and emotional stimuli.

William Butler Yeats said: "*Education is not the filling of a pail, but the lighting of a fire*". For too long, the classroom has been considered as a pail: a place to transfer notions.

In the twenty-first century, the classroom is no longer a closed box, one learns at home, at school, sitting at one's own desk, sitting at someone else's desk, standing or playing. One learns outside and all these different methods not only concern teaching, but also involve the design of environments.

The UNESCO Report '*Reimagining our futures together: a new social contract for education*' (UNESCO 2021) dedicates a chapter to '*Education across different times and spaces*', highlighting the need to foster opportunities for inclusion and sustainability. Schools are a central component of larger educational ecosystems, which needs to be updated as a platform for cooperation, care and also for change.

In the context of the new Contract for Education, it is written that in the face of emergencies related to climate change, persistent poverty and the risks of inequality, pedagogy must be oriented according to the principles of dialogue and action, cooperation and solidarity. The teachers' work will have to be transformative, as written in Chap. 5 '*The transformative work of teachers*':

> *The teacher does not necessarily have privileged access to the sagely truth. Like his students, he is in the process of becoming what he ought to be.*

The report does not provide rules or instructions for the school's architecture, minimum spaces, sizing standards, and so forth. The terms it uses are inclusion and sustainability.

In architecture, sustainability is a notion that has been known since Vitruvius and Leon Battista Alberti (Fabbri 2023), it consists of designing in relation to the context, and both architectural forms and construction technologies must be referred to the context. I would add that thermal comfort should also be referred to the context, trying to avoid situations where the air conditioning system provides internal temperatures that are not consistent with the external ones (e.g., 20 °C indoors when it's 32 °C outside).

12.3 Pedagogy of Architecture, Learning from the Classroom Space

As an architect, I think that the UNESCO Report provides useful stimuli for architects and for those involved in the design and management of that part of the '*school*' place that is the school building.

In studying and writing the first edition of this book, thanks to my experience in the field, I revised and updated my convictions on the relationship between thermal comfort and children. Thermal comfort is measurable, thanks to the laws of thermodynamics and the theory developed by Fanger (1970) and subsequent studies, but this measure does *not* take into consideration the stories, narration, and emotions of children. In the previous chapters we have seen how important the narrative is in the very definition of the predicted mean vote scale.

In this new, updated edition, I start from this point by delving into what emotion is and how and where it can be a moment of play and learning, always keeping in mind what the heart of this book is: thermal comfort, which includes the relationship with the senses and emotions, described in this paragraph, and the relationship between Outdoor Education and Outdoor Thermal Comfort, which constitutes the current area of my research.

The Finnish architect Pallasmaa (1994, 2005) is among the architects who have most deeply emphasized the predominance of the eye in architecture, both in the practice of building construction and in the design, study and history of architecture, compared to the other senses such as the ear and the acoustics of environments. Architecture engages all the senses without privileging any of them. The body is immersed in the experience of the perception of architecture, whether it is inside or outside the building. The architectural form, the materials, the light, the air, the temperature, the reverberation of sound, the smell, the color, all engage the senses.

Architecture is not only about space, but about the environment and involves all the senses:

- *proprioception* or kinesthesia, defined as "knowledge of the position and movement of parts of your body, which comes from sense organs in the muscles and joints" (Oxford Dictionary online) described by Edward T. Hall in The Hidden Dimension (Hall 1966), constitutes the sense that positions our body in space. In the case of children, this little-known sense is the main vector for learning about one's being-in-the-world. The position of the child in the classroom, often constantly the same, influences the relationship with others, the hierarchical relationship with the teacher and with one's peers, in a single sentence: proprioception expresses '*one's place in the world*';
- the *eye* and sight, which allows one to see the spatial-geometric elements that delimit the interior space, floors, walls, ceiling, and also the exterior thanks to the windows, as well as the furniture, the moldings, the colors of the rooms, the relationship of light, shadow, light, dark, shading, and so forth. The eye is the predominant organ in humans. Unlike other animals, in our case sight is the predominant sense of knowledge of the world, particularly in architecture. The training of architects always refers to sight, starting from the project: from the drawing of the floor plan or elevations, to perspective, to 3D models, renderings, etc. However, it is not the only sense in which we know architecture and, in our case, the classroom.

Colors interact with the emotional state of the students, the color red stimulates excitement, motor activities, it is the color of activities, it can affect the rhythm of breathing or heartbeats, it resembles the color of the fatigued or embarrassed face. Blue stimulates relaxation, calm, stillness, slows down the heartbeat and breathing, similar to green, even if the latter is more dynamic, it stimulates collective activities. Yellow stimulates sensations similar to the sun, reminiscent of warmth and feelings of joy, purple or brown are non-primary colors to which children can attribute different meanings, as is gray, an intermediate condition between black, which corresponds to darkness, and white, which corresponds to light.

Beyond this brief excursus that attempts to make a succinct summary of the theories of Goethe (1810) and Itten (1973), color is able to describe the world as Deutscher (2010) writes. Moreover, as in the case studies described in Part Two, the child also associates a thermal sensation with color;

- the *ear* and hearing, which allows one to understand sounds and noises, music, speech and dangerous conditions, and so forth. The acoustics of a room are able to provide pleasure, such as listening to music or to a lesson. It also can provide disturbance, as in the case of noisy environments. In addition, the reverberation of a sound is able to make us understand the size of the environment, depending on if it is large or small, the sound of the voice will be reflected differently. This sense is heightened in those who are visually impaired, but anyone is able to perceive the size and construction characteristics of a room, thanks to the ear;

- the *nose* and smell, which allows one to feel the Indoor Air Quality of an environment, the presence of pleasant or unpleasant smells, produced internally or coming from outside. This happens even before these smells can be associated with the level of pollution. The sense of smell makes it possible to attribute an emotional quality to the air, especially by children, for whom smell is an element of knowledge and play, even in teasing one another. Pallasma refers to the 'space of scent';
- the *palate*, and the sense of taste, even if it is not possible to lick the buildings to taste them (except in the fairy tale of Hansel and Gretel), the taste of architecture is perceived through the air: dry or humid environments, through the breath, give back to the mouth a taste, good or bad. The very dryness of the mouth can correspond to the sensation of a taste;
- *the skin* and *touch*, which concern both the organ (the skin), that allows thermoregulation, and the sense of touch itself, what one perceives by touching things, for architecture one touches the floor with or without shoes, one touches doors, windows and furnishings with one's hands, and so forth. The epidermis is the organ through which we perceive the sensation of heat or cold, lukewarm or cool, and thanks to the thermoregulation of the skin, allows one to "feel the air and the climate" (hot and cold, damp, dry, shivers, sweat, etc.). These sensations correspond, sometimes unconsciously, to an emotional sensation, for example, lukewarm is a thermal sensation AND a state of mind.

In the case of children, the '*tactile value*' and the '*thermal sensations*', as we have described in the previous chapters, correspond to an '*emotional experience*' and to one's own narration of the world. It is a narrative that can be internal (how the child remembers things) or towards others (how he or she reacts or tells them to others). For example, the experience of snow can be positive if the first snowfall of one's life is associated with a positive experience of play, discovery, sharing with friends or family, on the contrary it can be negative if associated with a feeling of cold, loneliness or no experience. In the same way, the sun, solar radiation and the feeling of heat can correspond to emotional or negative experiences.

In the human experience, the body is at the center of sensory experience, and it is through the senses that we experience the world. The eye is the sense that observes and knows everything, but it is also the sense of separation and distance. On the contrary, the skin, with thermoregulation, is the sense of union and interpenetration with the sensitive world.

The arrangement of the desks affects the senses, the senses influence sensations and emotions, emotions influence learning. The position of the body within the space and the characteristics of the space itself, including the air and therefore, the systems and thermal comfort, are architectural issues even before they are pedagogical ones.

Space educates: the position in space, the presence of drafts, known as air-leakage, or the presence of bad or good smells, of light or heat from the sun, of things in the classroom: blackboard, desks, basket, the teacher's desk, paintings, maps, posters, drawings, and so forth. When the child's body is inside the classroom he or she learns from the eye, from the skin, from the nose.

The classroom is the environment in which, or rather, within which the child is immersed with *all* his or her senses. The didactic and pedagogical activities, together with the architecture and furnishings of the rooms, must take into account this simultaneity of the classroom experience. In the classroom the eye does not dominate, the sense of sight allows one to read on the blackboard, draw, etc., but learning takes place by listening to the lesson in a comfortable and stimulating condition, i.e., one that is not boring for all the senses. This does not mean that the environment must be homogeneous, without colors, or aseptic, without odors or neutral, without temperature variations, on the contrary: every difference constitutes a possible new experience. Teacher and designer must include the senses in the design of the lesson and/or classroom.

The school is not the school building just as.

> *The soufflè isn't the soufflè. The soufflè is the recipe*
>
> (said by Clara Oswald, in the Doctor Who series, Season 7).

Thermodynamics is the science that studies the processes and exchanges of energy and mass, these exchanges concern: the body with the environment (e.g., comfort), between bodies (e.g., relationships between peers, classmates or with teachers, i.e., the hierarchies), and between these and the outdoor context, be it the city, nature or society.

The architectural design from the individual classroom (including the furniture, the colors, the windows, the arrangement of the desks), to the layout of the classrooms, from the common areas, internal and external, to the corridors, from the office services, to the distribution and hierarchical paths among the various functions, reflect the educational and pedagogical model.

The body, with all its senses, crosses the environments, constantly exchanging relationships.

The classroom and the corridor are the solid translation of the dominant pedagogical and educational model and affect both teaching and learning, as well as sensory and emotional aspects. The sensation of heat or cold is comparable to that of anger or joy, they are always emotions that run through the body.

The stimuli present in the classroom act on the child's emotional state, whether they are of a physical nature (the classroom), or relational (with other children or with the teacher(s) or experiential (physical activity, play, laboratory, artistic or musical activity). Thermal Comfort acts on the emotional state in all three cases since it depends on the physical characteristics of the environment (see Chap. 3), on the metabolism which depends both on the emotional state and on the teaching activity carried out.

Educational activities outside the classroom, in spaces adjacent to the school such as courtyards or urban parks, or in nature and, in general, all contexts in which Outdoor Education is practiced (see Chap. 13) multiply the number of sensory, relational and experiential stimuli. This is thanks to the relationship with nature, the play relationship with classmates that takes place outside and, at the same time, the autonomy of the individual child.

Interaction with nature has benefits for the mental health of children and teenagers, as reported in the literature (Tillmann et al. 2018). Mental health includes their emotional, psychological and social well-being and affects, and scientific literature also includes research on hyperactivity, inattention, depression, ADHD and stress. The natural elements include green space, water/blue space (rivers, sea, lake), outdoor adventure, horticulture, vegetation, wilderness therapy, generic outdoor space, schoolyard, playground, grass, garden, yard, park, forest school.

It is interesting to study the relationship between outdoor education and thermal comfort, also in order to know when the emotional state is such as to be detrimental to thermal comfort and lead to dangerous situations.

References

Amoatey P, Al-Jabri K, Al-Saadi S, Al-Harthy I, Al-Khuzairi M (2023) Impact of indoor environmental quality on students' comfort in high school buildings during the summer season in an extreme climate. J Architectural Eng 29(3). https://doi.org/10.1061/JAEIED.AEENG-1468

Cabanac M (1969) Plaisir ou déplaisir de la sensation thermique et homeothermie. Physiol Behav 4(3):359–364. https://doi.org/10.1016/0031-9384(69)90189-9

Cabanac M (1979) Sensory pleasure. Q Rev Biol 54(1):1–29. https://doi.org/10.1086/410981

Cabanac M (2002) What is emotion? Behav Proc 60(2):69–83. https://doi.org/10.1016/S0376-6357(02)00078-5

Cabanac M (2020) Alliesthesia. Update of the word and concept. Am J Biomed Sci Res 8(4):313–320. https://doi.org/10.34297/AJBSR.2020.08.001293

Deutscher G (2010) Through the language glass: why the world looks different in other languages. Henry Holt and Company

Hall ET (1966) The hidden dimension. Garden City

Fabbri K (2023) La forma e la formula. La sostenibile leggerezza ovvero cosa intendi per sostenibilità. Libria Editore

Fanger PO (1970). Thermal comfort. Analysis and applications in environmental engineering

Fanger PO, Breum NO, Jerking E (1977) Can colour and noise influence man's thermal comfort? Ergonomics 20(1):11–18. https://doi.org/10.1080/00140137708931596

Gagge AP, Stolwijk JAJ, Hardy JD (1967) Comfort and thermal sensations and associated physiological responses at various ambient temperatures. Environ Res 1(1):1–20. https://doi.org/10.1016/0013-9351(67)90002-3

Gierse A (1842) Quaemiam sit ratio caloris organii

Gluck L (1996) Meadowlands. Harper Collins

von Goethe, JW (1810) Zur Farbenlehre (Theory of colours) (John Murray)

Itten J (1973) The art of color: the subjective experience and objective rationale of color. Van Nostrand Reinhold

Kleinginna PR, Kleinginna AM (1981) A categorized list of emotion definitions, with suggestions for a consensual definition. Motiv Emot 5(4):345–379. https://doi.org/10.1007/BF00992553

Koyré A (1957) From the closed world to the infinite universe. John Hopkins Press

OECD (2020) Back to the future(s) of education. https://doi.org/10.1787/178ef527-en

Pallasmaa J (1994) An architecture of the seven senses. In: Holl S, Pallasmaa J, Pèrez-Gomes A (eds) Question of perception: phenomenology of architecture. A+U PUblishing, pp 29–37

Pallasmaa J (2005) The eyes of the skin. Architecture and the senses. John Wiley & Sons Ltd

Perlovsky LI, Bonniot-Cabanac M-C, Cabanac M (2010) Curiosity and pleasure. In: The 2010 International Joint Conference on Neural Networks (IJCNN), pp 1–3. https://doi.org/10.1109/IJCNN.2010.5596867

Tillmann S, Tobin D, Avison W, Gilliland J (2018) Mental health benefits of interactions with nature in children and teenagers: a systematic review. J Epidemiol Community Health 72(10):958–966. https://doi.org/10.1136/jech-2018-210436

UNESCO (2020) Learning to become with the world: education for future survival (Background paper for the Futures of Education initiative)

UNESCO. (2021). Reimagining our futures together: a new social contract for education. UNESCO. https://doi.org/10.54675/ASRB4722

Vellei M, Chinazzo G, Zitting K-M, Hubbard J (2021) Human thermal perception and time of day: a review. Temperature 8(4):320–341. https://doi.org/10.1080/23328940.2021.1976004

Vellei M, Pigliautile I, Pisello AL (2023) Effect of time-of-day on human dynamic thermal perception. Sci Rep 13(1):2367. https://doi.org/10.1038/s41598-023-29615-8

Chapter 13
Outdoor Education and Thermal Comfort

Abstract The chapter considers the role of Outdoor Thermal Comfort during Outdoor Education. Outdoor Education consists of carrying out educational activities or experiences with children outside the classroom, in the schoolyard, in open spaces or in nature, even including forests or the wilderness. The climate, in particular the study of the outdoor microclimate, makes it possible to evaluate the thermal sensation, expressed as Outdoor Thermal Comfort, determined by the energy exchanges between the human body, the sun and the external environment, with the aim of expressing indexes of heat stress or thermal discomfort. Finally, the relationships and connections between Outdoor Education and Outdoor Thermal Comfort are addressed, as a future research path in the framework of the various actions underway for the reduction of Urban Heat Islands and the implementation of Nature Based Solutions.

13.1 Outdoor Education: A Proven Pedagogical Way

Outdoor Education includes educational and pedagogical activities carried out outside the school building: from the schoolyard to the forest, even in the wilderness. It is a system that has been in use for decades with a consolidated practice and literature both in a pedagogical and didactic context, of which, here, a summary of the main contents and methods of development is reported. As far as children's perception of thermal comfort is concerned, the main difference lies in the fact that they are outside a confined environment: *the class is outside the classroom.*

In an outdoor environment, the system of relationships with the physical space of the classroom and with others, classmates, teachers, etc., is disrupted, we can even say it is reinvented and "up for grabs", with the result of new relationships being established with the context, with the people and with the activities that are carried out. The pedagogy of Outdoor Education focuses on learning and getting to know the world through relationships and experience. The diverse experiences tend to combine outdoor activities with educational paths related to the physical or natural sciences, or with the learning and development of one's own skills, through the willingness

K. Fabbri, *Thermal Comfort Perception*, https://doi.org/10.1007/978-3-031-52610-7_13

of children to rise to the occasion or the overcoming of challenges, whether they be individual or group challenges. Although these are the most common methods, in the outdoor environment it is also possible to learn abstract subjects such as mathematics, literature, art, and so forth. The only thing that changes is the context and the way the lesson is carried out.

Outdoor Education, according to the Oxford Dictionary, is defined as education.

> *that takes place in the natural environment. It includes conventional field studies, outdoor pursuits, and any other educational activity which takes place in the open air.*

In reality, Outdoor Education can be expressed as 'staying in nature' or as 'staying outside'.

Outdoor Play and Learning (OPAL, by Wallers et al. (2017)) is a consolidated international experience, in a theoretical and conceptual framework where outdoor activity gives preference to play, in the double meaning of playing and acting, i.e., the child/young person must '*take on a challenge*' by establishing a relationship with the environment, understood as nature and open space.

Another definition considers Outdoor Education as an educational strategy that provides explorative experiences connected with knowledge as in open-ended play (Cutter-Mackenzie and Edwards 2013). Articles published in the journal *Pedagogy of place*, points out that outdoor education has often been understood as an activity carried out in nature through risky or adventurous activities (Mullins 2013; Wattchow and Brown 2011).

In this chapter, Outdoor Education refers to being outdoors, carrying out educational activities in open spaces, which can be:

(a) open spaces in the vicinity of the school such as courtyards, schoolyards, playgrounds, with or without vegetation;
(b) urban spaces such as squares or urban parks;
(c) local parks or natural areas in the vicinity of the school or in any case close to the school;
(d) forests, sea, river, lake (also called Blue Education) or other adventures in the Wilderness.

The levels of '*naturalness*' can be different and each one of them interesting, since each context provides different stimuli and relationships with the environment: for example, in the schoolyard it is possible to learn by looking at the school building itself, to discover the windows (to see from the outside what is usually seen from the inside), see the regularity or irregularity of the design of the façade, the height of the building compared to the other surrounding buildings, the decorations and all the architectural elements that can be part of an outdoor lesson as well as the activities carried out in direct contact with nature (Fig. 13.1).

Outdoor space is a learning environment capable of promoting interconnected knowledge. The 2030 AGENDA, drafted by the UN and UNESCO (UNESCO 2011) promotes Outdoor Education, a well-established practice in pedagogy and teaching (Smith 2002; Waite 2020).

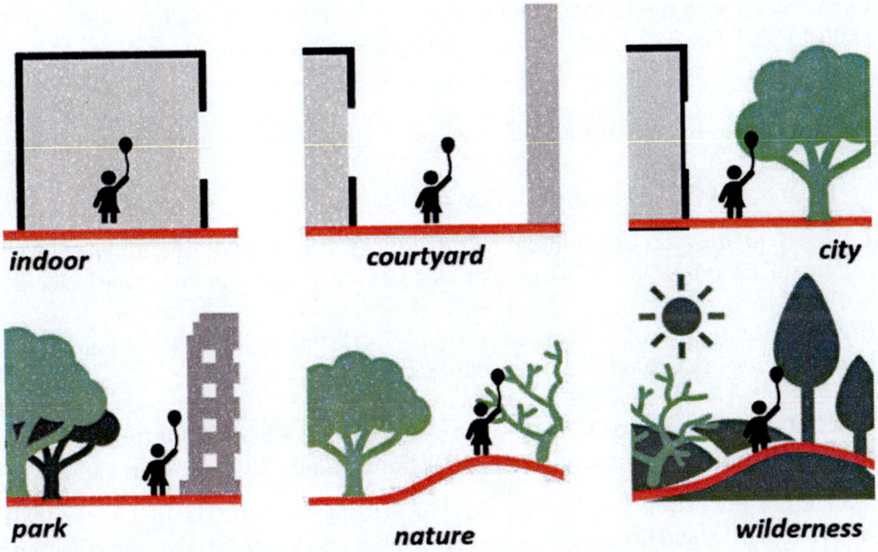

Fig. 13.1 Representation of the different kinds of open spaces for Outdoor Education

From a pedagogical and didactic point of view, the main distinction consists between doing school '*inside*' and doing school '*outside*'. In this second case, the literature reports that it is necessary to change the pedagogical approach because we have different instruments: the space is different, just as the experience of being in space is different; the equipment is different, there are no blackboards, no desks, there are natural elements and climatic variations. In addition, there are continuous external stimuli and the teacher is not in a hierarchically superior position in the space (unlike when, in the classroom, he or she is at the desk/blackboard in front of the students). Authority gives way to authoritativeness and the teacher's way of relating with the students.

Outdoor activity stimulates emotion, which is a determining element for learning, adventure, risk and, therefore, self-esteem; it promotes creativity, imagination and problem solving, requires listening skills and respect for others, allows people to learn to be together, promoting sociability and, at the same time, enhancing children's autonomy.

Outdoor activity allows the development of non-abstract knowledge, transversal skills, soft skills and life skills: '*mens sana in corpore sano'*, which means 'a healthy mind in a healthy body'.

The class climate, given by the relationship of the people involved, must interface with different spatial relationships. The child is no longer motionless and always sitting in the same place, but constantly changes position in a context of peer relationships. The outdoor context can disrupt the dynamics of relationships between pupils such as, for example, bullying. The relationship with the teacher is no longer hierarchical, with the role of command, on the contrary, the teacher must acquire skills to play the role of guide and *mentor*, and encourage new experiences.

It should be remembered that teachers, if not trained on how to carry out outdoor activities, tend to consider outdoor activity as a recreation for themselves, and leave children free to roam in the open spaces, especially in the schoolyard. This attitude is flawed.

The place of the educational context can be:

(a) *next to* the school building (courtyard, playground) in an urban environment (Ward 1978, 1979) or natural (e.g., vegetable garden or agri-crèches, agri-kindergartens);
(b) *close to* the school, such as an urban park or neighboring meadows or fields if the school is located in a rural setting, or
(c) *distant*, when the activity takes place in natural places such as woods (Constable 2014) or through activities such as scouts, adventure education, German Waldkindergarten or English Forest Schools (Claire 2012; Knight 2009).

Outdoor Education was born in the 1930s thanks to Kurt Hahn (James 1990), who fled to Scotland because of Nazi persecution and where, in 1941 together with Marina Ewald, he founded '*The Foundation of the Outward Bound Organisation*'. The curriculum studiorium for children includes outdoor activities, distinguishing between Outdoor Learning (OL), which is part of the formal school curriculum, and Outdoor Adventure Education (OAE) which is part of the extra-curricular, non-formal pathway.

More information can be found in the Australian Journal of Outdoor Education, Journal of Adventure Education and Outdoor Learning, Journal of Experiential Education and Journal of Leisure Research.

The pedagogical experiences and research in the following years would expand the themes of outdoor activities, both in terms of their role in the psycho-pedagogical development of the child, in relationships with others and in leadership (Pelchat and Karp 2012), and in terms of the modalities and contents of these activities, in what is called Place-Based Education (Sobel 2004).

Outdoor Education takes the outdoor space as a *learning environment* and a place of normal life for children. Farnè (Farnè et al. 2022) argues that there is no contradiction between the internal and external space of the school, since the two spaces are in a seamless relationship with each other. The doubts that are usually raised, such as the difficulty of controlling children (doubt, on the part of the teachers), the risk of getting sick (the doubt held by the parents), the adverse weather conditions, especially in winter, can be solved and faced, indeed the ways in which they are resolved are actually elements of the learning experience.

Outdoor Education acts on the mental, physical and social well-being of children and adolescents, particularly when combined with immersive experiences, especially in nature. The influence of the environment on children's development is known in the literature (Brooks-Gunn et al. 1993) the presence of parks near the home, for example, or other places to play where nature is present, allows children to carry out activities, explore, receive stimuli, discover the world and discover themselves.

On the contrary, the scarcity of opportunities for children to go to places to play and have social interactions negatively impacts their social-emotional competences (Frost et al. 2007; Pellegrini and Smith 1998).

The issue may seem, at some level, to be stating the obvious, but children are denied the benefits of these experiences in school curricula and educational programs where outdoor activities are limited to the moment of recess, entering and leaving the classroom or school. As a consequence, the child associates the open space and outdoor activities only with play, freedom, euphoria and de-stressing activities compared to the 'classroom' phase.

Experience shows that in schools where Outdoor Education is adopted for the first time, when children leave the classroom and are in the open space for the first time, it is a moment of pure euphoria and uncontrolled excitement: children may leave the classroom or school as if they were animals freed from captivity, and vent all their frustration, compressed by remaining motionless in the classroom, in an intense physical activity such as running, screaming, jumping, in a disorganized and excessive way, which does not allow the relationship with other children and with educators.

The experience of Outdoor Education, especially when scheduled, allows one to avoid these situations of euphoria that involve stress for the body, the psyche and the well-being of the child. And it can also allieviate the stress of the educator who, if not adequately trained, has the task of supervision alone.

Outdoor activities (in the garden, on the grass, in the courtyard), even in the presence of natural elements (urban park, green space) or directly in nature (forest, wilderness, blue education) educate the child to "be in the environment", to live in the moment, a condition that includes all the senses, even thermal sensations and the consequent emotions and inner motions. In addition, outdoor activity conducted according to the criteria of outdoor education improves physical and emotional development, as well as enhancement of language and communication skills (Hayley et al. 2015).

While outdoor play and education stimulate the children's imagination and creativity in different settings, on the contrary, a poor, repetitive design of urban environments can have a negative impact on children, and instead promote conditions of degradation.

The interactions and activities carried out with nature, and in nature, can be:

(a) *accessibility* to green space, which is the minimum level;
(b) *passive* or incidental exposure to nature; and
(c) *active use and engagement with nature*, as in forest school or blue education activities, and so forth.

The benefit is such that there is no need to identify a minimum or maximum time spent outside, especially if immersed in nature.

The scientific literature (Mygind et al. 2019) is also broad in this case because it acts on mental health, resilience (O'Brien and Lomas 2017) in problem-solving and self-concept skills, on mood, cognitive performance, skills and behavioral indicators, even in the case of specific conditions such as Attention Deficit Hyperactivity Disorder (ADHD).

'*Nature pedagogy*' (Waller et al. 2017) consists of carrying out outdoor educational activities in a natural setting by connecting children with nature through practices to learn the concepts of ecology, biology or sustainability. Teaching in nature follows design approaches that enhance children's skills in learning even through the unexpected, which become an opportunity for making new discoveries.

In general, the pedagogical and didactic approach can be expressed in various ways, as 'learning-*the*-nature' (e.g., the naturalist approach, ecology, geography, orientation, etc.), 'learning-*in*-nature' (e.g., outdoor play activities, proprioception, relationship dynamics, etc.) or 'learning-*from*-nature' (e.g., biology, physical mechanisms, chemistry and the like).

Whatever the approach, the relationship between sensory experience, learning and physical space is central. Compared to indoor activities, outdoor activities make it possible to increase opportunities of simulation related to negotiation skills, cooperative play, social and communicative skills, both in school activities (Liu and Green 2023; Moll et al. 2022; Taylor et al. 2002) and pre-scholastic ones such as in nursery schools (Fjørtoft 2001; Monti et al. 2019). In addition, it is possible to see a relationship between effects of physical interaction with use of the capacity for symbolization and imagination (O'Brien and Murray 2007). This aspect will have reverberations on the construction of the child's inner world and the experimentation of his or her own sensations, including those related to thermal comfort.

The role of parents, teachers and the '*educational setting*'

With respect to the activities of outdoor education, *parents* show an overestimated perception of the social risk involved and the potential damage to children's health when compared to the *children*'s abilities to react and understand actions, difficulties and their tolerance of thermal comfort (Hallam et al. 2021; Prezza et al. 2005; Rotaris et al. 2023).

In particular, taking into consideration the issue dealt with in this book, the weather and climate conditions, the risk of getting sick and the well-being of children are directly linked to outdoor thermal comfort and depend on clothing, climatic conditions and the design of the open space, on the conditions of greater activity of the child which affects both the metabolism and perception of the environment (sun or rain may not be a problem) as well as the motivation and interest in the activity and in learning.

The educational skills of *teachers* to carry out and guide outdoor activities concern the ability to rethink their own position as an adult in a situation where authority is not given by one's position and it is instead one's actions serving as an example, where everything that is present in the open space, habitual or unknown, can be a reminder of education and learning. The adult educator must know the characteristics of the outdoor spaces and acquire the relative expertise on the sites. In general, the educator must get used to being outside and in nature by seeking his or her own well-being in the relationship with nature and outdoor space in general (Chawla et al. 2014; Dadvand et al. 2015; Li et al. 2022; Liu and Green 2023; Moll et al. 2022).

The *educational setting* must define the scenario of space–time-material (objects), i.e., what is the outdoor space identified for outdoor education, the time during which to carry out the activities and what are the objects and the rules of the context; what are the forms of communication, when to speak, when to be silent and to listen; what are the rules that determine the climate and relationships within the group.

The group may have a collaborative or competitive, contractual or decision-making orientation, with or without leadership. These elements depend on the pedagogical models implemented by educators and on processes of attributing meaning to the educational experience carried out outdoors.

In the outdoor environment, unlike the 'traditional' indoor lesson, the pedagogical path may not have precise didactic/educational or training paths. Outdoor activity can be, at the same time, a risk, a limit and a challenge to change. Children's personalities can change, even considerably, during outdoor activities, improving interpersonal skills, favoring jokes or spontaneous play.

Experiences lived in an open-air context, urban or natural, can also affect the existential state of the person experiencing them, when they settle in the memory. Everything related to the outdoor experience, as an activity that takes place through the body, can be fixed in the child's memory as a lasting memory, whether positive or negative, related to the context, clothing, relationships, etc., including the response of one's body with outdoor thermal comfort.

Children are embodied with their experiences during the outdoor educational path that involves (a) *research*, (b) *thematization*, also present indoors, and (c) *problematization*, in the focus of attention (Freire 1970, 1998; Mahler et al. 2018; Menolascino 1972).

In the outdoors, the educational relationship tends to be more spontaneous, and the sensorial awareness increase as the senses are stimulated in a rich and simultaneous way. When materials, furnishings, objects, etc., are not structured, the unexpected or the change can come about easier. We realise that the same tool can be used for various things, outside there is '*more body*' and greater active attention that involves an emotional processing that subsequently becomes a cognitive category. The construction of outdoor teaching spaces can be a solution to turn the city into a classroom (arki_lab, 2019).

The educational setting may require that the teaching materials (chairs, tables, writing materials, etc.) are *brought* to the site, or already *set up* on site (outdoor place already equipped and furnished, street furniture), or *present* on site (e.g., sitting on the ground, on a log, using nature to learn, not taking notes, etc.) or even *absent* and *improvised*.

The scientific literature and experiences on Outdoor Education refer to the relationship with nature and include educational activities to learn about sustainability, biology, ethology, and so forth. As mentioned, outdoor activities are not limited to this kind of content: outdoors one can also learn traditional subjects such as mathematics, language, literature, and so forth.

In particular, i want to point out a notable lack in outdoor education, namely, the absence of education regarding architecture and the city.

Lewis Mumford in *The City in History'* (Mumford 1961), argues that the city is humanity's greatest invention and claims that "*Cities are a product of the earth*" (Mumford 1938). The city has been with humanity since the beginning, from the first human settlements, up to the construction of Aleppo, one of the world's oldest cities, founded at least in 4000 BC, and the founding of the first settlements in the Middle East. The problems that have emerged in the late twentieth century and the arguments related to sustainable development cannot make us forget this fact.

Outdoor Education, in my opinion, should also include understanding the grammar of urban spaces and architectures. Children should be able to recognize the space and environment that surrounds them, even in the horror or emptiness of the ghetto; without knowledge, it is impossible to think of improving open spaces in urban areas, those in which the community lives. Being outdoors in an outdoor space does not mean being in a new isotropic space without identity. We could talk about an Urban Outdoor Education.

Having clarified this aspect, which is particularly close to my heart, in this chapter I would like to address, in summary, some aspects that the pedagogy of Outdoor Education does not consider, in that it is extraneous to the discipline itself.

In particular, I'd like to focus on:

(a) what the outdoor microclimate is and how one studies it, what the physical/environmental factors that characterize it are and what the models to study them are;
(b) how to assess outdoor thermal comfort and outdoor comfort/discomfort indicators;
(c) the architectural and furnishing elements of outdoor spaces (pavement, grass, etc.); and
(d) the relationship between the emotional states of thermal comfort with experience and learning, to conclude with a proposal of considering thermal comfort in outdoor education.

13.2 What the Outdoor Microclimate is and How It Can Be Evaluated and Simulated

The microclimate in outdoor space is different than that of indoor space, not only due to a space dimension but also concerning the equation to define physical variables, body energy balance, solar radiation, and so forth. In this section we report a short summary of the research on the outdoor microclimate (e.g., environment simulation, outdoor microclimate mapping, urban heat island).

The confined or indoor environment is characterized, in fact, by having a boundary defined by fixed vertical partitions, such as walls, or mobile partitions such as doors and windows, and by horizontal partitions, such as the floor at the bottom and the roof or ceiling at the top.

The boundary forms a physical separation from the outside, even if a visual continuity remains, for example, through windows. The confined space separates a volume of air from the outside, so the physical parameters of the air, such as air temperature, relative humidity, mean radiant temperature, etc., do not depend directly on weather or climate conditions, but on the building envelope and the HVAC. The microclimatic characteristics of a confined environment are, generally, constant or with a reduced range of variations, as described in the previous chapters.

In the outdoor environment, the microclimate does not have a confined space. In the open space, in the outdoors, the only boundary is the ground on which one stands and the celestial sphere, so the phenomena and the study of the exchanges of energy and mass between the human body and the environment require different modeling criteria from those described in the previous chapters, just as there are different indexes to evaluate the thermal comfort of people. For example, in a confined space, the human body exchanges energy by radiation with the six surfaces that delimit the space: four vertical walls, the floor and ceiling. Outside, on the other hand, the human body exchanges energy by radiation with the soil and the sky (or celestial sphere) (Fig. 13.2).

Outdoor microclimate modeling

The virtual model is a tool we use so as to understand reality. Climate modeling involves different scales, and each scale requires different input data and returns different outputs. The first distinction is between climate and weather, where *climate* refers to phenomena on a global/world scale and refers to long periods, centuries or millennia, while *weather* concerns phenomena of a territorial nature and refers to short modeling periods, on the order of a few days and, usually, with a predictive character. The climate looks to the past, back centuries and millennia, to understand the trend of future years, while the weather predicts the next three or four days. The distinction is due both to the different dimensional and temporal scales, and to the number of variables and calculation models adopted.

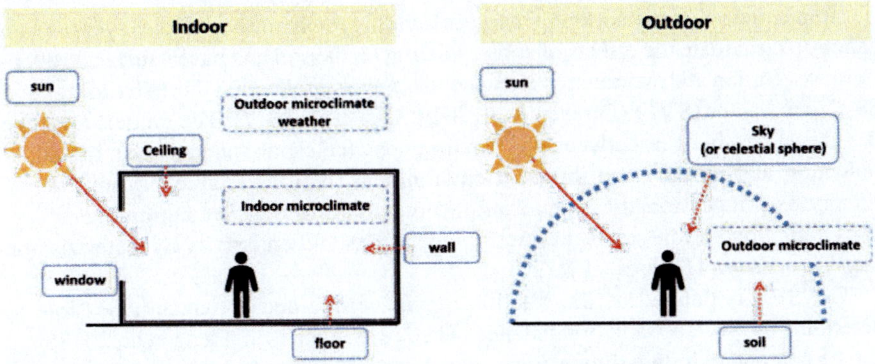

Fig. 13.2 Comparison between indoor and outdoor thermal comfort models

Going down the scale, in the territorial context, a distinction is made between: *macro-scale*, which is the continental or national territorial scale, for example that of weather forecasts; the *meso-scale*, that of the regional or urban and rural territory, the city and the surrounding territory, adopted in the study of the phenomenon of Urban Heat Islands, and the *micro-scale*, referring to portions of the city: the neighborhood, the square, the streets, where local phenomena due to the architectural conformation of the urban context are described, such as the reflection of solar radiation due to the façades of buildings, the effect of evotranspiration of lawns or trees, the role of the sun's path and shadows, and so forth. The micro-scale refers to the portion of the urban building. Some phenomena, such as those related to turbulence and fluid dynamics, and those related to the reflection and radiation of the solar component, behave differently in the three different scales.

To clarify:

- the *Urban Heat Island* (UHI) (Chui et al. 2018; Kim and Brown 2021; Marando et al. 2022; Noro and Lazzarin 2015; Oke 1973, 1982; Ward et al. 2016) is a structural condition of the territory, where a part of the territory, the city, is at a constantly higher average temperature, of at least 0.5–1 °C compared to the rest of the surrounding territory, urban and rural;
- *Heat Waves*, on the other hand, are summer and transient phenomena during which, for meteorological reasons, the air temperature is significantly higher, even by + 2 °C or +4 °C compared to the average and persists for several days (Ballester et al. 2023). The combination of the two factors has repercussions on people's health, which have also been studied in the epidemic context (Condemi and Gestro 2019; Gestro et al. 2020), such as the increase in people in conditions of malaise, metabolic imbalances, heat stroke and mortality.

Epidemiological literature related to heat waves reports cases of summers with excess mortality, such as the summer of 2003 (over 70,000 heat-related deaths) and of 2022 (62,862 heat-related deaths) in Europe (Ballester et al. 2023) and Klinenberg's milestone research as a social autopsy of the heat wave in Chicago during the summer of 1995 (Klinenberg 1999, 2002).

Urban microclimate modeling involves dividing reality into multiple layers (urban canopy) to evaluate the effects of solar radiation on the soil and paved surface (reflection); vegetation and evotranspiration phenomena, and reflection due to buildings and Sky View Factor (SVF) (Dirksen et al. 2019; Oke et al. 2017). The models describe the energy exchanges between the environment (celestial sphere, solar radiation, reflection and albedo) and the built environment (buildings, greenery and trees). Energy exchanges mainly depend on phenomena related to irradiation and blackbody radiation and the relevant laws, temperatures and emissivity of materials and Sky View Factor (Dirksen et al. 2019).

The SVF is defined as the fraction of visible sky and it makes it possible to determine the surface radiation balance (Theeuwes et al. 2017; Zeng et al. 2018) and can be calculated in two dimensions:

$$\mathrm{SVF}_{2\mathrm{D}} = \cos\left(\arctan\frac{H}{0.5W}\right) \quad (-) \tag{13.1}$$

where H is the height of the building and W is the distance between the center point where SVF is calculated and the building facing it;

or in three dimensions:

$$\mathrm{SVF}_{3\mathrm{D}} = \int_{\theta=0}^{2\pi} \cos^2(\beta(R,\theta))\mathrm{d}\theta \quad (-) \tag{13.2}$$

where β is the angle from the center point to the maximum obstacle height at a maximum distance equal to the constant search radius R, θ is sky hemisphere angles.

Urban Climate or Urban Climatology is the discipline that studies energy exchanges in urban areas in relation to:

(a) direct solar radiation from the sun and sky;
(b) the reflection of radiation from the ground and buildings, which depends on the characteristics of sidewalks and streets, façades and roofs, in particular the values of emissivity, reflection and albedo. The albedo is defined as the ratio between the reflected energy and the incident energy over a unit area, from which it can be inferred that there is some similarity between the albedo and the reflectance;
(c) the presence of shadows that affect energy exchanges due to irradiation;
(d) to mass exchanges, of the water vapor, due to the evotranspiration of plant elements, meadows, lawns and trees. (Oke et al. 2017).

In Outdoor Thermal Comfort, the *mean radiant temperature* influenced by the solar shortwave radiation (direct, diffuse and reflected) and, although to a lesser extent, by terrestrial longwave radiation, both atmospheric and environmental, plays a crucial role (Kántor et al. 2012).

13.3 Outdoor Thermal Comfort Indexes

The Predicted Mean Vote (see Chap. 4) and other indexes for the evaluation of the indoor thermal comfort standard are not applicable to the outdoor setting (Höppe 2002).

In the outdoor environment, thermal comfort depends on different phenomena than the indoor environment does. Firstly, because of the geometrical characteristics of physical phenomena, which do not take place in an environment delimited by six confine surfaces, and secondly, because the activities carried out, the clothing worn, the health conditions, gender and age, as well as the cultural aspects related to being outdoors, act in a different way.

Well-being conditions can be assessed according to three approaches:

(a) *psychological*, when the person expresses satisfaction with thermal comfort, an aspect that is difficult to evaluate except through questionnaires, even if in the outdoor environment the psychological aspects have a greater influence than they do in indoor environments, largely because they are linked to pleasant activities, sports, walks, can involve social interactions or collective activities, concerts, chats, and so forth;
(b) *thermophysiological*, which expresses thermal comfort according to thermal receptors in the skin and the hypothalamus, and comfort is defined as the minimum rate of nervous signals (Mayer 1993);
(c) *heat balance of the human body*, following laws of thermodynamics, for which comfort is a condition of equilibrium of energy exchanges (e.g., skin temperature, metabolism) and mass exchanges (e.g., sweat rate, breath) between the human body and the environment, an approach also followed by P.O. Fanger as described in the previous chapters.

The difference with respect to indoor comfort is that in the outdoors, psychological expectations and *thermal history*, individual and collective, as well as cultural habits, can influence the person's judgement. Thermal history refers to cultural expectations, habits and all cultural and social aspects upon which the person's judgement depends. In addition, the characteristics of outdoor environments are very different from one another and cannot be simplified in a box, as they can be in indoor environments.

The outdoor environment can be urban, rural or in nature, in a park, in a forest or near the sea, it can be experienced as a place of passage, play or relaxation, or a place of vacation, as in the case of summer holidays in seaside resorts (Hoppe and Seidl 1991).

Outdoor Education itself is a didactic activity that cannot be compared with that which takes place inside the classroom. The Subjective Assessment of Outdoor Thermal Comfort is important and includes the history of the individual, in the case of the child, the educational activity in the open air, the experience he or she is carrying out, the interactions with his or her classmates, the child's own (new) role in the group; all of these change the judgement with respect to the thermal sensation.

Heat Balance, on its own, is not able to explain or predict the child's judgement and behavior, regarding, for example, the clothing worn or the child's level of concentration/participation. The child tells himself or herself something about the world, including the weather conditions and the thermal sensations he or she experiences during an outdoor educational experience.

In the case of children's outdoor thermal comfort, the psychological, relational and emotional aspects are difficult to investigate and standardize, in addition to the little or almost no scientific literature available on the subject.

Urban Climatology tries to bring order between measurable variables (heat balance, energy exchange) and subjective evaluations. The indexes adopted in bioclimatology and outdoor thermal comfort indicate an equivalent temperature that corresponds to a thermal perception and grade of physiological stress.

The indexes for assessing outdoor human comfort and thermal stress are divided into three categories (Coccolo et al. 2016):

- *Empirical indexes*, based on monitoring or questionnaires, fall into this category: *Actual Sensation Vote* (ASV) used in the European project RUROS, Thermal Sensation (TS) and Thermal Sensation Vote (TVS), always with the use of monitored data and questionnaires;
- Based on *Linear Equation*, such as Thom Discomfort Index (TDI), Apparent Temperature (AT), Heat Index (HI), Relative Strain Index (RSI), Environmental Stress Index (ESI) and Humidex (H);
- *Thermal indexes*, which consider the interactions between metabolic activities of clothing and environmental parameters, derive from indoor indexes such as the PMV and the Standard Effective Temperature, but adapted to the outdoor environment and referring to a subject: a pedestrian thermal perception. This category includes: *COMFA model* and *COMFA+*, *Universal Effective Temperature (ETU)*, *Index of Thermal Stress (ITS)*, *Perceived Temperature (PT)*, *New Standard Effective Temperature (SET*)*, ***Physiological Equivalent Temperature (PET)* and *Urban Thermal Comfort Index (UTCI)***.

The ***Empirical Indexes*** express an empirical evaluation of the thermal sensation starting from some local climatic parameters such as air temperature, solar radiation, wind speed and relative humidity. The values are measured directly on site, using probes or monitoring stations, or through satellite surveys.

The *Actual Sensation Vote* (ASV) is an empirical index developed within the framework of the European project RUROS in Europe (ASV_{Europe}) and in the Mediterranean area (Chen et al. 2022; Nikolopoulou et al. 2001; Nikolopoulou and Lykoudis 2006), and calculated with the formula:

$$AVS = 0.0678 \cdot T_a + 0.0006 \cdot R_g - 0.107 \cdot \nu - 0.002 \cdot RH - 0.69 \quad (-) \tag{13.3}$$

where T_a is air temperature (°C), R_g is the global solar radiation (Wm^{-2}), ν is air speed (ms^{-1}) and RH is relative humidity (%).

The ASV value = 0 corresponds to the condition of neutrality and therefore of well-being (absence of discomfort).

The *Thermal Sensation* (TS) (ASHRAE 2017) is an empirical equation generated by multiple regression obtained through on-site monitoring and questionnaire, with the formula:

$$TS = 0.1185 \cdot T_a + 0.6919 \cdot \nu + 0.1155 \cdot RH - 4.77 \quad (-) \tag{13.4}$$

The *Thermal Sensation Vote* (TSV) is derived from the previous one but refers only to the results obtained during on-site surveys.

$$TSV = (a \cdot T_a) + (b \cdot R_g) + (c \cdot \nu) + (d \cdot RH) + e \quad (-) \tag{13.5}$$

where

a, *b*, *c*, *d* and *e* are regression constants obtained by survey.

The vote is expressed according to the PMV criteria of Cold (−3) and Hot (+3) or according to similar criteria based on the current research.

Other indexes of thermal stress and thermal perception in the outdoors, particularly for pedestrians, can be found in Patanvou et al. (2013).

Indexes ***based on linear equation*** are similar to empirical indexes, in that they are based on data measured on site or are predictive (e.g., weather forecasts), and express conditions of discomfort. Some indexes, such as the *Thom Discomfort Index*, are used as a Climate Comfort Index and for the prediction of potential hazards in weather forecasting, also for the purpose of weather warnings in the event of Heat Waves.

The Thom Discomfort Index (TDI or DI, it is also called the Temperature-Humidity Index THI), measured in °C proposed by Thom (1959), is a physiological thermal stress indicator for people based on dry-bulb and wet-bulb temperature calculated with the formula:

$$\mathrm{DI} = 0.4(T_a + T_{wb}) + 4.8 \tag{13.6}$$

where T_a is air temperature (°C) and T_{wb} is wet-bulb temperature. Literature reports several TDI formulas, this is from (Angouridakis and Makrogiannis 1982) another formula, reported in Coccolo et al. (2016) is:

$$\mathrm{DI} = T_a - 0.55 \cdot (1 - 0.01 \cdot \mathrm{RH}) \cdot (T_a - 14.5) \quad (-) \tag{13.7}$$

where RH is Relative Humidity (%).

The Thom index is very common in warning systems of summer weather conditions, such as heat wave risks and intense heat stress (Table 13.1).

The *Apparent Temperature* (AT), measured in °C (Steadman 1984), is used by the Australian Bureau of Meteorology to evaluate pedestrian exposition, with the formula:

$$\mathrm{AT} = T_a + 0.348 p_a - 0.7\nu + \frac{0.7 \cdot Q}{\nu + 10} - 4.25 \tag{13.8}$$

where p_a is the vapour pressure of the air (hPa), ν is the wind speed (m s^{-1}) at 10 m a.s.l., Q is the net radiation absorbed by the human body (W m^{-2}).

Table 13.1 Categories of the Discomfort Index (DI) (Angouridakis and Makrogiannis 1982)

DI	Categories
DI < 21	Comfort
21 ≤ DI < 24	10% of the total population feels discomfort
24 ≤ DI < 26	50% of the total population feels discomfort
DI ≥ 26	100% of the total population feels discomfort
> 26.7	the discomfort is very strong and dangerous

The *Heat Index* (HI), measured in °C (Rothfusz 1990; Steadman 1979), is a multi-regression derivative of the Apparent Temperature calculated with the formula (Table 13.2):

$$\begin{aligned} \mathrm{HI} = & -8.784695 + 1.61139411 \cdot T_a + 2.338549 \cdot \mathrm{RH} \\ & -0.14611605 \cdot T_a \cdot \mathrm{RH} - 1.2308094 \cdot 10^{-2} \cdot \mathrm{T}^2 \\ & -1.6424828 \cdot 10^{-2} \cdot \mathrm{RH}^2 + 2.211732 \cdot 10^{-3} \cdot \mathrm{T}^2 \cdot \mathrm{RH}^2 \\ & + 7.2546 \cdot 10^{-4} \cdot \mathrm{T}^2 \cdot \mathrm{RH}^2 - 3.582 \cdot 10^{-6} \cdot \mathrm{T}^2 \cdot \mathrm{RH}^2 \quad (^\circ\mathrm{C}) \end{aligned} \tag{13.9}$$

The *Relative Strain Index* (RSI) (Emmanuel 2005) is a simplified method to express the thermal comfort of a standard pedestrian with a metabolic activity equal to 100 Wm-2 and clothing resistance of 0.9 clo, calculated with the formula:

$$\mathrm{RSI} = \frac{T_a - 21}{58 - p_a} \quad (^\circ\mathrm{C}) \tag{13.10}$$

where T is the air temperature (1C) and p_a is the vapor pressure (hPa). The RSI scale is:

- if RSI less than 0.1 with 100%, unstressed i.e., everyone is comfortable;
- RSI equal to 0.2 with 75% unstressed;
- RSI equal to 0.3 with 50% unstressed i.e., upper limit of comfort;
- RSI equal to 0.4 with 75% distressed;
- RSI equal to 0.5 with 100% distressed.

The *Environmental Stress Index* (ESI), measured in °C (Moran et al. 2001) is the evolution of the *Wet Bulb Globe Temperature* (WBGT) a screening method for evaluating the heat stress to which a person is exposed and for establishing the presence or absence of heat stress (ISO 7243 2017). The ESI index was developed for military exercises in hot-dry or hot-wet climates, and calculated with the formula:

$$\mathrm{ESI} = 0.63 \cdot T_a - 0.03 \cdot \mathrm{RH} + 0.002 \cdot \mathrm{SR} + 0.0054 \cdot (T_a \cdot \mathrm{RH}) - 0.073 \cdot (0.1 + \mathrm{SR})^{-1} \tag{13.11}$$

Table 13.2 Effect of the Heat Index

HI in °C	Heat stress
HI < 27 °C	Comfort
27 °C ≤ HI < 32 °C	Caution: fatigue for prolonged exposure and activity
32 °C ≤ HI < 41 °C	Extreme Caution: heat cramps and heat exhaustion are possible. Continuing activity could result in heat stroke
41 °C ≤ HI < 54 °C	Danger: heat cramps and heat exhaustion are likely. Heat stroke is probable with continued activity
HI ≥ 54 °C	Extreme danger: heat stroke is imminent

Table 13.3 Humidex scale of comfort

H	Scale of comfort
H < 20 °C	Comfort
20 °C < H < 29 °C	Little discomfort
30 °C < H < 49 °C	Some discomfort
40 °C < H < 45 °C	Great discomfort
H > 45 °C	Dangerous: heat stroke possible

where SR is the solar radiation (W·m^{-2}).

Humidex (*H*) defined for the Canadian climate is a method of quantifying human discomfort due to excessive heat and humidity (Masterton and Richardson 1979) with the formula:

$$H = T_a + \frac{5}{9} \cdot (p_{as} - 10) \quad (-) \tag{13.12}$$

where p_{as} is calculated following the formula:

$$p_{as} = 6.112 \cdot \frac{RH}{100} \cdot 10^{\left(\frac{7.5 \cdot T_a}{237.7 + T_a}\right)} \tag{13.13}$$

The ***Thermal indexes*** return the equivalent temperature, the calculation formulas take into account the variables related to the person (e.g., metabolism or body energy budget), energy exchanges (e.g., evaporative phenomenon) and environmental conditions (e.g., air temperature, solar radiation), and can be either modeled by calculation software or measured in the field by weather data (Table 13.3).

The Computer Model for Adaptive Comfort *COMFA model* (Brown and Gillespie 1995; Kenny et al. 2009a, 2009b; Vanos et al. 2012) is described by the equation:

$$B = M + R_{RT} - C - E - L \quad (\mathrm{W\,m^{-2}}) \tag{13.14}$$

where B is the body energy budget (W m^{-2}); M is the metabolic rate (Vanos et al. 2012); R_{RT} is the short-wave and long-wave radiations absorbed (W m^{-2}) (Monteith and Unsworth 2013); C is the sensible heat loss by convection (W m^{-2}); E is the evaporative heat loss by breath; L is the long-wave radiation emitted by a person (W m^{-2}).

The COMFA model applies the range of the PMV sensation scale (from −3 cold, to +3 hot) to the outdoors. The evaluation of thermal sensation (from Cold to Hot) follows the seven-point scale (Table 13.4).

The *Index of Thermal Stress (ITS)* (Erell et al. 2014; Pearlmutter et al. 2007) expresses the overall thermal exchange between the body and its surroundings under warm conditions, with evaporation accounted for in a detailed manner, and is described with the equation:

Table 13.4 The seven-point scale for the predicted thermal sensation ratings that was applied to the COMFA budget output (Kenny et al. 2009a, b)

Sensation rating	Budget (W m^{-2})	Protection measures
−3	≤ 201	Cold
−2	−201 to −121	Cool
−1	−51 to −120	Slightly cool
0	−50 to +50	Neutral
+1	+51 to +120	Slightly warm
+2	+121 to +200	Warm
+3	≥ 201	Hot

$$\text{ITS} = E \cdot \left(\frac{1}{f}\right) \quad (\text{W}) \tag{13.15}$$

where E is the cooling rate (W) produced by sweat and f (Burton et al. 2009) is the dimensionless cooling efficiency of sweating that depends on the evaporative capacity of the air according to wind speed. The cooling rate is calculated with the formula:

$$E = (M - W) \pm R_n \pm C \quad (\text{W}) \tag{13.16}$$

where M is the body's metabolic rate, W mechanical work, Rn energy exchange by radiation and C energy exchange by convection (W m^{-2}) for body surface area.

The factor 1/f is evaluated by the empirical equation:

$$\frac{1}{f} = \exp\left[0.6 \cdot \left(\frac{E}{E_{\max}} - 0.12\right)\right] \tag{13.17}$$

where E is the evaporative capacity of the air (W) and Emax is evaluated by

$$E_{\max} = c_{\text{clo}} \cdot V^{0.3}(42 - Pv_a) \tag{13.18}$$

where c_{clo} is clothing coefficient, V is wind speed (m s^{-1}) and Pv_a is the vapor pressure of the air (mmHg).

The scale of ITS judgement following Fig. 13.3 graphics.

The *Perceived Temperature* (PT) is based on the calculation of the PMV for a *Standard Pedestrian*, called "Clima Michel", who is walking ($M + W = 135$ W m^{-2}) and is wearing clothing associated with the winter season (I_{cl} equal to 1.75 clo) or summer season (I_{cl} equal to 0.5 clo), is given by the formula:

$$\text{PT} = \begin{cases} 5.805 + 12.6784\text{PMV}, \text{PMV} < 0 \text{ and } I_{\text{cl}} = 1.75\,\text{clo} \\ 21.258 - 9.558 \cdot I_{\text{cl}}, PMV = 0 \text{ and } 1.75\text{cl} > I_{\text{cl}} > 0.5\,\text{clo} \\ 16.826 + 6.183 \cdot \text{PMV}, \text{PMV} > 0 \text{ and } I_{\text{cl}} = 0.5\,\text{clo} \end{cases} \tag{13.19}$$

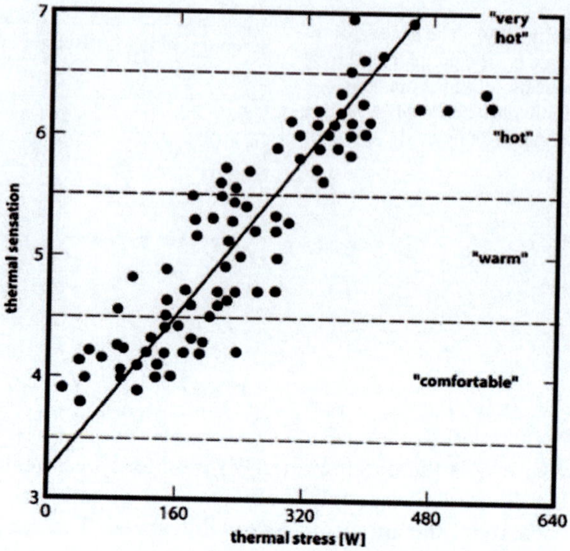

Fig. 13.3 Relation between index of thermal stress (*W*) and thermal sensation by (Pearlmutter et al. 2007)

The *Standard Effective Temperature* (SET) defined as the temperature of an imaginary environment at 50% RH, < 0.1 m/s air speed, and tr = ta, in which the total heat loss from the skin of an imaginary occupant with an activity level of 1.0 met and a clothing level of 0.6 clo is the same as that of a person in the actual environment, with actual clothing and activity level, following ASHRAE 55 "*Appendix D - Procedure for evaluating cooling effect of elevated air speed using SET*" (ASHRAE 2017).

PET and UTCI are the *Thermal Indexes* used both in on-site measurement and in the modeling of the outdoor microclimate, as they make it possible to obtain a description of outdoor thermal comfort that considers all energy exchanges with the environment and the celestial sphere, and return a perceived temperature related to thermal stress. For this reason, a detailed description is provided.

Physiological Equivalent Temperature (PET)

The Physiological Equivalent Temperature (PET) is a human biometeorological parameter that describes the thermal perception of an individual. It is defined as the air temperature at which, in a typical indoor setting, the heat exchange of the human body is balanced with the same body core and skin temperature (Höppe 1999; Mayer and Hoppe 1987).

The PET adopts the equations of the Two-Node Model and the Effective Temperature developed by Gagge (Gagge et al. 1971) and the Heat-Balance of the MEMI model (Munich Energy-balance Model for Individuals) applying the following equations:

(a) *heat-balance equation for the human body* (refer to m^2 of body area)

$$M + W + R + C + E_D + E_{Re} + E_{Sw} + S = 0 \ (\mathrm{W\,m^{-2}}) \tag{13.20}$$

where

M is metabolic rate, W physical work, R the net radiation for the body, C the convective heat flow, E_D the latent heat flow to evaporative water, E_{Re} the sum of heat flow for heating and humidity inspired air, E_{Sw} the heat flow due to evaporation of sweat and S the storage flow for heating or cooling the body mass;

(b) *heat flow from body-core to skin surface* ($F_{\text{Core-Skin}}$), with the formula:

$$F_{CS} = v_b \cdot \rho_b \cdot c_b \cdot (T_c - T_{sk}) \quad (\text{W m}^{-2}) \tag{13.21}$$

where

v_b is blood flow from body core to skin (l s^{-1} m^{-2}), ρ_b is blood density (kg l^{-1}), c_b is specific heat (W s K^{-1} kg^{-1}), T_c is temperature of body core (°C), T_{sk} is temperature of body skin (°C);

(c) *heat flow from skin surface to clothing surface* ($F_{\text{Skin-Clothes}}$), with the formula:

$$F_{SC} = \left(\frac{1}{I_{cl}}\right) \cdot (T_c - T_{sk}) \quad (\text{W m}^{-2}) \tag{13.22}$$

where

I_{cl} is heat resistance of the clothing (measured in kg m^2 W^{-1} or in *clo*).

Following the Equivalent Temperature approach, the PET is defined as the physiological equivalent temperature at any given space and is equivalent to the air temperature at which, in a typical indoor setting, the heat balance of the human body is maintained with the same core and skin temperature as those under the conditions being assessed. The human body is assumed with a work metabolism 80 W and heat resistance of clothing 0.9 clo. The indoor reference climate assumed with a mean radiant temperature equal to air temperature ($T_{mrt} = T_a$), air velocity is 0.1 m/s, water vapor pressure is 12 hPa (RH 50% with T_a 20 °C).

Table 13.5 shows the PET values and the perceived thermal sensation. Figure 13.4 report a scheme to explain how to determine the equivalent temperature.

Table 13.5 Physiological Equivalent Temperature (PET)

PET		Thermal perception	Grade of physiological stress
	< 4 °C	Very Cold	Extreme cold stress
4 °C	8 °C	Cold	Strong cold stress
8 °C	13 °C	Cool	Moderate cold stress
13 °C	18 °C	Slightly cool	Slight cold stress
18 °C	23 °C	Comfortable	No thermal stress
23 °C	29 °C	Slightly warm	Slight heat stress
29 °C	35 °C	Warm	Moderate heat stress
35 °C	41 °C	Hot	Strong heat stress
	> 41 °C	Very hot	Extreme heat stress

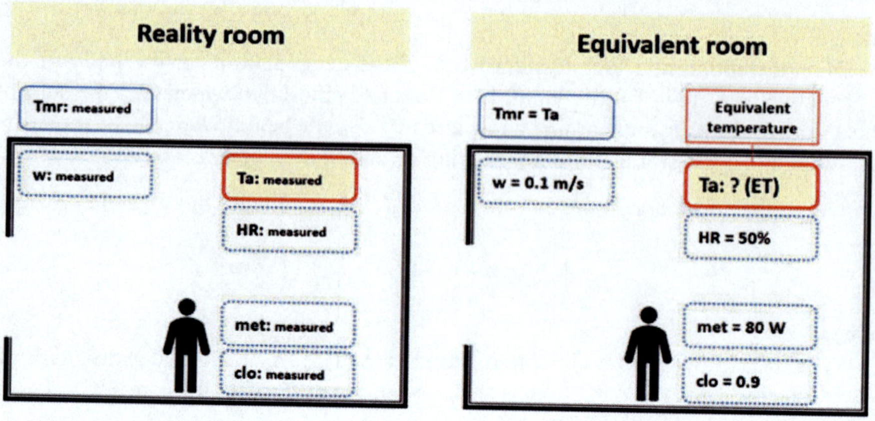

Fig. 13.4 Schematic representation of reality room and virtual room (equivalent room) to determine the equivalent temperature (T_a)

13.4 Universal Thermal Climate Index (UTCI)

The Universal Thermal Climate Index was created a few years after the PET, in 1999 as part of the COST Action 730 project (Jendritzky and Nübler 1981) (website www.utci.org).

The UTCI is based on a Multi-Node dynamic thermophysiological model that evaluates the energy exchanges of the human body, referring to the whole body and individual parts. The human body is divided into 12 body elements: head, face, neck, shoulders, thorax, abdomen, upper and lower arms, hands, upper and lower legs and feet and internal organs for a total of 187 nodes.

The Universal Thermal Climate Index (UTCI) is expressed as an equivalent ambient temperature (°C) of a reference environment providing the same physiological response of a reference person as the actual environment (Blazejczyk et al. 2012; Weihs et al. 2012).

The reference data are: air temperature (T_a, in °C) equal to mean radiant temperature ($T_{mrt} = T_a$, in °C), wind speed equal to 0.5 m s^{-1} at 10 m, Relative Humidity equal to 50% up to a constant water vapor pressure of 20 hPa, and metabolic rate (M) equal to 150 W m^{-2}.

The equation for calculating UTCI (Bröde et al. 2012) is as follows:

$$\text{UTCI}\,(T_a, T_r, v_a, p_a) = T_a + \text{Offset}\,(T_a, T_r, v_a, p_a) \quad (°\text{C}) \tag{13.23}$$

where

T_a is the air temperature (°C), T_r is the mean radiant temperature (T_{mrt}, measured in °C), v_a is the wind speed (m s^{-1}) p_a is the water vapor pressure (hPa) and *Offset* is the deviation from air temperature. Table 13.6 shows the UTCI values and the thermal stress category.

Table 13.6 Urban Thermal Climate Index (UTCI)

UTCI		Thermal stress category	Protection measures
	> 46 °C	Extreme heat stress	Temporary body cooling, Drinking > 0.5 L hr^{-1} No physical activity
38 °C	46 °C	Very strong heat stress	Temporary use of air conditioning, Shaded places necessary, Drinking > 0.5 L hr^{-1} Reduce physical activity
32 °C	38 °C	Strong heat stress	Use shaded place, Drinking > 0.25 L hr^{-1} Temporarily reduce physical activity
26 °C	32 °C	Moderate heat stress	Drinking > 0.25 L hr^{-1}
9 °C	26 °C	No heat stress	Physiological thermoregulation sufficient to keep thermal comfort
0 °C	9 °C	Slight cold stress	Use gloves and hat
−13 °C	0 °C	Moderate cold stress	Intensify activity and protect face and extremities against cold
−13 °C	−27 °C	Strong cold stress	Intensify activity and protect face and extremities against cold. Use warmer clothing
−27 °C	−40 °C	Very strong cold stress	Intensify activity and protect face and extremities against cold. Use warmer clothing. Reduce outdoor exposure time
	< −40 °C	Extreme cold stress	Stay at home. If outdoor exposure is necessary, use heavy and wind protected clothing

13.5 Thermal Comfort in Outdoor Education

How to evaluate if it is possible to do outdoor education. Who can evaluate it? The choice depends on several factors such as teaching volunteers, outdoor space around the school (if present, or if there is none, garden, park, green spaces, etc.), climate, seasons or weather.

The parameters to consider are:

(a) *local climate, weather data* and/or *weather* based on the season of the year, in relation to which the duration and outdoor activities are defined, as well as clothing, and the weather conditions of the day itself to assess any contextual and specific conditions. The objective of these analyses, as far as thermal comfort is concerned, is to assess whether there are conditions to ensure the health and comfort of children, the presence of any risks related to the weather-climatic context (rain, wind, storm, heat wave, etc.) with possible thermal stress conditions (cold or hot) that may lead to the suspension of the activity for some or all of the children;

(b) the *activities* carried out in relation to the metabolism of children, taking into account that activities with greater physical effort (running, jumping, etc.) correspond to a higher value of metabolism and mechanical work of the body (M-W, see Chap. 3);
(c) the *children's clothing*, to be divided into clothing related to the season and any protection systems such as coats, jackets, etc., for the winter or hats and sunscreen for the summer;
(d) the characteristics of the *experience* and of the pedagogical activities envisaged, based on motion (activity) and emotion (senses and experience);
(e) the *Architecture role* in outdoor thermal comfort and outdoor education, namely:
    (i) the characteristics of the context that depend on whether it is the *natural* and/or *built environment* (e.g., courtyard, city), and how these affect the on-site microclimate (radiation, air temperature, relative humidity, etc.);
    (ii) the presence or absence of furniture to be used as a support for outdoor activities.

Children's response to thermal stress during outdoor education activity

Children respond to heat stress differently than adults do (see Chaps. 3 and 12), in particular, there is different thermoregulation during exposure to hot or cold environments. The differences are both physical and *physiological*, so educators must take into account that their own heat stress may not correspond to that of the children.

The main physical difference between the adult and child thermoregulation system is the much higher surface-area-to-mass ratio for children, and the higher rate of heat absorption or heat loss. The lower amount of blood in the child's body limits heat transfer during heat exposure, and influences the activities performed. From a *physiological* point of view, the sweating mechanism of children is different from that of adults. Children have a lower sweating rate, which depends on the smaller number and size of the sweat glands.

Table 13.7 summarizes the physical and physiological difference between children and adults affecting thermoregulation, from the article by Falk (1998).

The *physical differences* concern the *Surface-Area-to-Mass (S/M)* ratio, which can be 40% higher in children than in adults. In an 8 or 9-year-old child, 130 cm tall and weighing 20 kg, the body surface is 0.87m^2 (Dubois body area, see Chap. 3) with an S/M ratio of 0.043, while an adult of 175 cm in height, 67 kg in weight, with a body surface of 1.81 m^2, the S/M ratio is 0.27.

The heat exchange by radiation, convection and conduction is greater in the child. When the difference between the temperature of the air and that of the skin is high (e.g., in case of 'hot' or 'cold' condition, there may be a greater heat exchange between the body and the environment. The lower adiposity of the human body influences the thermal insulation of the body in cold exposure conditions, especially for girls and adult women.

The *body composition* of children has a lower fat-free mass, they are 'less fat' than adults, so the specific heat is reduced (1.67 kJ/kg°C for children and 3.35 kJ/kg°C

Table 13.7 Physical and physiological difference between children and adults affecting thermoregulation

Difference	Effect on thermoregulation
Physical	
Greater surface-area-to-mass ratio	Increase heat gain in hot environments. Increase heat loss in warm and cold environments
Greater body density (in girls vs women)	Higher specific heat of the body
Lower blood volume per body surface area	Possible compromise in perfusion of periphery, muscles or central nervous system
Smaller sweat glands	Lower sweat gland output
Physiological	
Greater oxygen cost of locomotion	Greater metabolic heat production per kg body mass
Lower maximal cardiac output	Possible compromise in perfusion of periphery, muscles or central nervous system
Lower sweat gland sensitivity	Higher sweating threshold; longer time to sweating onset; lower sweating rate
Lower sweat gland anaerobic metabolism	Lower sweating rate
Lower prolactin response to exercise in the heat	Possible effect on sweat electrolyte composition

for adults). Finally, a child's *blood volume* is smaller, which may affect peripheral vasoconstriction (Astrand 1952).

The *physiological differences* concern aspects related to the stage of growth towards adolescence and the adult body. Some characteristics are peculiar in children such as *Metabolism*: the oxygen cost of walking or running per unit of mass is higher in children than in adults. For example, a 7-year-old consumes 20% more energy on an 8 km/h run than an adolescent does (Bar-Or 1983). The hormonal status of children affects two hormones implicated in thermoregulation: testosterone, which acts on sweating, and prolactin which is implicated in osmoregulation and affects the electrolyte composition of sweat.

Sweating plays an essential role in *Human Body Thermoregulation*, thanks to which it is possible to exchange heat with the warm or hot environment. The *sweating mechanism* allows body heat dissipation thanks to sweat evaporation, by two types of sweat glands: eccrine and apocrine. The sweat glands of Children generally are smaller than those of adults (Wolfe et al. 1970), a factor that affects their ability to dissipate energy through sweat. Studies report higher skin temperatures in children compared to adults.

Children's physiological response to heat stress leads to an increase in the rate of oxygen consumption, in order to increase the metabolic response. When the ambient temperature is high, the average temperature of the child's body is higher than that of an adult. The cardiovascular system promotes a more rapid decrease in skin temperature in children compared to adults (Falk and Dotan n.d.; McCullough et al. 2009). During physical activities, sweating ensures the distribution of fluids within the body and affects hypo-hydration.

The acclimation rate is slower in children compared to that of adults (Inbar et al. 1981). The heat acclimation process is slower in children than it is in adults. In children, the change in body temperature is accompanied by an increase in metabolic cost, especially in cold environments. In extreme environmental conditions or thermal stress, heat or cold, children's thermoregulation appears to be less effective than it is in adults (Falk 1998).

Children's physiology, clothes and thermal comfort in outdoor space

Children's physiology, clothes and thermal comfort in outdoor space: how to guarantee their comfort in the outdoors?

A child's clothing has multiple functions in the school environment:

(a) the thermal protection provided by '*ordinary*' clothing, from the cold and heat, and the thermal protection provided by '*auxiliary*' clothing and their relative '*practicality*' depending on the season, for example the use of thermal jumpsuits and padded jackets to protect oneself from cold days, or hats, visors and sunscreen to protect oneself from the sun on hot summer days;
(b) the symbolic meaning of the clothing including affection, comfort or pleasure in wearing particular garments; and
(c) practicality with respect to the activities carried out, particularly in the case of outdoor activities.

Thermal protection and clothing insulation have been described (see Chap. 3) and it is possible to attribute thermal resistance to the clothing and, consequently, based on the metabolic activity and physical conditions of the classroom, we are able to evaluate thermal comfort in the indoor environment.

Compared to what has been said in the previous chapters, in the case of outdoor activities, it should be remembered that the level of clothing insulation depends on the outdoor temperature and the difference with respect to the indoor operating temperatures. The child will tend to adapt his or her behavior or clothing in relation to the indoor temperature, the weather, the activities to be carried out, and so forth (Mors et al. 2011; Nam et al. 2015; Teli et al. 2017).

The symbolic meaning concerns the taste of the child or of the person who dresses him or her and chooses/buys the clothes, usually the mother, who also decides which clothes the child should wear based on the day's weather. The choice does not always take into account the metabolism of the child, which is different than that of the adult: 1 met, as 48.8 W/m^2 for 1.14 m^2 of body area (Parsons 2002). In addition, the school may require children to wear the same uniform or smock to avoid discrimination based on income (at school every child is equal the others) or other factors.

The practicality of clothing mainly concerns outdoor activities, whether they take place in the schoolyard, in an urban environment, in the park or in other outdoor contexts. The clothing must be suitable for the environmental and climatic context, following the Baden Powell maxim:

«There is no such thing as good or bad weather, but good or bad equipment.»

The clothing must allow the child to be able to run without falling or tripping, it must be able to allow movement to the arms, legs (avoid long clothes) and hands (if gloves are used, they must have fingers). In Outdoor Education a dress-code is not always required, waterproof jackets with hood, raincoats or similar items are not particularly recommended and the clothes must guarantee protection from the various conditions that may occur during outdoor activities by protecting from outdoor atmospheric agents, acting on thermal insulation preferably through the use of several layers of clothing. They must guarantee protection from water with clothing having waterproof or washable materials, including shoes (cold or wet feet negatively affect the experience). In spring or summer, it is necessary to avoid excessive thermal insulation, it is preferable to use clothes made of cotton or breathable material, a hat with a visor, comfortable shoes and, if necessary, socks, without the risk of chafing, sunscreen with adequate protection for children's skin, mosquito repellent, and always provide a sweater or other protection systems in case of sudden temperature changes during the day.

In addition, children can have a bag or backpack with them to carry a bottle of water or a snack, or they can bring along notebooks, sheets of paper, pens and pencils and any other equipment and material useful for the activity, e.g., bags to collect stones of various shapes, leaves, and so forth.

Clothing must guarantee protection from external climatic conditions (clothes resistance, measured in *clo*); the ability to adapt itself during the course of the activities, adding or removing layers; protect from rain, including protecting the feet; protect from the sun, especially the head; ensure functionality with respect to the physical activities carried out, i.e., it must not be an obstacle to the child's movements.

Clothing that 'makes you sweat' or 'makes you feel cold' diverts children's attention and participation, maintaining the condition of discomfort given by the personal condition of thermal comfort. This condition may not be obvious to the adult, in the sense that the child may prefer not to complain so as not to be different from others, or the child himself or herself may not notice the discomfort if he or she is very busy with the activity together with others. The shoes must be suitable for the place, with solid soles that have a grip (avoid smooth or leather soles), preferably sneakers or water-resistant shoes. A good outfit is a completely waterproof jumpsuit, with fleece material in the padding, to be worn over clothes, like a sort of dungarees with the strap under the shoes.

The adult educator must be able to predict, prevent, assess and act in the event that such conditions of discomfort occur for the child due to clothing, including the physiological needs of children, in particular urinating (e.g., if the child '*has to pee*' a large part of the child's attention is directed to his or her own body).

In winter it is important to have layered clothing, for footwear to use rubber boots, thermal socks of the right thickness, gloves suitable for the context (snow, rain) avoiding woolen gloves that get wet easily, for the head use hats, beanies, baseball caps, ear covers and scarves.

Children must have the opportunity to undress even if it is January and it is cold: their metabolism during activities is different from that of adults, as mentioned in the previous paragraphs, adults perceive thermal comfort in a very different way than children do. The child must be able to observe, contemplate, dig or run, without having to worry about his or her body.

Lastly, the educator (adult) can consider the weather conditions, rain or snow, as learning conditions: in other words, it is not always necessary that activities be interrupted in case of light rain.

Emotion and Motion

Emotional and motivational factors that influence children's perception of thermal comfort. Self-awareness of children and in the social context of the classroom, teaching awareness of children and classroom in the outdoors in relation to outdoor weather.

The adaptation strategies implemented by children in outdoor environments, depending on the seasons, are similar to those adopted by adults but with some peculiarities related to the emotional aspects, the motion of the activities carried out, or the acclimatization strategies adopted.

Outdoor Thermal Comfort requires *physical*, *physiological* and *psychological* adaptation strategies (Nikolopoulou and Steemers 2003). Adaptation consists of the set of activities that the child performs in space to achieve a condition of well-being, i.e., thermal comfort understood as a condition of balance with respect to the context. In the case of outdoor activities and Outdoor Education, it is a condition of unstable equilibrium, continuously subject to change, therefore adaptation can concern both specific actions, physical or physiological, implemented by the individual, as well as psychological or social interpretations that occur in interactions with other persons.

Unlike thermal comfort in indoor environments, the social context in which activities take place can slow down or speed up the adaptation processes. For example, during play, the feeling of cold or wanting to urinate can be attenuated or repressed, in order to continue to carry out the activity and get to the end. To this can be added situations given by peer relationships: leaving the game because one feels cold or tired from too much heat can mean that others see the child as giving up, or not having enough courage, or isolating himself or herself from the group.

A child's autonomous choice to '*leave*' the outdoor environment because he or she feels too hot or too cold does not follow rational reasoning, but emotional or relational reasoning. In other words, it is likely that the child will decide to leave or suspend activities, because the condition of personal discomfort does not allow him or her to carry out activities on an equal footing or in competition with others, or because it has made the child feel nervous (e.g., anger) and has altered the relationship with others (e.g., quarrelling) rather than only for the feeling of discomfort.

For this reason, the supervision of an adult can be helpful in noticing situations or risks of thermal suffering in order to adopt those strategies that make it possible to mitigate and adapt peer relationships and allow the adoption of solutions to bring equilibrium back to the situation. For example, in the case of a competitive game that takes place under the sun, the educator may notice that a child is fatigued due to excessive heat and may decide to stop the game to allow him or her to don protection (hat, sunscreen) or to move into the shade or to drink something, all actions that the heated child, in competitive play conditions, may not autonomously adopt, given that emotions have taken the upper hand as well as the interest in the play activity.

The *physical adaptation strategy* concerns the behavior of the person in the climatic and meteorological context, whether urban or natural. Adaptation can be active and *reactive*, if conscious strategies are adopted to respond to an external stimulus (e.g., clothing or motor activity are reactive adaptations) or *interactive* or *adaptive*, when it is the context that is adapted, and not the body, or the activity performed in the context (Nikolopoulou et al. 2001). For example, the presence/absence of furniture elements or the use of natural elements for other needs, such as a log for sitting, fall under this second strategy.

The *physiological adaptation strategy* concerns the body's reaction (stimuli) to the climatic and meteorological context, to clothing and to the physical activity that is carried out. Stimuli and concentration are physiological (body) responses and are influenced by the child's emotions and also by fatigue.

The *psychological adaptation strategy* includes the emotional aspects (see Chap. 12) which include the characteristics of the places and the activities carried out. In turn, the adaptation process depends on:

- the *genuineness* of the relationship with nature, the child in the natural context discovers the sensation of wonder, of discovery, given by the large number of stimuli with which he or she comes into contact, alone and/or together with others;
- the *expectations* of the activity carried out outdoors, also in relation to the number of hours spent indoors and the ratio of outdoor/indoor time in school and extracurricular contexts;
- the *duration* of the experience, which can be short-term, long-term or very-long-term in the case of school trips lasting one or more days,
- from the *time of exposure* to outdoor weather conditions, rain, sun, wind, snow, high/low temperatures, etc. The time of exposure is decided by the educator, generally extreme weather conditions (e.g., wind or heavy rain) determine the start or end of the activities;
- the *perception* of the external context, urban or natural, and one's reaction and relationship to it. The outdoor context, from the city to the wilderness, can be more or less known by the child, based on his or her personal experience and having already carried out similar activities in the past. For example, children who live in the countryside have a different relationship with nature, with animals and with the unexpected, compared to children who live in cities and who know nature as it is 'tamed' in urban parks or holiday destinations.

Architecture of the Outdoor Space

The characteristics of the built environment can have a significant impact on children's physical activities and health (Powell 2005). In general, natural elements, in indoor environments (e.g., decorative plants) or outdoors (e.g., trees or lawns in the courtyard, school garden), promotes the well-being of students (van den Bogerd et al.

2020) at any level of education. The presence of natural elements, plants in particular, makes it possible to:

(a) *mitigate damage to health*, for example, trees are able to reduce particular matter (PM), volatile organic compounds (VOC) and carbon dioxide (CO_2), reduce noise, acting as a screen, and reduce solar radiation, thanks to their shade;
(b) It promotes physical and psychological *refreshment*, resting from stress and mental fatigue, an effect that is also obtained during outdoor educational activities. The mere view of nature, for example, from the windows of a school, is able to bring benefits (Guo et al. 2019; Li and Sullivan 2016);
(c) it promotes '*building capacities*', socialization, physical and non-physical group activities, play, social cohesion, collaborative activities (e.g., horticulture, classroom vegetable gardens or gardening activities) (Bikomeye et al. 2021).

Outdoor Education takes place in different types of '*spaces*': courtyard, garden, yard, park, woods, forest, skypark, urban parks, playground, sport fields, backyards, etc., each of which has its own architectural and landscape characteristics.

The development of outdoor education activities in these places depends on their design: the ground surface can be made up of concrete pavement for the courtyards, or gravel, bricks, wood or bare ground, there can be a lawn, shrubs, fruit trees or tall trees, with different aggregation configurations, more or less dense. There may be fixed or mobile furnishing elements, playground equipment or benches. In the case of activities in nature the space can be prepared and cleaned before or during the activity, there may be landscaped elements or paths, beaten trails, bridges, and so forth. All these are elements that can determine the microclimate of the open spaces and can influence the children's experience (Ciucci et al. 2011, 2013; Vanos 2015).

The elements that influence the microclimate are (i) the sky view factor, (ii) the land covers (surface pavement), (iii) vegetation.

The *Sky View Factor*, whose value varies from 1, without shade, to 0 (zero), totally overcast, describes the amount of solar radiation that hits the area. For example, if the activity takes place during the summer, a totally uncovered area (SVF = 0) is too sunny and can more easily lead to situations of heat stress. SVF values between 0.2 and 0.4 make it possible to reduce solar radiation and ensure good illuminance. The design strategies can be different, from the presence of existing buildings, vertical elements, pergolas or canopies, hedges, to the planting of plants, rows or curtains of trees.

The land covers influence the reflection of solar radiation, in the case of urban environments, it is necessary to avoid mineral pavements such as asphalt and always prefer the soil or lawn, with the presence of any partially permeable surfaces, such as wooden flooring or surfaces that can be wetted with jets, sprays or water blades, to ensure summer comfort.

Vegetation, especially shrubs and trees, reduces the air temperature thanks to evotranspiration (water vapor). In addition, the foliage of the trees, thanks to the tree's shaded surface, make it possible to reduce direct shortwave radiation by limiting direct solar radiation and lowering the surface temperature of the soil.

The outdoor spaces where outdoor education takes place, whether it is the school courtyard or the city park or the woods, can be 'furnished' with shelves and seats to allow reading, drawing or taking notes, with fixed or mobile elements.

13.6 A Proposal to Consider Thermal Comfort in Outdoor Education

A proposal to consider thermal comfort in outdoor education—how to observe. proposal to design/manage outdoor space around school to promote outdoor education

Outdoor Education makes it possible to incorporate the relationship between children and nature into the educational journey. Scientific literature shows the impact of nature in learning processes and in situations in which students present stress, burnout, depression and other mental health problems. Nature improves the health of people, promotes mental and physical well-being, as well as the exchange of emotions and social relationships.

The activities carried out outside the school building, from the courtyard to the wilderness, connect the child with diverse worlds:

(a) the *outside world*, the city, the park, the forest, etc.,
(b) the *world of one's peers*, through social activities, and
(c) one's *inner world*, with regard to sensations and emotions.

The research on Outdoor Thermal Comfort, as described in the previous paragraphs, has focused on the human body and environmental energy exchanges and the related indicators. The next step consists of including the sensory experience of well-being in the research and architectural design of outdoor environments.

The activity outside the school must be equipped with the instruments that allow it to adapt to the pedagogical experience, both material instruments, furnishings or equipment, and conceptual ones. Reading the outdoor space, by children, can (and must) be part of the pedagogical and didactic activity. Nature, even when it is wilderness, provides adaptation strategies on account of the mere presence of human beings. The tent, the camp oven, the clothing, the trekking equipment are all 'architectures' that we carry with us, instruments to adapt the environment to us, and vice versa.

Below I will try to outline some elements that, in my opinion, can contribute to a debate on the relationship between Outdoor Education and Outdoor Thermal Comfort.

Outdoor Thermal Comfort adopts the conceptual pattern and equations of Indoor Thermal Comfort, adapting it to the context and energy exchanges that take place in open spaces. The first issue related to Outdoor Thermal Comfort is: *to avoid stress and health risks.*

The various indicators refer to situations of thermal stress, considering two limit conditions: avoiding stressful conditions and damage to health due to cool/cold environments that can lead to flu diseases, or those due to warm/hot environments or very hot ones that can lead to heat stroke, malaise, and extreme discomfort. During the outdoor activity, the educator plans, manages and supervises the conditions of the children, individually and in groups, to prevent uncomfortable conditions. In addition, as mentioned above, conditions of physical stress, such as holding back urine, feeling cold/hot, thirsty, etc., de-concentrate the child from teaching activities or play, because he or she continuously feels the physiological needs dictated by his or her body. Thermal comfort and the external environment are seen as a potential detriment and must be 'accommodated'.

The second issue that constitutes the proposal of this chapter concerns *how to observe* and *how to learn to feel*, Outdoor Thermal Comfort in Outdoor Education. The brief history of thermal comfort (see Chap. 2) has been developed by trying to describe and obtain the condition of balance of the heat exchange between the human body and the environment. The condition of balance and neutrality is rightly considered the condition in which most people express a favorable judgement. Adaptive thermal comfort and studies on the comfort of subjects other than the standard human have shown that the condition of balance is not always the ideal condition, on the contrary, a continuous series of procedures are put in place to adapt to energy exchanges, from metabolism to clothing, to meet the needs of the individual or the community. The role of collective thermal comfort has not yet been investigated in literature.

In Outdoor Education, children, alone and together with the educator, can learn to observe how their body reacts to the conditions of the external environment: the air and its temperature or smell, solar radiation, the heat/cold of surfaces, the humidity of the lawn, trees, the color of the sky, the grass, the streets, the buildings, etc., are all learning opportunities, teaching and pedagogical tools for teachers, as well as being experiences, emotions and feelings for children.

Seen not as stress but as a sensation, Thermal Comfort is a way to get to know the environment around oneself.

Depending on the relationship between architecture and wilderness, the didactic-pedagogical experience of Outdoor Education can vary as well as the levels of 'adaptation' to the environment. For example, in the most 'unfortunate' case of a closed courtyard without vegetation, during the outdoor activity the child can follow the path of the shadows cast, the solar path and feel hot/cold, the child can also see his or her school from the outside, learn to read its height, the size of the windows, the architectural details, etc., and from these invent stories, make drawings and replicas.

The plant elements, the lawn, school garden, shrubs and trees, have a shape, a color that varies according to the season, they cast a shadow that follows the path of the sun, they can promote heat and mass exchanges (water vapor, humid or dry environment); in addition, they can be touched, manipulated, counted, drawn by children. Experiences outside the schoolyard, for example in an urban park or in city squares or streets, can include a walk during which one's metabolic activity varies.

The child can discover the city from different perspectives, going from narrow and shady streets, to open and sunny squares, and so forth.

Outdoor Education can also be carried out in the city to learn how to read the city and architecture, otherwise children learn it on their own and, in a trivial way, they may not be able to recognize simple concepts such as the proportions of a building, the architectural orders, the sequence of shapes, the relationship between empty and full, between nature and artifice, as well as the beauty of architecture and art.

L'urbanista Kevin Andrew Lynch, in '*The Image of the City*' (Lynch 1960), uses on-site observation, perspectives, as an instrument of investigation, hence, strolling means learning by looking at the outside of buildings and is an integral part of learning about architecture. In conclusion, Outdoor Education can influence and be influenced by recent research on the redesign of open spaces in cities to avoid the effects of heat waves and mitigate the phenomena of urban heat islands (Gaspari and Fabbri 2019; Huang et al. 2016; Ibraheem and Hassan 2020; Tsitoura et al. 2017).

Nature Based Solutions (Eggermont et al. n.d.; European Commission 2021; Wild et al. n.d.) can be one of the educational contexts where Outdoor Education can be carried out in the city but, nevertheless in nature.

Acknowledgements A Special Thanks to Silvia Magnani for the Chats and the Discussion on Outdoor Education and Children's Clothing.

References

Angouridakis VE, Makrogiannis TJ (1982) The discomfort-index in Thessaloniki, Greece. Int J Biometeorol 26(1):53–59. https://doi.org/10.1007/BF02187617

arki_lab (ed) (2019) The city as a classroom. Designing for outdoor learning. arki@0332lab

ASHRAE (2017) ASHRAE 55 - 2017, Thermal environmental condition for human occupancy

Astrand PO (1952) Experimental studioes of physical working capacity in relation to sex an age. Ejnar Munksgaard

Ballester J, Quijal-Zamorano M, Méndez Turrubiates RF, Pegenaute F, Herrmann FR, Robine JM, Basagaña X, Tonne C, Antó JM, Achebak H (2023) Heat-related mortality in Europe during the summer of 2022. Nat Med. https://doi.org/10.1038/s41591-023-02419-z

Bar-Or O (1983) Pediatric sports medicine for the practitioner. Springer, New York. https://doi.org/10.1007/978-1-4612-5593-2

Bikomeye J, Balza J, Beyer K (2021) The impact of schoolyard greening on children's physical activity and socioemotional health: a systematic review of experimental studies. Int J Environ Res Public Health 18(2):535. https://doi.org/10.3390/ijerph18020535

Blazejczyk K, Epstein Y, Jendritzky G, Staiger H, Tinz B (2012) Comparison of UTCI to selected thermal indices. Int J Biometeorol 56(3):515–535. https://doi.org/10.1007/s00484-011-0453-2

Bröde P, Fiala D, Błażejczyk K, Holmér I, Jendritzky G, Kampmann B, Tinz B, Havenith G (2012) Deriving the operational procedure for the Universal Thermal Climate Index (UTCI). Int J Biometeorol 56(3):481–494. https://doi.org/10.1007/s00484-011-0454-1

Brooks-Gunn J, Duncan GJ, Klebanov PK, Sealand N (1993) Do neighborhoods influence child and adolescent development? Am J Sociol 99(2):353395

Brown RD, Gillespie TJ (1995) Microclimatic landscape design: creating thermal comfort and energy efficiency. Wiley, Hoboken

Burton I, Ebi KL, McGregor G (2009) Biometeorology for adaptation to climate variability and change. In: Biometeorology for adaptation to climate variability and change. Springer Netherlands, pp 1–5. https://doi.org/10.1007/978-1-4020-8921-3_1

Chawla L, Keena K, Pevec I, Stanley E (2014) Green schoolyards as ha-vens from stress and resources for resilience in childhood and adolescence. Health Place 28:1–13. https://doi.org/10.1016/j.healthplace.2014.03.001

Chen L, Kántor N, Nikolopoulou M (2022) Meta-analysis of outdoor thermal comfort surveys in different European cities using the RUROS database: the role of background climate and gender. Energy Build 256:111757. https://doi.org/10.1016/j.enbuild.2021.111757

Hayley C, Zubrick SR, Foster S, Giles-Corti B, Bull F, Wood L, Knui-man M, Brinkman S, Houghton S, Boruff B (2015) The influence of the neighborhood physical environment on early child health and devel-opment: a review and call for research. Health Place 33:25–36. https://doi.org/10.1016/j.healthplace.2015.01.005

Chui AC, Gittelson A, Sebastian E, Stamler N, Gaffin SR (2018) Urban heat islands and cooler infrastructure—measuring near-surface tempera-tures with hand-held infrared cameras. Urban Climate 24:51–62. https://doi.org/10.1016/j.uclim.2017.12.009

Ciucci E, Calussi P, Menesini E, Mattei A, Petralli M, Orlandini S (2011) Weather daily variation in winter and its effect on behavior and af-fective states in day-care children. Int J Biometeorol 55(3):327–337. https://doi.org/10.1007/s00484-010-0340-2

Ciucci E, Calussi P, Menesini E, Mattei A, Petralli M, Orlandini S (2013) Seasonal variation, weather and behavior in day-care children: a multilevel approach. Int J Biometeorol 57(6):845–856. https://doi.org/10.1007/s00484-012-0612-0

Claire W (2012) Nature Kindergartens and forest schools: an exploration of naturalistic learning within nature kindergartens and forest schools. Mindstretchers

Coccolo S, Kämpf J, Scartezzini JL, Pearlmutter D (2016) Outdoor hu-man comfort and thermal stress: a comprehensive review on models and standards. Urban Climate 18. https://doi.org/10.1016/j.uclim.2016.08.004

Condemi V, Gestro M (2019) Health impact of fuel poverty. Urban Fuel Poverty 105–123. https://doi.org/10.1016/B978-0-12-816952-0.00003.X

Constable K. (2014) Bringing the forest school approach to your early years practice. Routledge, London

Cutter-Mackenzie A, Edwards S (2013) Toward a model for early child-hood environmental education: foregrounding, developing, and connect-ing knowledge through play-based learning. J Environ Educ 44(3):195–213. https://doi.org/10.1080/00958964.2012.751892

Dadvand P, Nieuwenhuijsen MJ, Esnaola M, Forns J, Basagaña X, Alva-rez-Pedrerol M, Rivas I, López-Vicente M, De Castro Pascual M, Su J, Jerrett M, Querol X, Sunyer J (2015) Green spaces and cognitive development in primary school children. Proc Natl Acad Sci 112(26):7937–7942. https://doi.org/10.1073/pnas.1503402112

Dirksen M, Ronda RJ, Theeuwes NE, Pagani GA (2019) Sky view factor calculations and its application in urban heat island studies. Urban Climate 30:100498. https://doi.org/10.1016/j.uclim.2019.100498

Eggermont H, Balian E, Azevedo MN, Beumer V, Brodin T, Claudet J, Fady B, Grube M, Keune H, Lamarque P (n.d.) Nature-based solutions: new influence for environmental management and research in Europe Nature-based solutions, an emerging term.

Emmanuel R (2005) Thermal comfort implications of urbanization in a warm-humid city: the Colombo Metropolitan Region (CMR), Srilanka. Build Environ 40(12):1591–1601. https://doi.org/10.1016/j.buildenv.2004.12.004

Erell E, Pearlmutter D, Boneh D, Kutiel PB (2014) Effect of high-albedo materials on pedestrian heat stress in urban street canyons. Urban Climate 10:367–386. https://doi.org/10.1016/j.uclim.2013.10.005

European Commission (2021) Evaluating the impact of the nature-based solutions. https://doi.org/10.2777/244577

Falk B (1998) Effects of thermal stress during rest and exercise in the paediatric population. Sports Med 25(4):221–240. https://doi.org/10.2165/00007256-199825040-00002

Falk B, Dotan R (n.d.) Physiologic and health aspects of exercise in hot and cold environments. In: The young athlete. Blackwell Publishing Ltd., pp 249–267 https://doi.org/10.1002/9780470696255.ch20

Farnè R, Bortolotti A, Terrusi M (2022) Outdoor education: prospettive teoriche e buone pratiche. Carocci Editore

Fjørtoft I (2001) The natural environment as a playground for children: the impact of outdoor play activities in pre-primary school children. Early Childhood Educ J 29(2):111–117. https://doi.org/10.1023/A:1012576913074

Freire P (1970) Pedagogy of the oppressed. Herder and Herder

Freire P (1998) Pedagogy of freedom: ethics, democracy and civic courage. Lanham, Rowman & Littlefield Publishers

Frost JL, Wortham SC, Reifel S (2007) Play and child development. Pearson College

Gagge AP, Stolwijk JAJ, Nishi Y (1971) An effective temperature scale based on a simple model of human physiological regulatory response. ASHRAE Trans 77(1)

Gaspari J, Fabbri K (2019) Greening actions and the related potential impacts on outdoor comfort in a dense built environment. IOP Conf Ser: Earth Environ Sci 290(1). https://doi.org/10.1088/1755-1315/290/1/012163

Gestro M, Condemi V, Bardi L, Tomaino L, Roveda E, Bruschetta A, So-limene U, Esposito F (2020) Short-term air pollution exposure is a risk factor for acute coronary syndromes in an urban area with low annual pollu-tion rates: results from a retrospective observational study (2011–2015). Arch Cardiovasc Dis 113(5):308–320. https://doi.org/10.1016/j.acvd.2020.03.013

Guo L-N, Zhao R-L, Ren A-H, Niu L-X, Zhang Y-L (2019) Stress recovery of campus street trees as visual stimuli on graduate students in autumn. Int J Environ Res Public Health 17(1):148. https://doi.org/10.3390/ijerph17010148

Hallam J, Gallagher L, Harvey C (2021) 'I don't wanna go. I'm staying. This is my home now.' Analysis of an intervention for connecting young people to urban nature. Urban Forestry Urban Greening 65:127341. https://doi.org/10.1016/j.ufug.2021.127341

Höppe P (1999) The physiological equivalent temperature—a universal index for the biometeorological assessment of the thermal environment. Int J Biometeorol 43(2):71–75. https://doi.org/10.1007/s004840050118

Höppe P (2002) Different aspects of assessing indoor and outdoor thermal comfort. Energy Build 34(6):661–665. https://doi.org/10.1016/S0378-7788(02)00017-8

Hoppe PR, Seidl HAJ (1991) Problems in the assessment of the biocli-mate for vacationists at the seaside. Int J Biometeorol 35(2):107–110. https://doi.org/10.1007/BF01087486

Huang J, Zhou C, Zhuo Y, Xu L, Jiang Y (2016) Outdoor thermal envi-ronments and activities in open space: an experiment study in humid sub-tropical climates. Build Environ 103:238–249. https://doi.org/10.1016/j.buildenv.2016.03.029

Ibraheem SQ, Hassan SA (2020) The effect of the characteristics for-mation of urban open space on thermal comfort for pedestrian. IOP Conf Ser: Mater Sci Eng 881(1). https://doi.org/10.1088/1757-899X/881/1/012016

Inbar O, Bar-Or O, Dotan R, Gutin B (1981) Conditioning versus exercise in heat as methods for acclimatizing 8- to 10-yr-old boys to dry heat. J Appl Physiol 50(2):406–411. https://doi.org/10.1152/jappl.1981.50.2.406

ISO 7243:2017 (2017) Ergonomics of the thermal environment-Assessment of heat stress using the WBGT (wet bulb globe temperature) index

James T (1990) Kurt Hahn and the aims of education. J Experiential Educ 13(1):6–13. https://doi.org/10.1177/105382599001300101

Jendritzky G, Nübler W (1981) A model analysing the urban thermal envi-ronment in physiologically significant terms. Arch Meteorol Geophys Bioclimatol Ser B 29(4):313–326. https://doi.org/10.1007/BF02263308

Kántor N, Unger J, Gulyás Á (2012) Subjective estimations of thermal environment in recreational urban spaces—part 2: international comparison. Int J Biometeorol 56(6):1089–1101. https://doi.org/10.1007/s00484-012-0564-4

Kenny NA, Warland JS, Brown RD, Gillespie TG (2009a) Part A: assessing the performance of the COMFA outdoor thermal comfort model on subjects performing physical activity. Int J Biometeorol 53(5):415–428. https://doi.org/10.1007/s00484-009-0226-3

Kenny NA, Warland JS, Brown RD, Gillespie TG (2009b) Part B: revisions to the COMFA outdoor thermal comfort model for application to subjects performing physical activity. Int J Biometeorol 53(5):429–441. https://doi.org/10.1007/s00484-009-0227-2

Kim SW, Brown RD (2021) Urban heat island (UHI) intensity and magni-tude estimations: a systematic literature review. Sci Total Environ 779:146389. https://doi.org/10.1016/j.scitotenv.2021.146389

Klinenberg E (1999) Denaturalizing disaster: a social autopsy of the 1995 Chicago heat wave. Theor Soc 28(2):239–295. https://doi.org/10.1023/A:1006995507723

Klinenberg E (2002) Heat wave. A social autopsy of disaster in Chicago. University of Chicago Press

Knight S (2009) Forest school and outdoor learning in the early years, play and learn. SAGE

Li D, Sullivan WC (2016) Impact of views to school landscapes on recovery from stress and mental fatigue. Landsc Urban Plan 148:149–158. https://doi.org/10.1016/j.landurbplan.2015.12.015

Li D, Zhai Y, Chang P-J, Merrill J, Browning MHEM, Sullivan WC (2022) Nature deficit and senses: relationships among childhood nature exposure and adulthood sensory profiles, creativity, and nature relatedness. Landsc Urban Plan 226:104489. https://doi.org/10.1016/j.landurbplan.2022.104489

Liu J, Green RJ (2023) The effect of exposure to nature on children's psychological well-being: a systematic review of the literature. Urban Forestry Urban Greening 81:127846. https://doi.org/10.1016/j.ufug.2023.127846

Lynch K (1960) The image of the city. MIT Press

Mahler MS, Pine F, Bergman A (2018) The psychological birth of the human infant. Routledge. https://doi.org/10.4324/9780429482915

Marando F, Heris MP, Zulian G, Udías A, Mentaschi L, Chrysoulakis N, Parastatidis D, Maes J (2022) Urban heat island mitigation by green in-frastructure in European functional urban areas. Sustain Cities Soc 77:103564. https://doi.org/10.1016/j.scs.2021.103564

Masterton JM, Richardson FA (1979) Humidex: a method of quantifying human discomfort due to excessive heat and humidity. Canada. Atmospheric Environment Service

Mayer E (1993) Objective criteria for thermal comfort. Build Environ 28(4):399–403. https://doi.org/10.1016/0360-1323(93)90016-V

Mayer H, Hoppe P (1987) Thermal comfort of man in different urban environments. Theoret Appl Climatol 38(1):43–49. https://doi.org/10.1007/BF00866252

McCullough EA, Eckels S, Harms C (2009) Determining temperature ratings for children's cold weather clothing. Appl Ergon 40(5):870–877. https://doi.org/10.1016/j.apergo.2008.12.004

Menolascino FJ (1972) Playing and reality. D. W. Winnicott London: Tavistock Publications Limited, 1971, pp 169. Price £2.10. Br J Psychiatry 120(554):106–106. https://doi.org/10.1192/bjp.120.554.106-a

Moll A, Collado S, Staats H, Corraliza JA (2022) Restorative effects of exposure to nature on children and adolescents: a systematic review. J Environ Psychol 84:101884. https://doi.org/10.1016/j.jenvp.2022.101884

Monteith JL, Unsworth MH (2013) Principles of environmental physics. Elsevier. https://doi.org/10.1016/C2010-0-66393-0

Monti F, Farné R, Crudeli F, Agostini F, Minelli M, Ceciliani A (2019) The role of outdoor education in child development in Italian nursery schools. Early Child Dev Care 189(6):867–882. https://doi.org/10.1080/03004430.2017.1345896

Moran DS, Pandolf KB, Shapiro Y, Heled Y, Shani Y, Mathew WT, Gonzalez RR (2001) An environmental stress index (ESI) as a substitute for the wet bulb globe temperature (WBGT). J Therm Biol 26(4–5):427–431. https://doi.org/10.1016/S0306-4565(01)00055-9

Mullins PM (2013) Book review: a pedagogy of place: outdoor education for a changing world. J Experiential Educ 36(1):78–79. https://doi.org/10.1177/1053825913484708

Mumford L (1938) The culture of cities. Harvest/HBJ

Mumford L (1961) The city in history: its origins, its transformations, and its prospects paperback. Harcourt, Brace and World

Mygind L, Kjeldsted E, Hartmeyer R, Mygind E, Bølling M, Bentsen P (2019) Mental, physical and social health benefits of immersive nature-experience for children and adolescents: a systematic review and quality assessment of the evidence. Health Place 58:102136. https://doi.org/10.1016/j.healthplace.2019.05.014

Nam I, Yang J, Lee D, Park E, Sohn J-R (2015) A study on the thermal comfort and clothing insulation characteristics of preschool children in Korea. Build Environ 92:724–733. https://doi.org/10.1016/j.buildenv.2015.05.041

Nikolopoulou M, Baker N, Steemers K (2001) Thermal comfort in out-door urban spaces: understanding the human parameter. Sol Energy 70(3):227–235. https://doi.org/10.1016/S0038-092X(00)00093-1

Nikolopoulou M, Lykoudis S (2006) Thermal comfort in outdoor urban spaces: analysis across different European countries. Build Environ 41(11):1455–1470. https://doi.org/10.1016/j.buildenv.2005.05.031

Nikolopoulou M, Steemers K (2003) Thermal comfort and psychological adaptation as a guide for designing urban spaces. Energy Build 35(1):95–101. https://doi.org/10.1016/S0378-7788(02)00084-1

Noro M, Lazzarin R (2015) Urban heat island in Padua, Italy: simulation analysis and mitigation strategies. Urban Climate 14:187–196. https://doi.org/10.1016/j.uclim.2015.04.004

O'Brien K, Lomas T (2017) Developing a growth mindset through outdoor personal development: can an intervention underpinned by psychology in-crease the impact of an outdoor learning course for young people? J Adventure Educ Outdoor Learn 17(2):133–147. https://doi.org/10.1080/14729679.2016.1232199

O'Brien L, Murray R (2007) Forest school and its impacts on young children: case studies in Britain. Urban Forestry Urban Greening 6(4):249–265. https://doi.org/10.1016/j.ufug.2007.03.006

Oke TR (1973) City size and the urban heat island. Atmos Environ 7(8):769–779

Oke TR (1982) The energetic basis of the urban heat island. Q J R Meteorol Soc 108(455):1–24. https://doi.org/10.1002/qj.49710845502

Oke TR, Mills G, Christen A, Voogt JA (2017) Urban climates. Cambridge University Press. https://doi.org/10.1017/9781139016476

Pantavou K, Santamouris M, Asimakopoulos D, Theoharatos G (2013) Evaluating the performance of bioclimatic indices on quantifying thermal sensation for pedestrians. Adv Build Energy Res 7(2):170–185. https://doi.org/10.1080/17512549.2013.865557

Parsons KC (2002) The effects of gender, acclimation state, the opportunity to adjust clothing and physical disability on requirements for thermal comfort. Energy Build 34(6):593–599. https://doi.org/10.1016/S0378-7788(02)00009-9

Pearlmutter D, Berliner P, Shaviv E (2007) Integrated modeling of pedestrian energy exchange and thermal comfort in urban street canyons. Build Environ 42(6):2396–2409. https://doi.org/10.1016/j.buildenv.2006.06.006

Pelchat C, Karp GG (2012) A historical review of outdoor leadership curricular development and the future with action research. SCHOLE: J Leisure Stud Recreation Educ 27(2):1–13. https://doi.org/10.1080/1937156X.2012.11949685

Pellegrini AD, Smith PK (1998) Physical activity play: the nature and function of a neglected aspect of playing. Child Dev 69(3):577–598

Powell KE (2005) Land use, the built environment, and physical activity. Am J Prev Med 28(2):216–217. https://doi.org/10.1016/j.amepre.2004.10.031

Prezza M, Alparone FR, Cristallo C, Luigi S (2005) Parental perception of social risk and of positive potentiality of outdoor autonomy for children: the development of two instruments. J Environ Psychol 25(4):437–453. https://doi.org/10.1016/j.jenvp.2005.12.002

Rotaris L, Del Missier F, Scorrano M (2023) Comparing children and parental preferences for active commuting to school. A focus on Italian mid-dle-school students. Res Transp Econ 97:101236. https://doi.org/10.1016/j.retrec.2022.101236

Rothfusz LP (1990) The heat index "Equation" (or, More Than You Ever Wanted to Know About Heat Index)

Smith GA (2002) Place-based education: learning to be where we are. Phi Delta Kappan 83(8):584–594. https://doi.org/10.1177/003172170208300806

Sobel D (2004) Place-based education: connecting classrooms & communities. Orion Society

Steadman RG (1979) The assessment of sultriness. Part I: a temperature-humidity index based on human physiology and clothing science. J Appl Meteorol 18(7):861–873. https://doi.org/10.1175/1520-0450(1979)018<0861:TAOSPI>2.0.CO;2

Steadman RG (1984) A universal scale of apparent temperature. J Climate Appl Meteorol 23(12):1674–1687. https://doi.org/10.1175/1520-0450(1984)023%3c1674:AUSOAT%3e2.0.CO;2

Taylor AF, Kuo FE, Sullivan WC (2002) Views of nature and self-discipline: evidence from inner city children. J Environ Psychol 22(1–2):49–63. https://doi.org/10.1006/jevp.2001.0241

Teli D, Bourikas L, James PAB, Bahaj AS (2017) Thermal performance evaluation of school buildings using a children-based adaptive comfort model. Proc Environ Sci 38:844–851. https://doi.org/10.1016/j.proenv.2017.03.170

ter Mors S, Hensen JLM, Loomans MGLC, Boerstra AC (2011) Adaptive thermal comfort in primary school classrooms: creating and vali-dating PMV-based comfort charts. Build Environ 46(12):2454–2461. https://doi.org/10.1016/j.buildenv.2011.05.025

Theeuwes NE, Steeneveld G-J, Ronda RJ, Holtslag AAM (2017) A diagnostic equation for the daily maximum urban heat island effect for cities in northwestern Europe. Int J Climatol 37(1):443–454. https://doi.org/10.1002/joc.4717

Thom EC (1959) The discomfort index. Weatherwise 12(2):57–61. https://doi.org/10.1080/00431672.1959.9926960

Tsitoura M, Michailidou M, Tsoutsos T (2017) A bioclimatic outdoor design tool in urban open space design. Energy Build 153. https://doi.org/10.1016/j.enbuild.2017.07.079

UNESCO (2011) EDUCATION 2030. Incheon Declaration Framework for Action for the implementation of Sustainable Development Goal 4 Ensure in-clusive and equitable quality education and promote lifelong learning opportunities for all

van den Bogerd N, Coosje Dijkstra S, Koole SL, Seidell JC, de Vries R, Maas J (2020) Nature in the indoor and outdoor study environment and secondary and tertiary education students' well-being, academic outcomes, and possible mediating pathways: a systematic review with recommenda-tions for science and practice. Health Place 66:102403. https://doi.org/10.1016/j.healthplace.2020.102403

Vanos JK (2015) Children's health and vulnerability in outdoor microclimates: a comprehensive review. Environ Int 76:1–15. https://doi.org/10.1016/j.envint.2014.11.016

Vanos JK, Warland JS, Gillespie TJ, Kenny NA (2012) Improved predictive ability of climate–human–behaviour interactions with modifica-tions to the COMFA outdoor energy budget model. Int J Biometeorol 56(6):1065–1074. https://doi.org/10.1007/s00484-012-0522-1

Waite S (ed) (2020) Outdoor learning research. Insight into forms and functions. Routledge

Waller T, Ärlemalm-Hagsér E, Sandseter E, Lee-Hammond L, Lekies K, Wyver S (2017) The SAGE handbook of outdoor play and learning. SAGE Publications Ltd. https://doi.org/10.4135/9781526402028

Ward C (1978) The child in the city. Society 15(4):84–91. https://doi.org/10.1007/BF02694719

Ward C (1979) The child in the city. Pantheon

Ward K, Lauf S, Kleinschmit B, Endlicher W (2016) Heat waves and urban heat islands in Europe: a review of relevant drivers. Sci Total Environ 569–570:527–539. https://doi.org/10.1016/j.scitotenv.2016.06.119

Wattchow B, Brown M (2011) A pedagogy of place. Outdoor education for a changing world. Monash University Publishing

Weihs P, Staiger H, Tinz B, Batchvarova E, Rieder H, Vuilleumier L, Ma-turilli M, Jendritzky G (2012) The uncertainty of UTCI due to uncer-tainties in the determination of radiation fluxes derived from measured and observed meteorological data. Int J Biometeorol 56(3):537–555. https://doi.org/10.1007/s00484-011-0416-7

Wild T, Freitas T, Vandewoestijne S, Bulkeley H, Naumann S, Vojinović Z, Calfapietra C, Whiteoak K. European Commission. Directorate-General for Research and Innovation (n.d.) Nature-based solutions: state of the art in EU-funded projects

Wolfe S, Cage G, Epstein M, Tice L, Miller H, Gordon RS (1970) Metabolic studies of isolated human eccrine sweat glands. J Clinical Investigation 49(10):1880–1884. https://doi.org/10.1172/JCI106407

Zeng L, Lu J, Li W, Li Y (2018) A fast approach for large-scale sky view factor estimation using street view images. Build Environ 135:74–84. https://doi.org/10.1016/j.buildenv.2018.03.009

Chapter 14
Conclusions

14.1 Why I Wrote This Book

The subject of this book is a methodology to evaluate the indoor thermal comfort assessment with subjective judgment and a questionnaire, especially in the case of children. The field research of indoor thermal comfort adopts, on the one hand, physical parameter measurements, comfort indexes (e.g. PMV or adaptive comfort), on the other hand, an ergonomic assessment by means of questionnaire. In the second case, not everybody understands the query about comfort, sometimes they do not understand the meaning of a word, or instead, in the case of children, they do not have the same words as an adult to talk about comfort or temperature. Children's judgment of energy, temperature, comfort and odour involves a personal and emotional connection.

The main benefits of the book are to discover how to ask to children about their sensations on environmental ergonomics. In my experience—about which I would like to write a book—I discovered several new points of view as regards architectural design, and how to measure comfort, e.g. children correlate temperature with outdoor climate, more than with their real perception inside a room, or the odours are correlated to food or smells. Furthermore, my book could be a useful guide for designing questionnaires judging comfort, so that this approach could be replicated in other subjects, e.g. the elderly, the ill, young people, etc. The keywords are: occupant Indoor Environmental Quality (IEQ), survey and building performance benchmarking.

The Special features concern a methodological approach to consider the psychological and pedagogical point of view on the judgment of thermal comfort, followed by architectural and environmental knowledge and including sustainability and sociality. All these issues can be useful in improving children's knowledge, the educational path and architectural design, including HVAC design.

K. Fabbri, *Thermal Comfort Perception*, https://doi.org/10.1007/978-3-031-52610-7_14

In conclusion, as an author I want to emphasise that the study of Thermal Comfort and Indoor Environmental Quality is transversal to multiple disciplines, as is architectural design.

14.2 And Now?

The book is a work open to multiple disciplines: Architecture, Environmental Psychology, Healthy and Medicine, Pedagogy, Engineering, Physiology, etc., including communication and teaching concepts regarding energy, environment, sustainability, knowledge of physics, chemistry, scientific method, etc., under programmes teaching children. Experience has shown that the story of experience is the preferred way for children to define their Thermal Cognitive Knowledge.

Experience also shows how crucial it is to listening to children, in this case, and in general to other users, so as to locate the physical, physiological and even behavioural parameters upon which to act in order to improve the conditions of Thermal Comfort and Indoor Environmental Quality.

Once defined, it is necessary to study the Thermal Cognitive Knowledge (TCK) that people can provide for other research to other entities (hospitals, nursing homes, workplaces, etc.).

14.3 Why I Updated This Book

The publisher's request, "*The book is clearly having a big influence on researchers in its subject area. I wonder if you would be interested in adding/updating some of the book for a second edition,*" had truly caught me, pleasantly, by surprise.

In updating the edition, I have asked myself in these (almost) ten years what has changed in the field of research on thermal comfort, referring in particular to children. The thought immediately went to the pandemic, the event that changed lives the world over, including the world of school, on which much has been discussed and will continue to be discussed. The class has left the classroom and this has allowed us to apply some experiences already in place on the new teaching methods and on Outdoor Education: being outside the classroom is no longer an "oddity" of some experimental pedagogical programme, but it can be one way, among others, of doing lessons.

That said, I wondered how these aspects affect children's perception of thermal comfort? Because changing the environmental context changes the response and the narrative of children. Following this reasoning, this edition expands with a new chapter that draws attention to the role of the senses and emotions as elements that contribute to defining thermal comfort and outdoor thermal comfort, currently focused on aspects related to environmental risks during the summer due to heat waves, climate change and effects on people's health.

The updates, especially Part III, have the objective of bringing fresh impetus to the research on thermal comfort and children.

These years have been an opportunity for me to reflect on the issues concerning thermal comfort in applications outside the confined space, and on the role that comfort has in architectural design, too often still focused on the eye, as Juhani Pallasmaa says, and I add, on Instagram, that is, on the image.

Architecture, on the other hand, concerns all the senses, and thermal comfort is the "sense" that sums them all up, given that when an environment is too hot or too cold, if not equipped with specific qualities, such as heritage buildings are, it is not used, it is abandoned. The reflection on thermal comfort and children is the "starting point", the first step for reflecting on the architectural project and architecture tout court, which, in the twenty-first century, will have to give itself a new definition.